Lecture Notes
in
LabVIEW & DATA ACQUISITION

Fadhil A. Ali

Oklahoma State University

Electrical and Computer Engineering

Acknowledgement

I would like to express my special appreciation and thanks to my advisor **James E. Stine**, he has been a tremendous mentor for me. I would like to thank you for encouraging my research and for allowing me to grow as a research scientist at Oklahoma State University / Electrical and Computer Engineering Department. Your advice on both research as well as on my career have been priceless.

I would also like to thank **Oklahoma State University** / Electrical and Computer Engineering Department, for hosting me. I also want to thank you for letting my researches be an enjoyable moment.. I would especially like to thank all staff at Oklahoma State University for helping my visiting period.

A special thanks to **Institute of International Education** (IIE) / **Iraq Scholar Rescue Project, Scholar Rescue Fund** . Words cannot express how grateful I am to all staff and administration, my program officer, Iraq project director, and for all of the helps that you've made on my behalf. Your efforts for me was what sustained me thus far.

At the end I would like express appreciation to my beloved wife who spent sleepless nights with and was always my support in the moments when there was no one to answer my queries.

Multi thanks to the national instruments for giving us this vast opportunity of great applications and open new era of technology in the future.

i

About the Author

 Currently, Fadhil A. Ali is visitor scholar and PostDoc Researcher at **Oklahoma State University.** The most of activities were focused on researches and an engineering academic strategy. This was taken place in different areas in computer engineering concerns, some projects and proposals have been written and presented.

- Neural Networks and Image processing units

- Pattern recognition design / data mining

- FPGA (Field Programmable Gate Array).

- LabVIEW presentations edited for undergrad and post grad students.

Previously, he taught many topics in computer engineering department in Basra University/ IRAQ;

Undergrad Students:

- ❖ Concepts of Operating Systems
- ❖ Software Engineering & Web Engineering
- ❖ Operating system lab's Supervisor (Red Hat | Ubuntu- Linux)
- ❖ Software Engineering lab's Supervisor (VB.Net)
- ❖ Software & Web Engineering lab's Supervisor (ASP.Net)

Artificial Intelligence – for (academic year 2010-2011) only.

Post grad Students (MSc in Engineering):

- Advanced Discrete Mathematics – (academic year 2008-2009)
- Advanced Artificial Intelligence – (academic year 2009-2010)
- Advanced Operating System - for (academic year 2010-2011)

Advanced Object Oriented Programming – for (academic year 2010-2011)

And many of projects in different engineering fields, and supervision for under grad students.

www.drfadhil.com

---

fadhil.ali@okstate.edu | drfadhil@drfadhil.com | fadhilcad@gmail.com
drfadhil@ieee.org | falial-sahlanee@acm.org

Contents at a Glance

Table of Contents

Preface

Lecture notes in LabVIEW and Data Acquisition is based on prelude that professional engineers , students in (electrical and computer) engineering college , and who are pioneers in technical works in both private and public sectors can afford to LabVIEW and its applications.

In today's electrical and computer engineering world , the LabVIEW has become essential for creating so many applications. It involves very large scale environment of researches, academic teaching , acquiring data and processing signals, automating test and validation systems, instrument control and embedded monitoring and control systems.

This book summaries LabVIEW principles and definitions, with some applications. Many of examples and reconstruction programming tools in electrical and computer engineering, controlling systems, database acquiring and data acquisition have been described in those chapters. The book has multi series as Lectures presented in the engineering colleges, which provides so many examples for both faculties / engineers and students. These are Lecture notes in very specific identifications and explanations into applications of Data acquisition deals into LabVIEW 2012.

Also, this book reflects a powerful trends that have recast the role of wireless data acquisition and database acquiring systems. Besides , it has so many graphical presentations in order to give more clarifications for the freshmen.

I had three main goals in writing these lecture notes:

o **The Simplicity :** These texts are the result of researching and training on LabVIEW and its applications at Oklahoma State University / Electrical and Computer Engineering department . It is very important to teach students the skills that are based in research from both the field of computer engineering and other related disciplines by using LabVIEW toolkits. This book places a clear simple emphasis on control and data acquisition skills by LabVIEW first, but also ensures that those skills are based on rigorous and current researches.

o **Simulation :** To describe and explain LabVIEW concepts, this book uses realistic examples to help students get inside what tools and applications are really like. In addition to examples, I have provided palettes of interacting terms; therefore many students and beginners can *follow* the details. As well as identify the practices they wish to adopt to improve their interaction with LabVIEW environment.

This book includes twelve chapters, and each chapter consists of sections and subsections. Also, these chapters have a lot of sketches and presentations to convey the technological details of the LabVIEW in simple way.

Chapter one has comprised six main sections, which explain the most LabVIEW concepts and principles. While chapter two included fourteen sections with subsections, who show the LabVIEW terminology in simple way.

In chapter three the main five sections presented some samples and examples for simple applications in LabVIEW , mostly used in students labs and projects. However, chapter four customized for explaining the LabVIEW as graphical programming environment, which mainly consisted of ten major sections.

In order to clarify the connectivity of LabVIEW and database , chapter five presented fourteen major sections figured out database systems and communications in LabVIEW and its applications.

Presenting the main features and applications of the data acquisition "DAQ" through LabVIEW, chapters six, seven, eight and nine have shown such technology as simple as possible. chapter six has included 11 sections and shows the DAQ principles and some properties and applications. While, chapter seven figured out 7 sections customized for explaining the wireless data acquisition, then chapter eight presented only 3 main sections with subsections for using the Wi-Fi DAQ systems in LabVIEW as very simple way. In the chapter nine one of the powerful technology used in wireless systems has presented , the wireless DAQ using ZigBee as simple as possible to clarify the role of this technology in LabVIEW environment.

Chapter ten has presented the basic concepts of using LabVIEW in controlling and simulation, this chapter includes twelve major sections with a lot of clarifications. In the way for explaining vision systems chapter eleven presented six sections declared the most features in vision systems as applied in LabVIEW.

At last chapter twelve has consisted 12 sections mainly showed some tips and tricks in LabVIEW in simple way for beginners and students, while gave extra view to the professionals.

Chapter 1

Introduction

Definition of LabVIEW : The LabVIEW stands for Laboratory Virtual Instrument Engineering Workbench; it is a system-design platform development environment for a visual programming language from National Instruments company. LabVIEW programs are called virtual instruments includes:

VIs (virtual instrument) that contains three main parts:

- Front Panel – How the user interacts with the VI.
- Block Diagram – The code that controls the program.
- Icon/Connector – Means of connecting a VI to other VIs.

   Front Panel consists of;

   - **Controls = Inputs**

   - **Indicators = Outputs**

The Front Panel is used to interact with the user when the program is running. Users can control the program, change inputs, and see data updated in real time. Indicators are used as outputs. These may include data, program states, and other information.

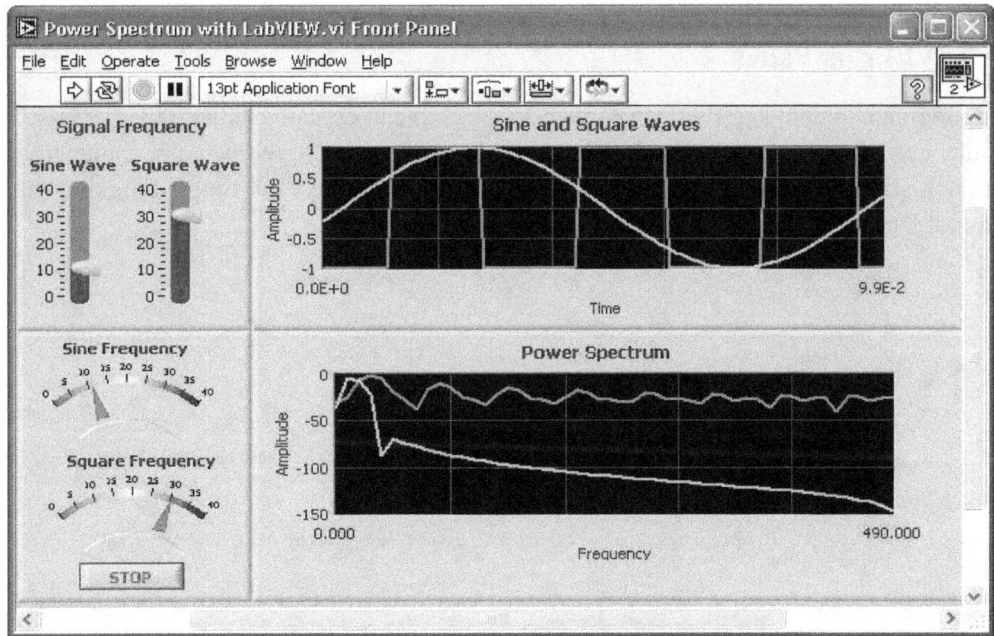

Figure 1-1 The front panel

Block Diagram consists of;

- o **Accompanying "program" for**

- o **Components " together**

Every front panel control has a corresponding terminal diagram. When a VI is run, values from controls flow through the block diagram, where they are used in the functions on the diagram, and the results are passed into other functions or indicators. Roughly, the following two examples will show some of these functions and tools been used through this environment.

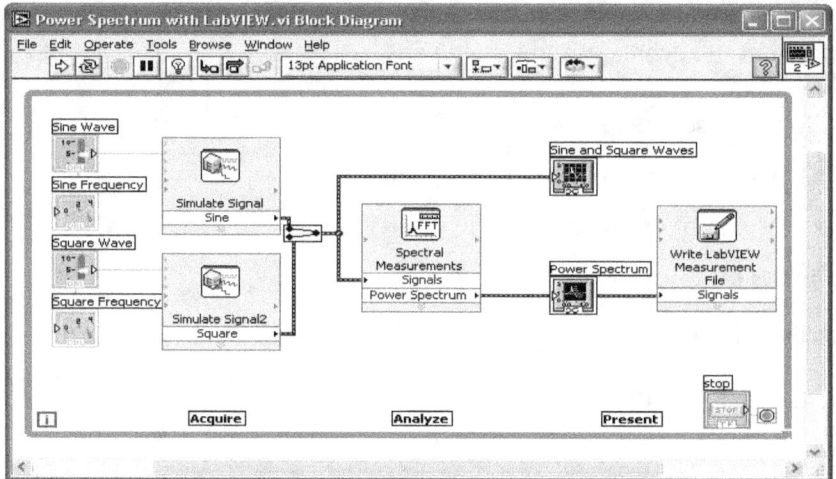

Figure 1-2 The Block diagram

Example 1- VI Front Panel:

The **Power switch** example is a Boolean control VI. A Boolean contains either a true or false values. The value is false until the switch is pressed. When the switch is pressed, the value becomes true. While the **temperature history** indicator is seen as waveform graph. It displays multiple numbers. In this case, the graph will plot Degree in (F) versus Time in (Seconds).

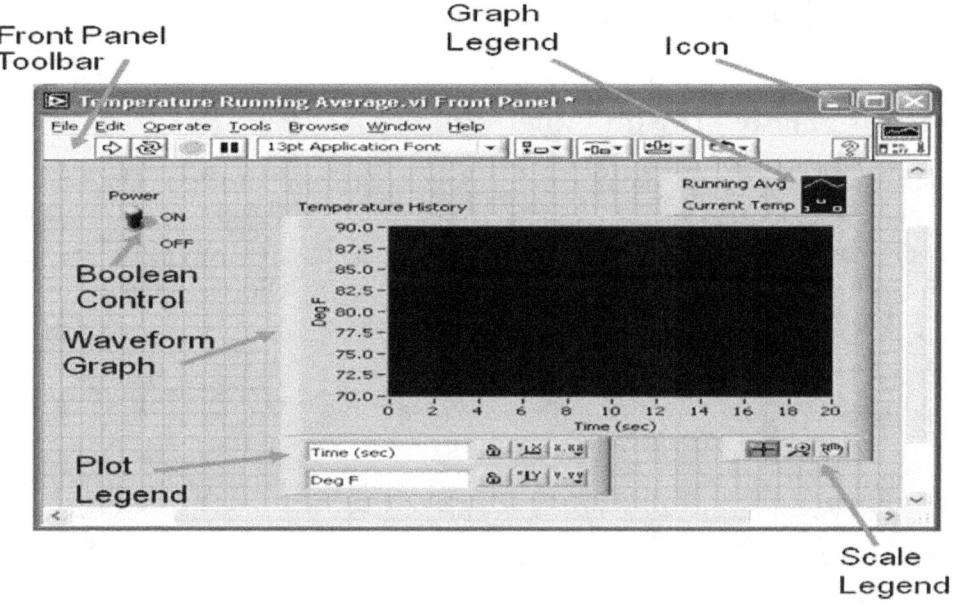

Figure 1-3 Power switch front panel

Example 2-VI Block Diagram:

The sub VI **Temp** calls the subroutine which retrieves a temperature from a Data Acquisition (DAQ) board. This temperature is plotted along with the running average temperature on the waveform graph **Temperature History**. The **Power** switch is a Boolean control on the Front Panel which will stop execution of the While Loop. The While Loop also contains a Timing Function to control how frequently the loop iterates.

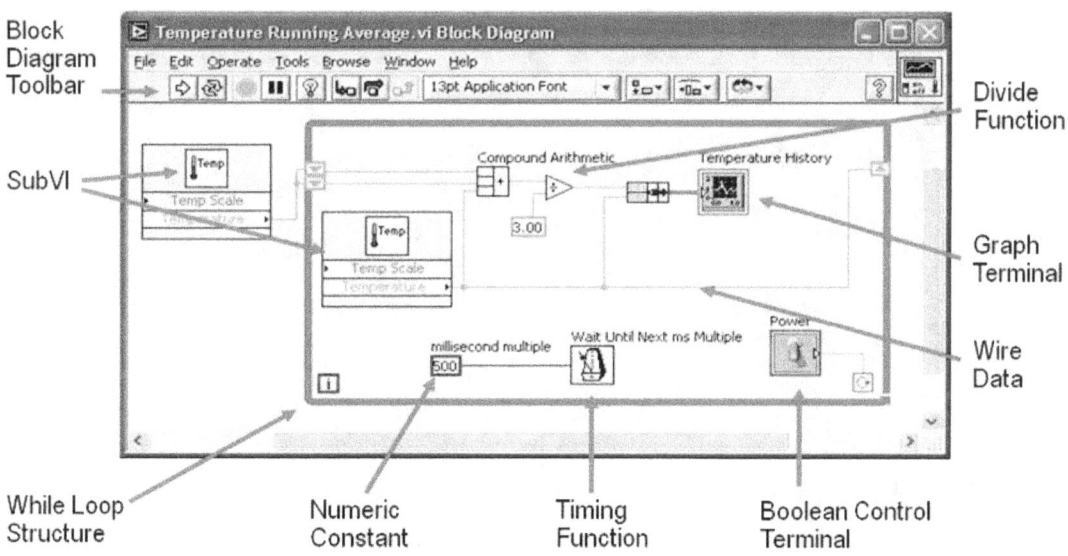

Figure 1-4 Waveform Graph Temperature Block diagram

1.1 Controls and Functions Palettes:

➤ Controls Palette - (Front Panel Window):

Use the **Controls** palette to place controls and indicators on the front panel. The **Controls** palette is available only on the front panel. Select **Window » Show Controls Palette** or right-click the front panel workspace to display the **Controls** palette. You also can display the **Controls** palette by right-clicking an open area on the front panel. Tack down the **Controls** palette by clicking the pushpin on the top left corner of the palette.

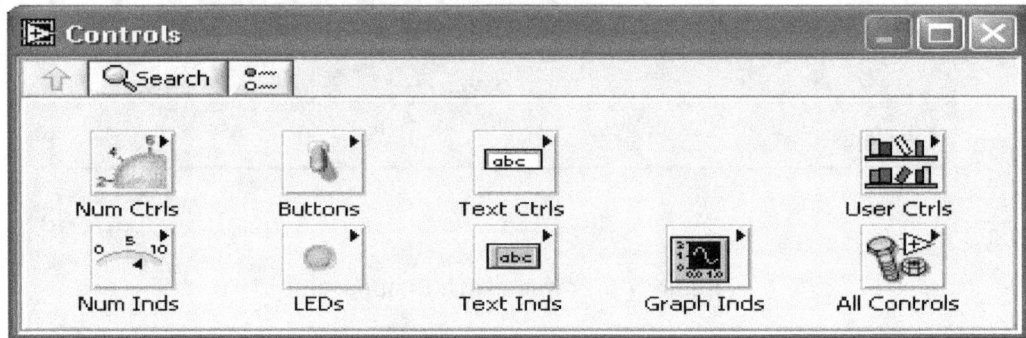

Figure 1-5 Controls Palette

➤ Functions Palette - (Block Diagram Window):

Use the Functions palette, to build the block diagram. The Functions palette is available only on the block diagram. Select Window » Show Functions Palette or right-click the block diagram workspace to display the Functions palette. You also can display the Functions palette by right-clicking an open area on the block diagram. Tack down the Functions palette by clicking the pushpin on the top left corner of the palette.

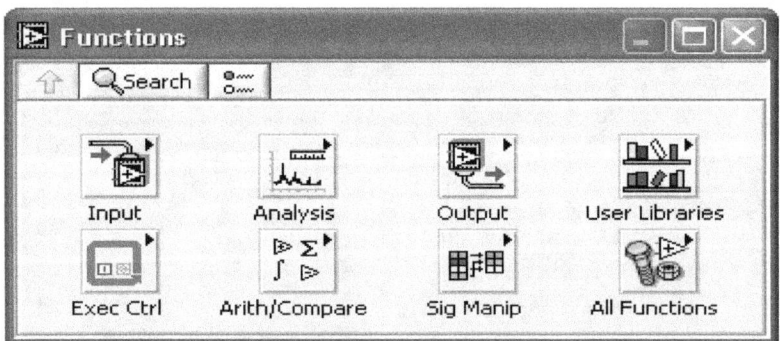

Figure 1-6 Functions Palette

➤ Tools Palette:

If automatic tool selection is enabled and you move the cursor over objects on the front panel or block diagram, Lab VIEW automatically selects the corresponding tool from the Tools palette. Toggle automatic tool selection by clicking the Automatic Tool Selection button in the Tools palette.

4

- Floating Palette
- Used to operate and modify front panel and block diagram objects.

Operating Tool	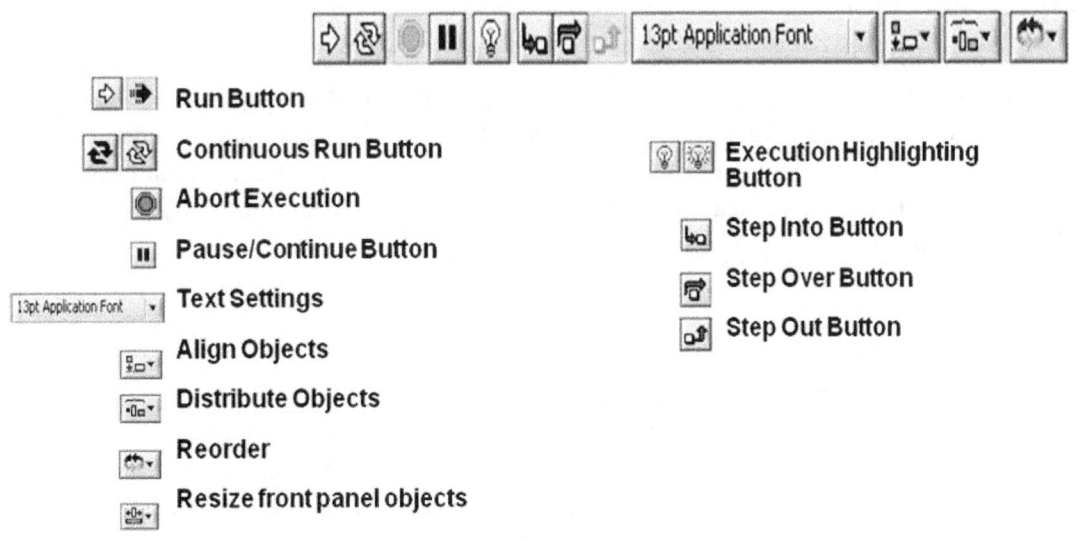
Positioning/Resizing Tool	
Labeling Tool	
Wiring Tool	
Shortcut Menu Tool	
Scrolling Tool	
Breakpoint Tool	
Color Copy Tool	

o Use the Operating tool to change the values of a control or select the text within a control.

o Use the Positioning tool to select, move, or resize objects. The Positioning tool changes shape when it moves over a corner of a resizable object.

o Use the Labeling tool to edit text and create free labels. The Labeling tool changes to a cursor when you create free labels.

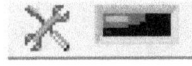

o Use the Wiring tool to wire objects together on the block diagram.
Automatic Selection Tool

➤ Status Toolbar:

Run Button

Continuous Run Button

Abort Execution

Pause/Continue Button

Text Settings

Align Objects

Distribute Objects

Reorder

Resize front panel objects

Execution Highlighting Button

Step Into Button

Step Over Button

Step Out Button

- Click the **Run** button to run the VI. While the VI runs, the **Run** button appears with a black arrow if the VI is a top-level VI, meaning it has no callers and therefore is not a sub VI.

- Click the **Continuous Run** button to run the VI until you abort or pause it. You also can click the button again to disable continuous running.

- While the VI runs, the **Abort Execution** button appears. Click this button to stop the VI immediately.

Note: Avoid using the **Abort Execution** button to stop a VI. Either let the VI complete its data flow or design a method to stop the VI programmatically. By doing so, the VI is at a known state. For example, place a button on the front panel that stops the VI when you click it.

- Click the **Pause** button to pause a running VI. When you click the **Pause** button, LabVIEW highlights on the block diagram the location where you paused execution. Click the **Pause** button again to continue running the VI.

- Select the **Text Settings** pull-down menu to change the font settings for the VI, including size, style, and color.

- Select The **Align Objects** pull-down menu to align objects along axes, including vertical, top edge, left, and so on.

- Select the **Distribute Objects** pull-down menu to space objects evenly, including gaps, compression, and so on.

- Select the **Resize Objects** pull-down menu to change the width and height of front panel objects.

1.2 Creating a VI :

When you create an object on the Front Panel, a terminal will be created on the Block Diagram. These terminals give you access to the Front Panel objects from the Block Diagram code. Each terminal contains useful information about the Front Panel object it corresponds to the color and symbols provide the data type. Double-precision, floating point numbers are represented with orange terminals and the letters DBL Boolean terminals are green with TF lettering.

In general, orange terminals should wire to orange terminals, green to green, and so on. This is not a hard-and-fast rule; LabVIEW will allow a user to connect a blue terminal (integer value) to an orange terminal (fractional value). But in most cases, look for a match in colors. Controls have an arrow on the right side and have a thick border. Indicators have an arrow on the left and a thin border. Logic rules apply to wiring in LabVIEW: Each wire must have one (but only one) source (or control), and each wire may have multiple destinations (or indicators). The program in this example shown in the following Figure 1-7, that takes data from A and B and passes the values to both an Add function and a subtract function. The results are displayed on the appropriate indicators.

Front Panel Window

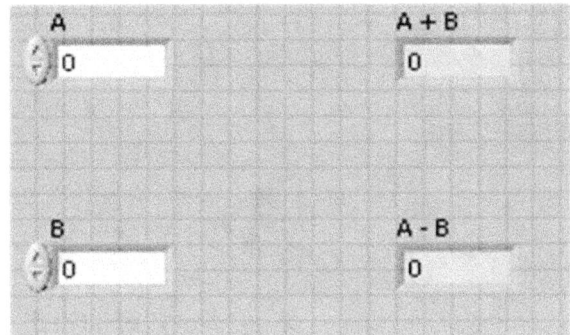

Block Diagram Window

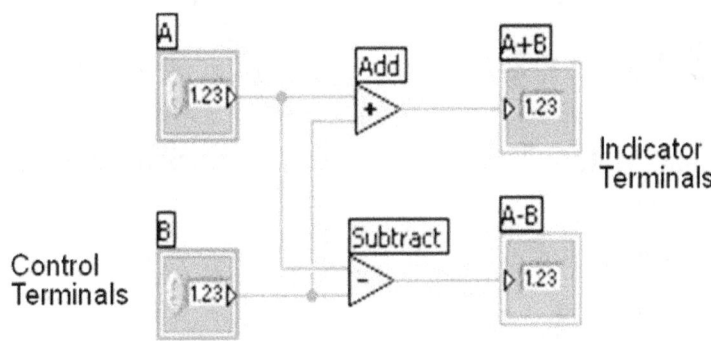

Figure 1-7

In addition to Front Panel terminals, the Block diagram contains functions, and each function may have multiple input and output terminals. Wiring to these terminals is an important part of LabVIEW programming. Once anyone has some experience programming in LabVIEW, the wiring will become easy. At first, one may need some assistance, here are some tips to get you started:

- The wiring tool is used to wire to the nodes of the functions. When you "aim" with the wiring tool, aim with the end of the wire hanging from the spool. This is where the wire will be placed.
- As you move the wiring tool over functions, watch for the yellow tip strip. This will tell you the name of the terminal you are wiring to.
- As you move the wiring tool over a terminal, it will flash. This will help you identify where the wire will attach.
- For more help with the terminals, right-click on the function and select **Visible Items>>Terminals**. The function's picture will be pulled back to reveal the connection terminals. Notice the colors- these match the data types used by the front panel terminals.
- For additional help, select **Help>>Show Context Help**, or press **CTRL+H**. This will bring up the context help window. As you move your mouse over the function, this window will show you the function, terminals, and a brief help description. Use this with the other tools to help you as you wire.

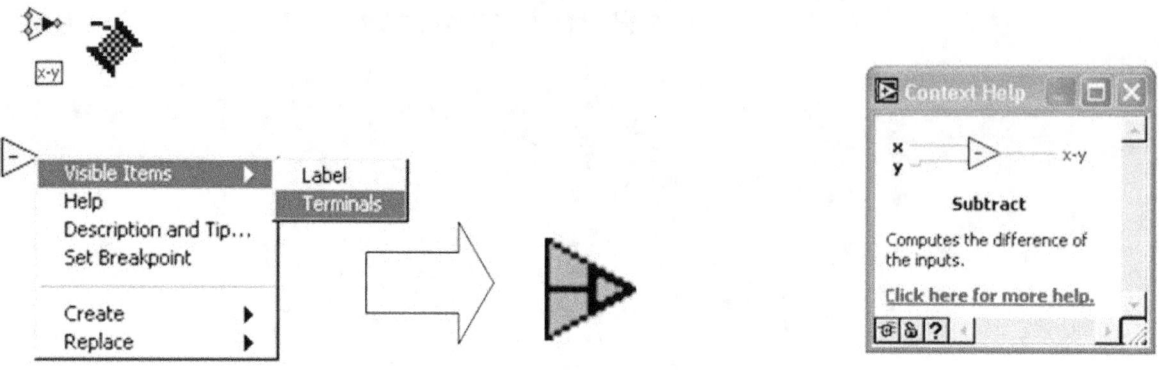

Wiring Tips – Block Diagram

Wiring "Hot Spot"

Click To Select Wires

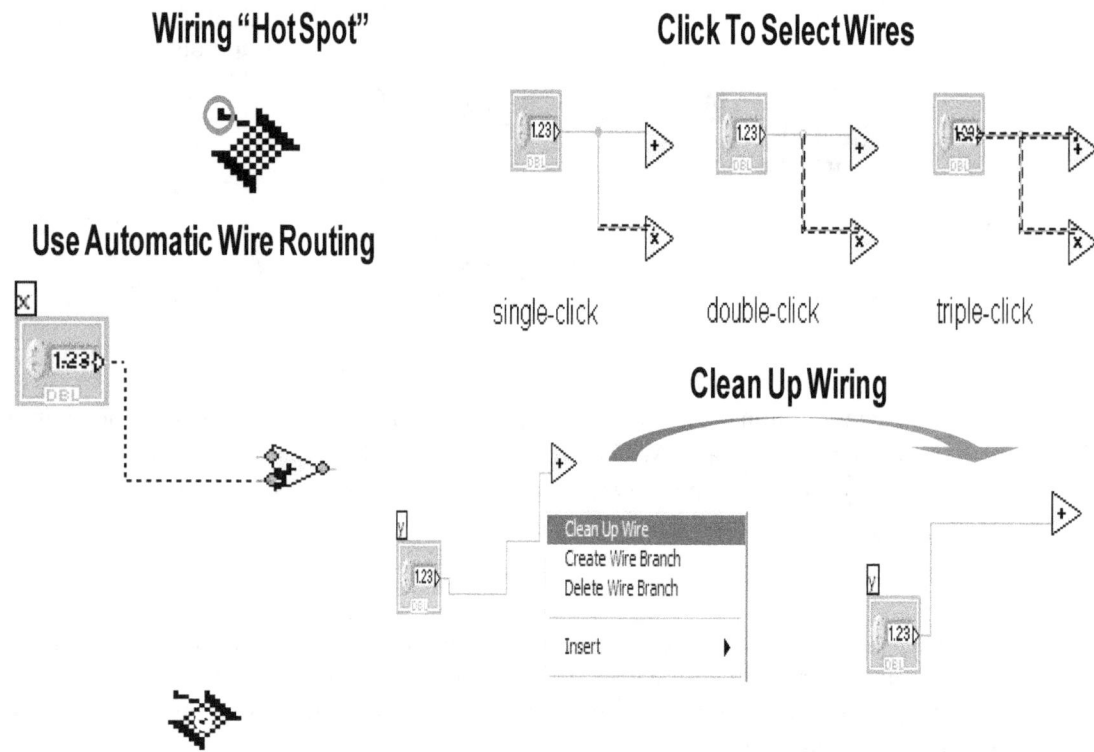

single-click   double-click   triple-click

Use Automatic Wire Routing

Clean Up Wiring

Clean Up Wire
Create Wire Branch
Delete Wire Branch

Insert

The following Figure 1-8 is shown the wire's types , style , thickness and colors are used in LabVIEW block diagrams.

The Color, Style, and Thickness of Common Wires

Wire Type	Scalar	1D Array	2D Array	Color
Floating Point				Orange
Integer				Blue
Boolean				Green
String				Pink
Error				Yellow

A "broken wire" represents a data type conflict that LabVIEW cannot automatically resolve. Fix it, or your code won't run!

Figure 1-8

In addition to Front Panel terminals, the Block diagram contains functions. Each function may have multiple input and output terminals. Wiring to these terminals is an important part of LabVIEW programming. Once you have some experience programming in LabVIEW, wiring will become easy. At first, you may need some assistance. Here are some tips to get you started:

- The wiring tool is used to wire to the nodes of the functions. When you "aim" with the wiring tool, aim with the end of the wire hanging from the spool. This is where the wire will be placed.
- As you move the wiring tool over functions, watch for the yellow tip strip. This will tell you the name of the terminal you are wiring to.
- As you move the wiring tool over a terminal, it will flash. This will help you identify where the wire will attach.
- For more help with the terminals, right-click on the function and select **Visible Items>>Terminals**. The function's picture will be pulled back to reveal the connection terminals. Notice the colors- these match the data types used by the front panel terminals.
- For additional help, select **Help>>Show Context Help**, or press **CTRL+H**. This will bring up the context help window. As you move your mouse over the function, this window will show you the function, terminals, and a brief help description. Use this with the other tools to help you as you wire.

- If your wiring becomes doesn't look very good, right-click on the particular wire in question and choose **Clean Up Wire** to automatically re-route that wire.

1.3 Dataflow Programming :

LabVIEW follows a dataflow model for running VIs. A block diagram node executes when all its inputs are available. When a node completes execution, it supplies data to its output terminals and passes the output data to the next node in the dataflow path. Visual Basic, C++, JAVA, and most other text-based programming languages follow a control flow model of program execution. In control flow, the sequential order of program elements determines the execution order of a program as following:

- Block diagram executes dependent on the flow of data; block diagram does NOT execute left to right

- Node executes when data is available to ALL input terminals

- Nodes supply data to all output terminals when done

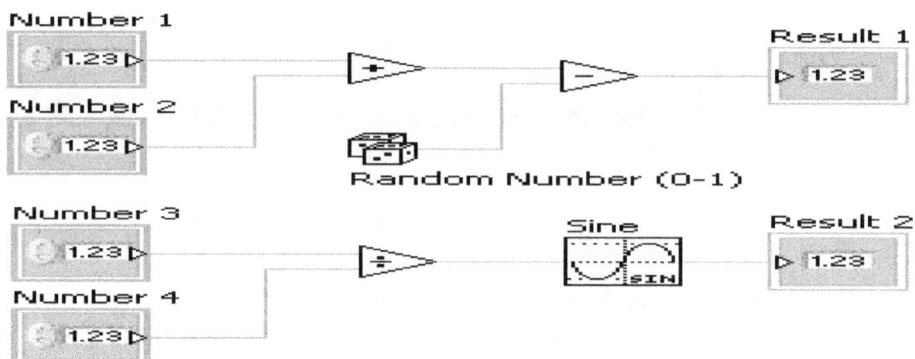

Consider the block diagram above; It adds two numbers and then subtracts 50.0 from the result of the addition. In this case, the block diagram executes from left to right, not because the objects are placed in that order, but because one of the inputs of the Subtract function is not valid until the Add function has finished executing and passed the data to the Subtract function. Remember that a node executes only when data are available at all of its input terminals, and it supplies data to its output terminals only when it finishes execution.

In the code to the right, consider which code segment would execute first—the Add, Random Number, or Divide function. You cannot know because inputs to the Add and Divide functions are available at the same time, and the Random Number function has no inputs. In a situation where one code segment must execute before another, and no data dependency exists between the functions, use a Sequence structure to force the order of execution.

Example 3 - Convert °C to °F:

This example is easy, but since it will be the first VI that's to create, it is good to allow ample time to explore the LabVIEW environment. Instructions: Build a VI that converts °C to °F. When run, the VI should take an input value (°C), multiply it by 1.8, add 32, and display the result (°F). The front panel should display both the input value and the result. Save the VI as Convert C to F.vi, Figure 1-9 shows this example.

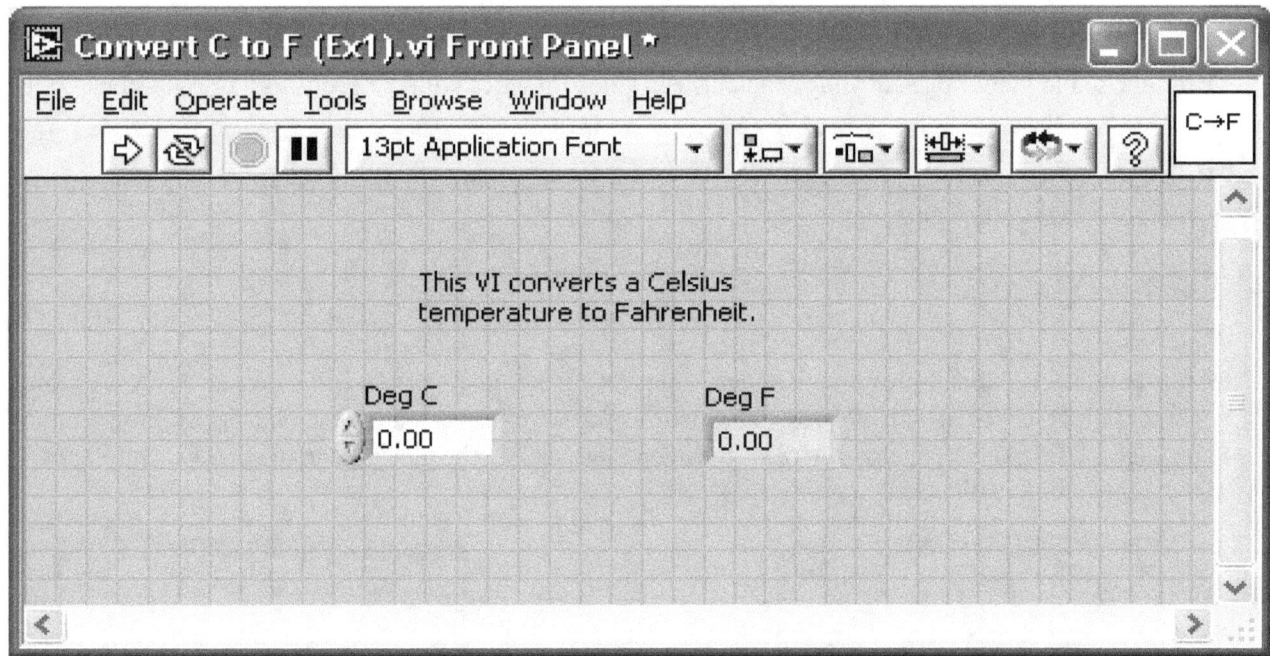

Figure 1-9 Convert C to F VI front panel

1.4 Debugging Techniques:

When your VI is not executable, a broken arrow is displayed in the Run button in the palette.

- **Finding Errors**

  Click on broken Run button, Window showing error appears

- **Execution Highlighting**

  Click on Execution Highlighting button; data flow is animated using bubbles. Values are displayed on wires.
- **Probe**

 Right-click on wire to display probe and it shows data as it flows through wire segment.

select Probe tool from Tools palette and click on wire

Finding Errors: To list errors, click on the broken arrow. To locate the bad object, click on the error message.

Execution Highlighting: Animates the diagram and traces the flow of the data, allowing you to view intermediate values, Click on the light bulb on the toolbar.

Probe: Used to view values in arrays and clusters, Click on wires with the Probe tool or right-click on the wire to set probes.

Breakpoint: Set pauses at different locations on the diagram, Click on wires or objects with the Breakpoint tool to set breakpoints.

1.5 Sub-VIs:

After building a VI and create its icon and connector pane, one can use it in another VI. A VI within another VI is called a sub-VI. A sub-VI corresponds to a subroutine in text-based programming languages. Using sub-VIs helps you manage changes and debug the block diagram quickly. The Figure 1-10 is shown the sub-VI configuration.

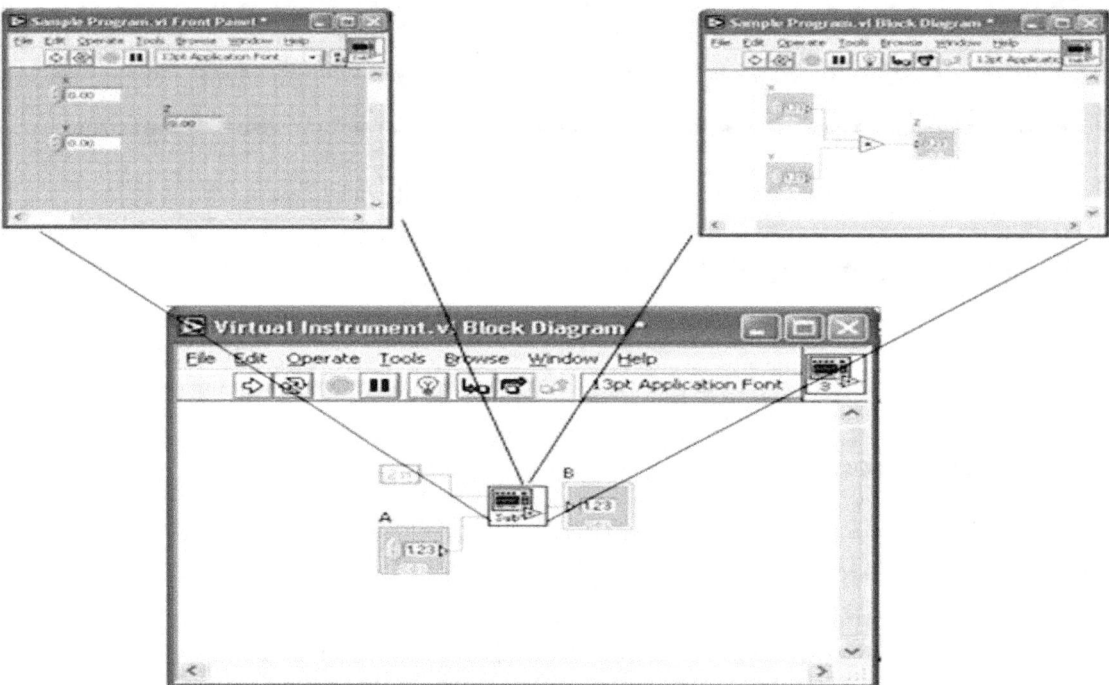

Figure 1-10 Sub VI

12

- A Sub-VI is a VI that can be used within another VI
- Similar to a subroutine
- Advantages
    - Modular

    - Easier to debug

    - Don't have to recreate code

    - Require less memory

The following Figure 1-11 shows a block diagram contains two sub-VIs. To see the front panel of a sub-VI, simply double click the sub-VI. You can also view the hierarchy of sub-VIs within a top level VI by clicking on Browse » Show VI Hierarchy.

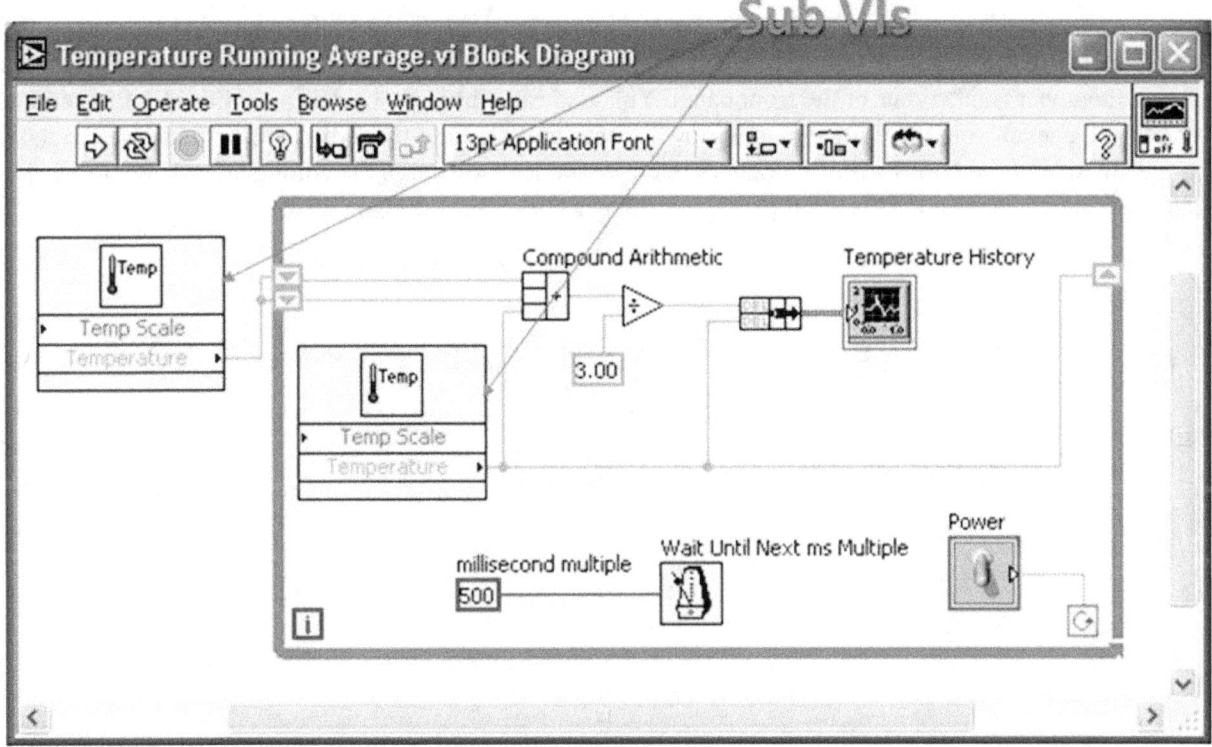

Figure 1-11

There are some steps to create Sub-VI , the following steps are showing in details for this important process.

> ## Steps to Create a Sub-VI:

- Create the Icon
- Create the Connector
- Assign Terminals

13

- Save the VI
- Insert the VI into a Top Level VI

Create the Icon

Right-click on the icon in the block diagram or front panel

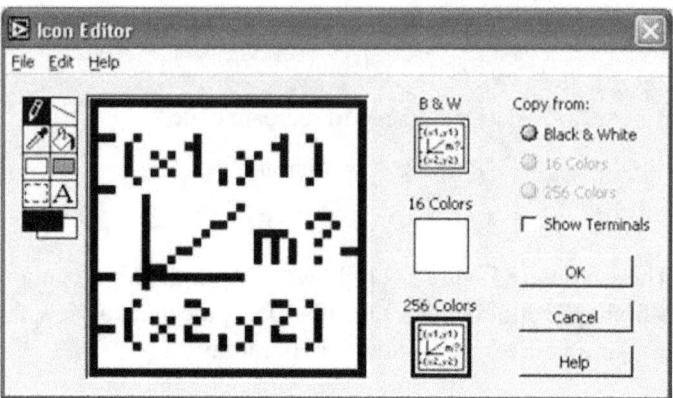

Create custom icons to replace the default icon by right-clicking the icon in the upper right corner of the front panel or block diagram and selecting **Edit Icon** from the shortcut menu or by double-clicking the icon in the upper right corner of the front panel. You also can edit icons by selecting **File»VI Properties**, selecting **General** from the **Category** pull-down menu, and clicking the **Edit Icon** button. Use the tools on the left side of the **Icon Editor** dialog box to create the icon design in the editing area. The normal size image of the icon appears in the appropriate box to the right of the editing area.

Create the Connector

Right click on the icon pane

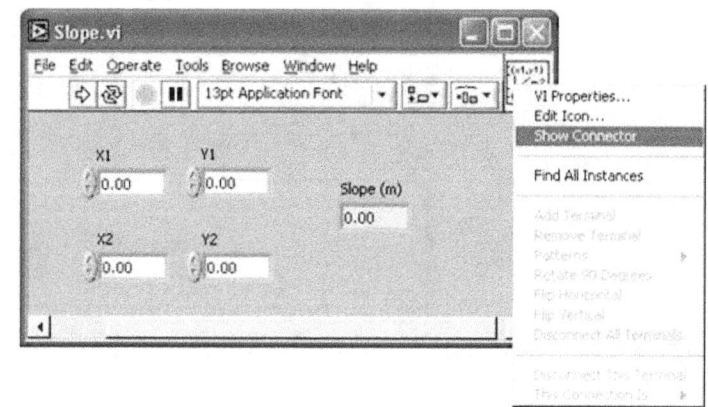

To use a VI as a sub-VI, you need to build a connector pane. The connector pane is a set of terminals that corresponds to the controls and indicators of that VI, similar to the parameter list of a function call in text-based programming languages. The connector pane defines the inputs and outputs you can wire to the VI so you can use it as a sub-VI. Define connections by assigning a front panel control or indicator to each of the connector pane terminals. To define a connector pane, right-click the icon in the upper right corner of the front panel window and select **Show Connector** from the shortcut menu. The connector pane replaces the icon. Each rectangle on the connector pane represents a terminal. Use the rectangles to assign inputs and outputs. The number of terminals LabVIEW displays on the connector pane depends on the number of controls and indicators on the front panel. The above front panel has four controls and one indicator, so LabVIEW displays four input terminals and one output terminal on the connector pane.

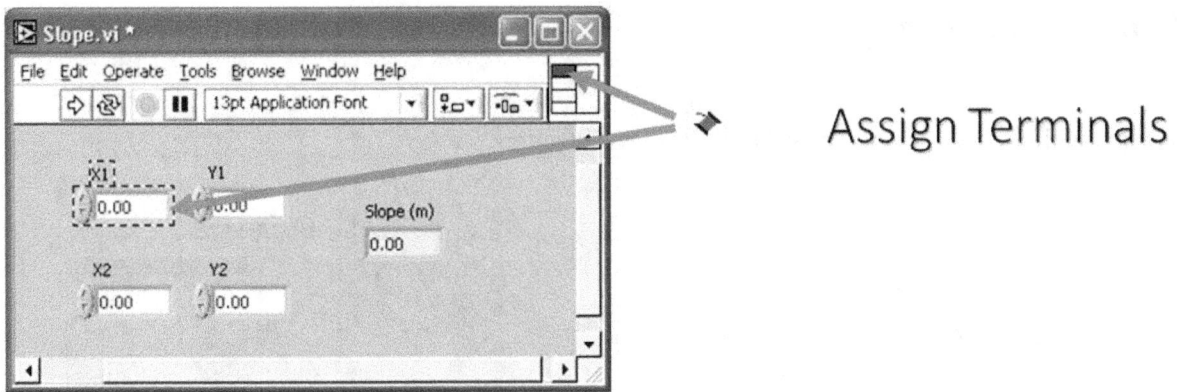

Assign Terminals

After you select a pattern to use for your connector pane, you must define connections by assigning a front panel control or indicator to each of the connector pane terminals. When you link controls and indicators to the connector pan, place inputs on the left and outputs on the right to prevent complicated, unclear wiring patterns in your VIs . To assign a terminal to a front panel control or indicator, click a terminal of the connector pane. Click the front panel control or indicator you want to assign to the terminal. Click an open area of the front panel. The terminal changes to the data type color of the control to indicate that you connected the terminal. You also can select the control or indicator first and then select the terminal. Make sure you save the VI after you have made the terminal assignments.

➤ Save The VI & Insert the Sub VI into a Top Level VI

There are several ways to organize your sub VIs. The most common way is to organize by application. In this case, all the VI's for a particular application are saved into the same directory or into a VI Library file. Saving into a library file allows you to transport an entire application within a single file. Saving into library is simple. After clicking Save As..., click New VI Library. This will allow you to name the library, and then save your VI into it. To add subsequent VI's, simply double click the **.llb** file from the standard Save window, and give the VI a name.

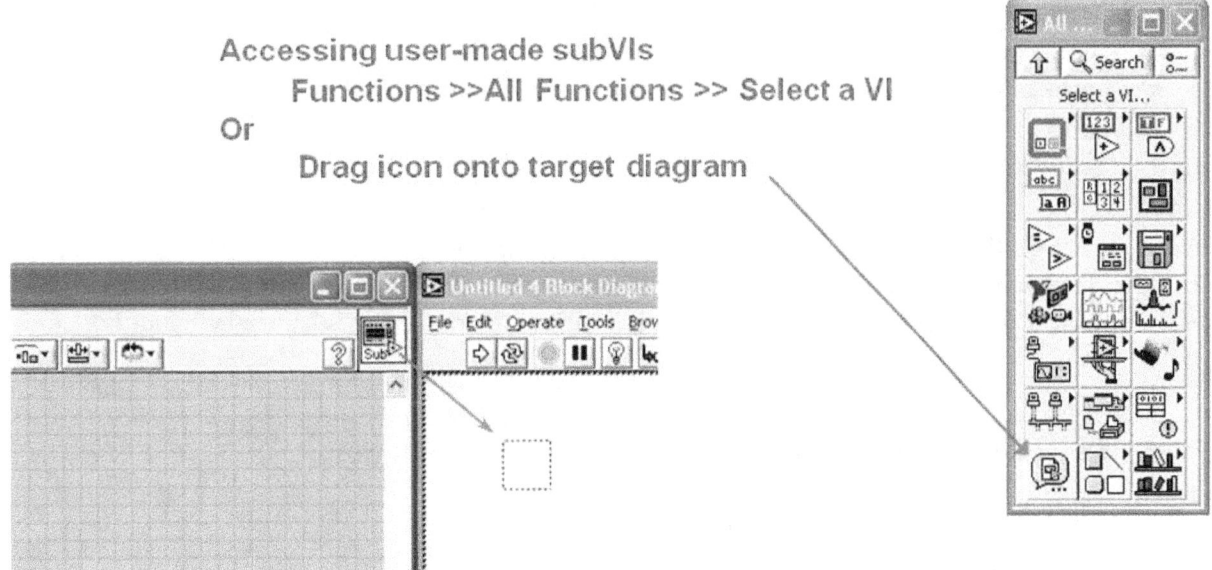

Accessing user-made subVIs
    Functions >>All Functions >> Select a VI
Or
    Drag icon onto target diagram

1.6 Data Acquisition:

This topic will discuss in more details later in the next chapters.

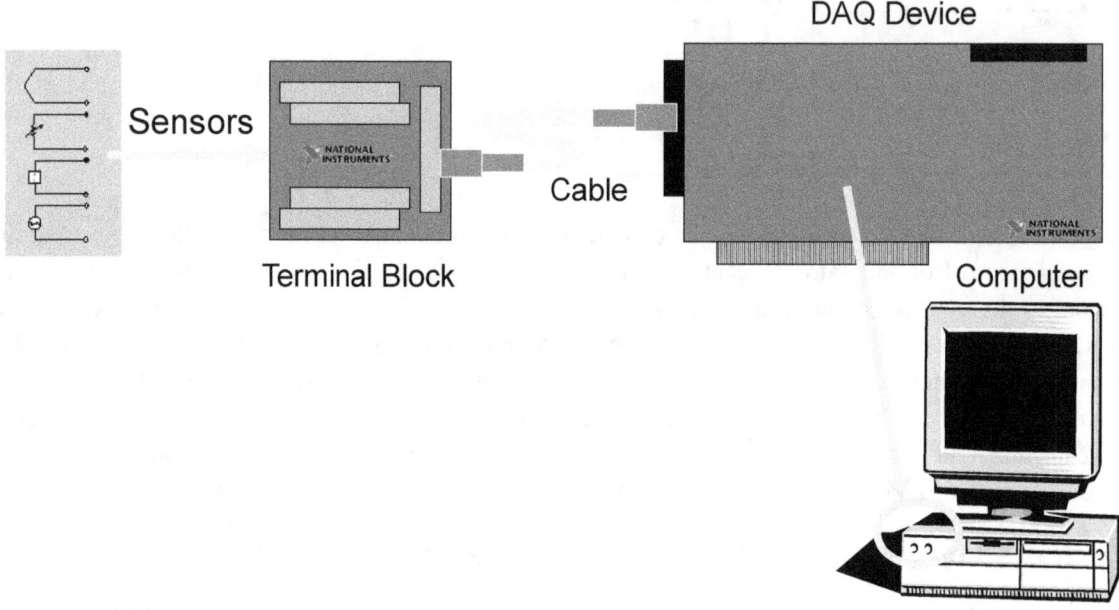

Data acquisition is the process of sampling signals that measure real world physical conditions and converting the resulting samples into digital numeric values that can be manipulated by a computer. Data acquisition systems (DAQ) typically convert analog waveforms into digital values for processing. The components of data acquisition systems include:

- Sensors that convert physical parameters to electrical signals.
- Signal conditioning circuitry to convert sensor signals into a form that can be converted to digital values.
- Analog-to-digital converters, which convert conditioned sensor signals to digital values.

The Data Acquisition palette in LabVIEW contains a palette for traditional NI-DAQ and one for NI-DAQmx. Traditional VIs are divided by the type of measurement; DAQmx VIs are divided by the type of task. The Data acquisition in LabVIEW has the following configurations , it is shown in Figure 1-12.

The devices should be configured for the computers in this class as following:

1. NI-DAQ software must be installed on the computer
2. It must have installed an E-series DAQ board and configured it using Measurement & Automation Explorer (MAX).

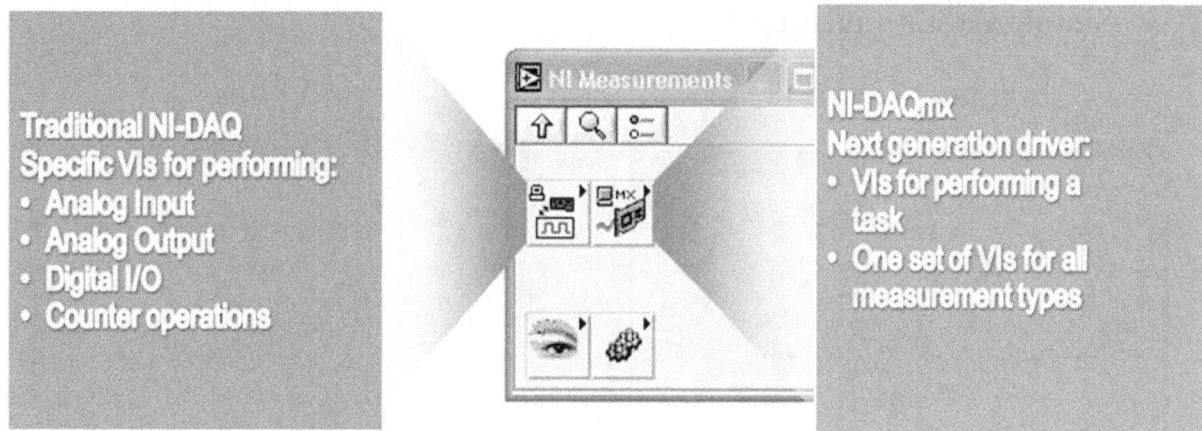

Figure 1-12 LabVIEW Data Acquisition configuration

➢ **Data Acquisition Terminology:**

➢ **Resolution** - Determines How Many Different Voltage Changes Can Be Measured

     – Larger Resolution → More Precise Representation of Signal

➢ **Range** - Minimum and Maximum Voltages

     – Smaller range → More Precise Representation of Signal

➢ **Gain** - Amplifies or Attenuates Signal for Best Fit in Range

Resolution: When acquiring data to a computer, an Analog-to-Digital Converter (ADC) takes an analog signal and turns it into a binary number. Therefore, each binary number from the ADC represents a certain voltage level. The ADC returns the highest possible level without going over the actual voltage level of the analog signal. Resolution refers to the number of binary levels the ADC can use to represent a signal. To Figure out the number of binary levels available based on the resolution you simply take $2^{Resolution}$. Therefore, the higher the resolution, the more levels you will have to represent your signal. For instance, an ADC with 3-bit resolution can measure 2^3 or 8 voltage levels, while an ADC with 12-bit resolution can measure 2^{12} or 4096 voltage levels.

Range: Unlike the resolution of the ADC, the range of the ADC is selectable. Most DAQ devices offer a range from 0 - +10 or -10 to +10. The range is chosen when you configure your device in NI-DAQ. Keep in mind that the resolution of the ADC will be spread over whatever range you choose. The larger the range, the more spread out your resolution will be, and you will get a worse representation of your signal. Thus it is important to pick your range to properly fit your input signal.

➢ Simple Data Acquisition:

Complete Convert C to F.vi, then create Thermometer

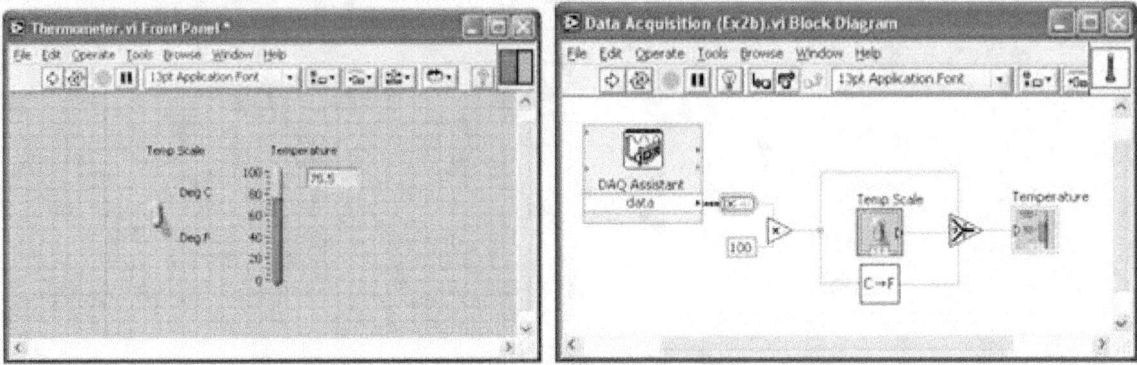

- **Create** an icon and connector for Convert C to F.vi. The icon should remind you of the functionality of the VI (e.g. C→F or C to F). The connector should have one input and one output, allowing a terminal for °C in, and °F out.

- **Create** a top level VI that acquires a data point from channel 0 (the temperature sensor) of your DAQ board and allows the user to display the temperature in Celsius or Fahrenheit. To do it this will need to acquire a single data point from your DAQ board and scale it by a factor of 100. This will give you °C. one should have a Boolean switch or button that allows the user to select Celsius or Fahrenheit. If the user selects Celsius, the scaled value should be displayed in a thermometer indicator. If the user selects Fahrenheit, the Celsius value should be passed into Convert C to F.vi (used as a sub-VI), and the output Fahrenheit value should be displayed.

Chapter 2

LabVIEW Terminology

➢ **Loops and Charts**

Both the While and For Loops are located on the **Functions » Structures** palette

- For Loops
  - Have Iteration Terminal
  - Run According to input **N** of Count Terminal

- While Loops
  - Have Iteration Terminal
  - Always Run at least Once
  - Run According to Conditional Terminal

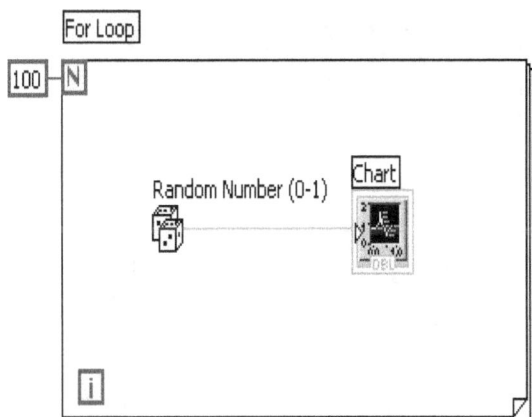

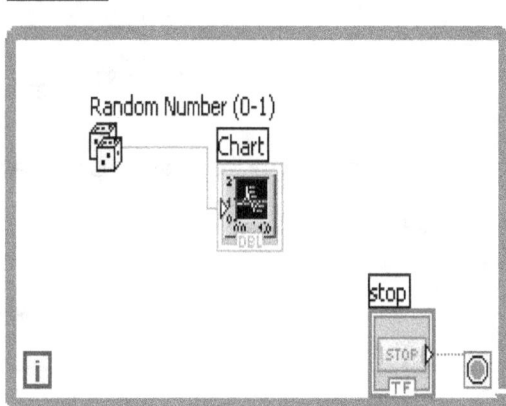

While Loops

Similar to a Do Loop or a Repeat-Until Loop in text-based programming languages, a While Loop, shown at the top right, executes a sub diagram until a condition is met. The While Loop executes the sub diagram until the conditional terminal, an input terminal, receives a specific Boolean value. The default behavior and appearance of the conditional terminal is **Continue If True**, shown at left. When a conditional terminal is **Continue If True**, the While Loop executes its sub diagram until the conditional terminal receives a FALSE value. The iteration terminal (an output terminal), shown at left, contains the number of completed iterations. The iteration count always starts at zero. During the first iteration, the iteration terminal returns 0.

For Loops

A For Loop, shown at left, executes a sub diagram a set number of times. The value in the count terminal (an input terminal) represented by the N, indicates how many times to repeat the sub diagram. The iteration terminal (an output terminal), shown at left, contains the number of completed iterations. The iteration count always starts at zero. During the first iteration, the iteration terminal returns 0.

2.1 Selection of Loops :

Place loops in your diagram by selecting them from the Structures palette of the Functions palette, Figure 2-1 shows the required steps:

- When selected, the mouse cursor becomes a special pointer that you use to enclose the section of code you want to repeat.
- Click the mouse button to define the top-left corner, click the mouse button again at the bottom-right corner, and the While Loop boundary is created around the selected code.
- Drag or drop additional nodes in the While Loop if needed.

1. Select the loop

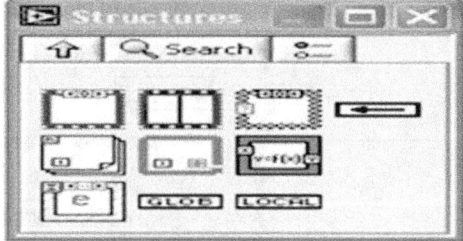

2. Enclose code to be repeated

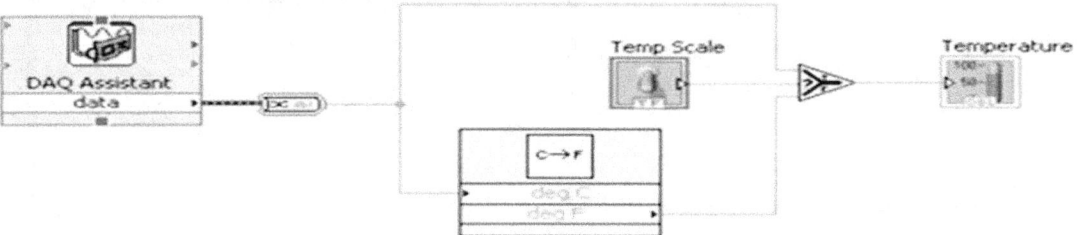

3. Drop or drag additional nodes and then wire

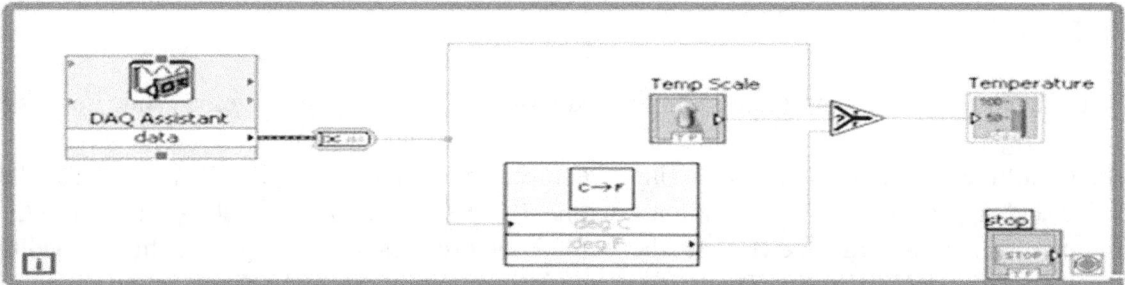

Figure 2-1

20

2.2 Charts :

Waveform charts can display single or multiple plots. The following Figure 2-2 of the front panel shows an example of a multi-plot waveform chart. It can change the min and max values of either the x or y axis by double clicking on the value with the labeling tool and typing the new value. Similarly, it can change the label of the axis, also right click the plot legend and change the style, shape, and color of the trace that is displayed on the chart.

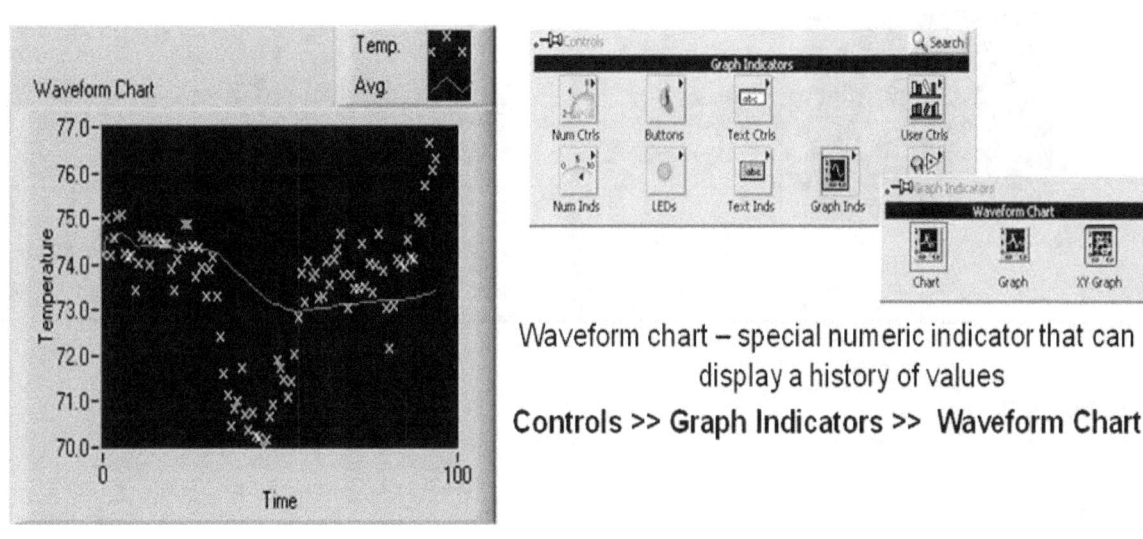

Figure 2-2

2.3 Wiring Data into Charts:

It can wire a scalar output directly to a waveform chart to display one plot. To display multiple plots on one chart, use the Merge Signals function found in the Functions >> Signal Manipulation palette. The Merge Signal function bundles multiple outputs to plot on the waveform chart. To add more plots, use the Positioning tool to resize the Merge Signal function.

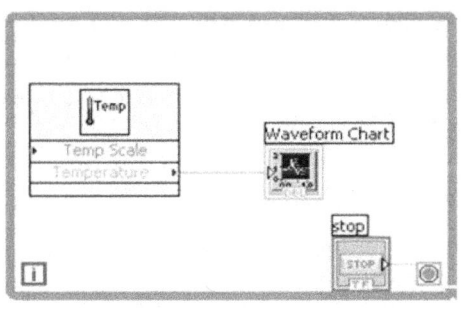

Single plot Charts

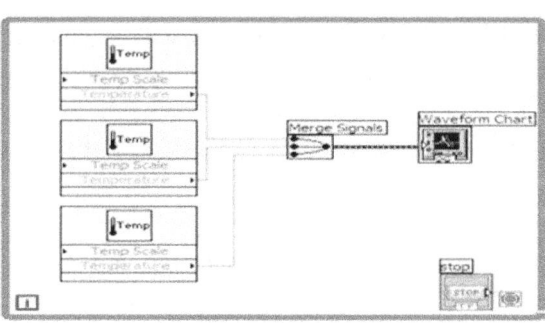

Multi plot Charts

Figure 2-3

21

Example 1 - Using loops:

Create a VI that generates a random number at a specified rate and displays the readings on a Waveform Chart until stopped by the user. Connect the termination terminal to a front panel stop button, and add a slider control to the front panel. The slider control should range from 0 to 2000 in value, and be connected to the Time Delay , Express VI function inside your while loop. Save the VI as Use a loop.vi.

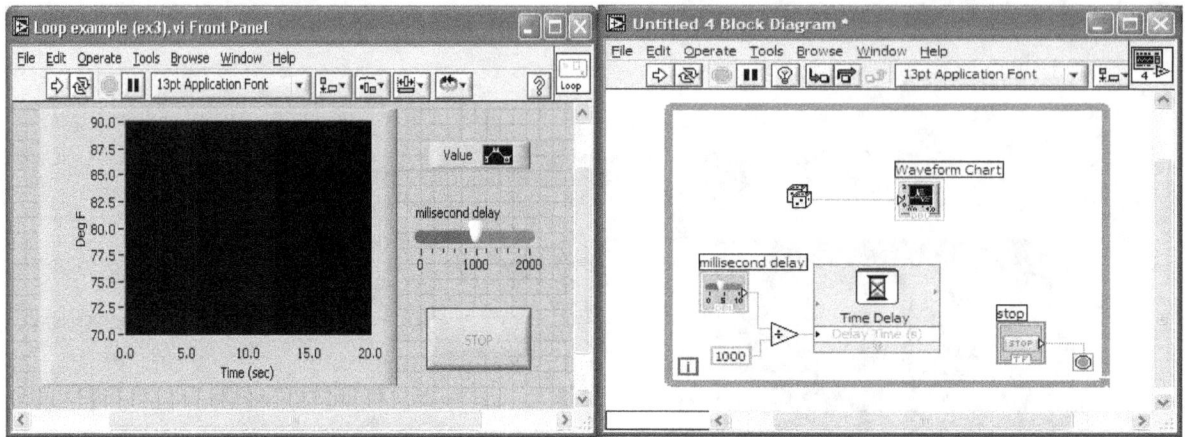

2.4 Arrays & File I/O:

- Build arrays manually
- Have LabVIEW build arrays automatically
- Write to a spreadsheet file
- Read from a spreadsheet file

Arrays are group of data elements of the same type. An array consists of elements and dimensions. Elements are the data that make up the array. A dimension is the length, height, or depth of an array. An array can have one or more dimensions and as many as $2^{31} - 1$ elements per dimension, memory permitting. It can build arrays of numeric, Boolean, path, string, waveform, and cluster data types.

Consider using arrays when one work with a collection of similar data and when one perform repetitive computations. Arrays are ideal for storing data that collects from waveforms or data generated in loops, where each iteration of a loop produces one element of the Array. Besides, an Array elements are ordered. An array uses an index so you can readily access any particular element.

The index is zero-based, which means it is in the range 0 to $n - 1$, where n is the number of elements in the array. For example, $n = 9$ for the nine planets, so the index ranges from 0 to 8. Earth is the third planet, so it has an index of 2. File I/O operations pass data to and from files. Use the File I/O VIs and functions located on the **Functions » File I/O** palette to handle all aspects of file I/O.

2.4.1 Adding an Array to the Front Panel:

To create an array control or indicator as shown, select an array on the **Controls » All Controls » Array & Cluster** palette, place it on the front panel, and drag a control or indicator into the array shell. If one attempt to drag an invalid control or indicator such as an XY graph into the array shell, you are unable to drop the control or indicator in the array shell.

It must insert an object in the array shell before it uses the array on the block diagram. Otherwise, the array terminal appears black with an empty bracket.

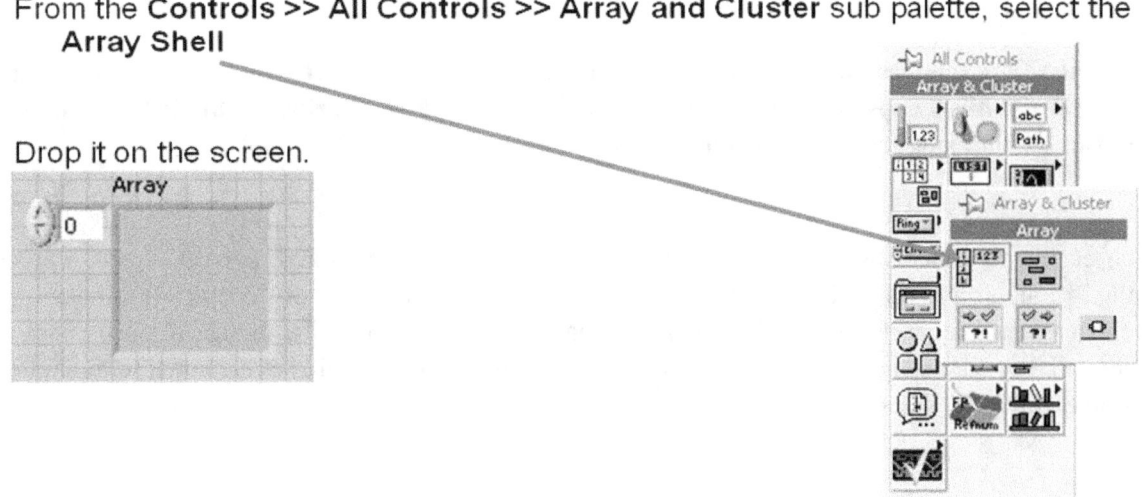

2.4.2 Place data object into shell (Numeric Control):

To add dimensions to an array one at a time, right-click the index display and select **Add Dimension** from the shortcut menu. Also , it can use the Positioning tool to resize the index display until you have as many dimensions as you want.

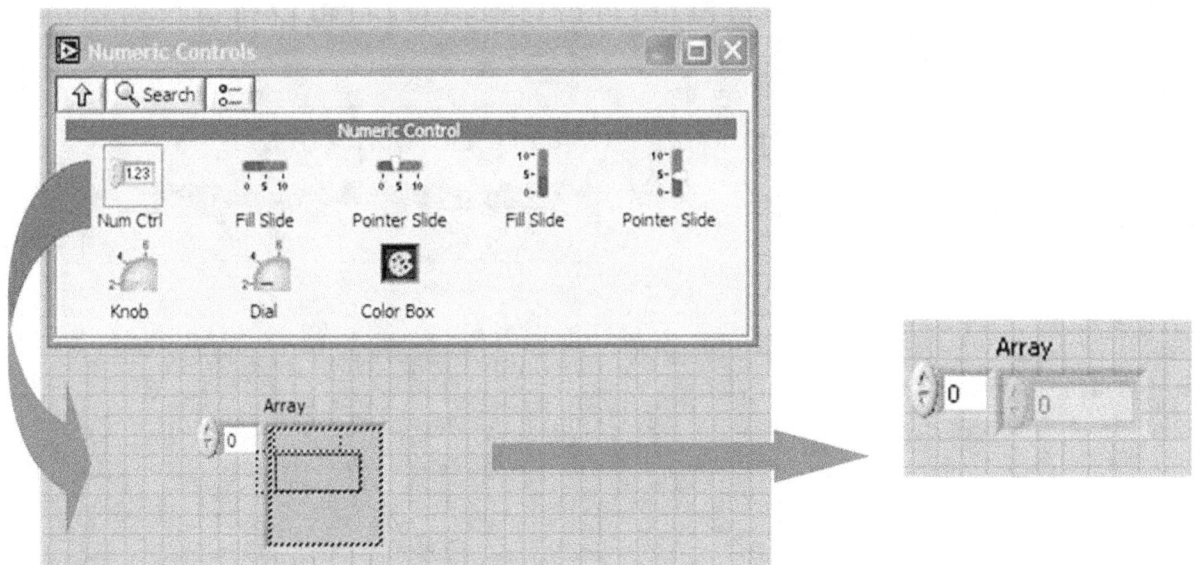

2.4.3 Creating an Array with a Loop:

If someone wire an array to a For Loop or While Loop input tunnel, it can read and process every element in that array by enabling auto-indexing. When an auto-index an array output tunnel, the output array receives a new element from every iteration of the loop. The wire from the output tunnel to the array indicator becomes thicker as it changes to an array at the loop border, and the output tunnel contains square brackets representing an array. Disable auto-indexing by right-clicking the tunnel and selecting **Disable Indexing** from the shortcut menu.

For example, disable auto-indexing if someone need only the last value passed to the tunnel, without creating an array.

Note: LabVIEW enables auto-indexing by default for every array that wire to a For Loop. Auto-indexing for While Loops is disabled by default. To enable auto-indexing, right-click a tunnel and select Enable Indexing from the shortcut menu.

If you enable auto-indexing on an array wired to a For Loop input terminal, LabVIEW sets the count terminal to the array size so it do not need to wire the count terminal. If one enable auto-indexing for more than one tunnel or if you wire the count terminal, the count becomes the smaller of the choices. For example, if you wire an array with 10 elements to a For Loop input tunnel , set the count terminal to 15, the loop executes 10 times.

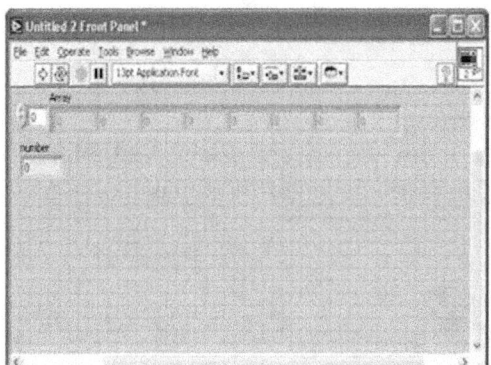 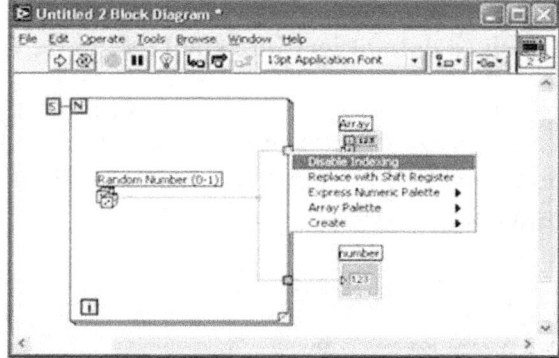

Loops accumulate arrays at their boundaries

2.4.4 Creating 2D Arrays:

It can use two For Loops, one inside the other, to create a 2D array. The outer For Loop creates the row elements, and the inner For Loop creates the column elements.

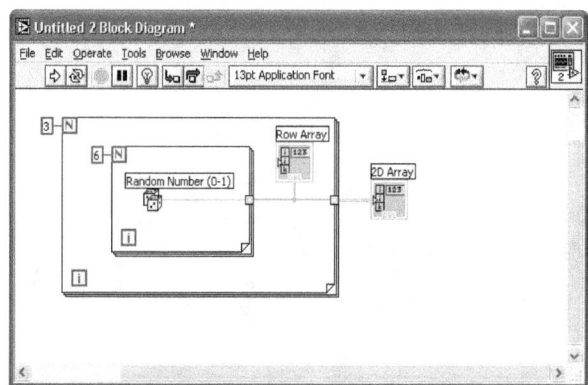

24

2.5 File I/O:

- Passing data to and from files
- Files can be binary, text, or spreadsheet
- Write/Read LabVIEW Measurements file (*.lvm)

File I/O operations pass data to and from files. In LabVIEW, it can use File I/O functions to:

- Open and close data files
- Read data from and write data to files
- Read from and writ to spreadsheet-formatted files
- Move and rename files and directories
- Change file characteristics
- Create, modify, and read a configuration file
- Write to or read from LabVIEW Measurements files

2.6 Write LabVIEW Measurement File:

The Write LVM file can write to spreadsheet files. However, its main purpose is for logging data, that will be used in LabVIEW. This VI creates a .lvm file which can be opened in a spreadsheet application. For simple spreadsheet files, use the Express VIs: Write LVM and Read LVM.

- Includes the open, write, close and error handling functions.
- Handles formatting the string with either a tab or comma delimiter.
- Merge Signals function is used to combine data into the dynamic data type.

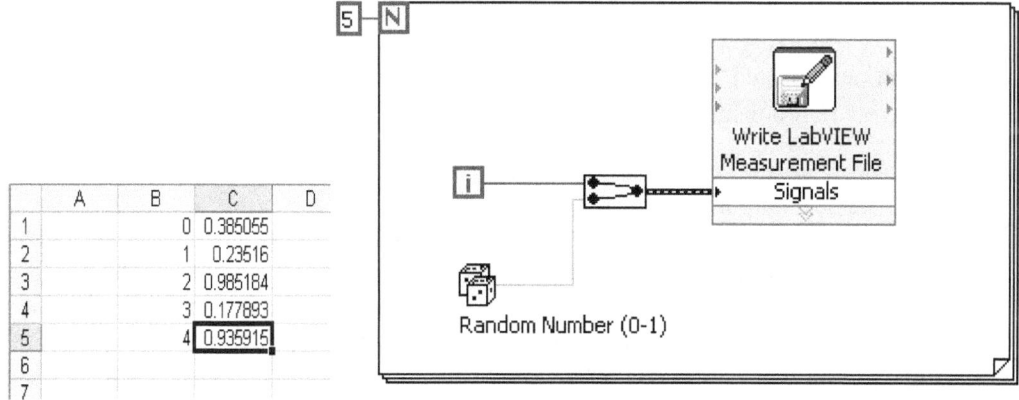

Example 2 – Analyzing and Logging Data:

Create a VI that acquires and displays temperature data at a fixed rate until stopped by the user. It can use the Digital Thermometer.vi from the Tutorial sub palette of the functions palette. Once stopped, the VI should perform analysis on the data it collected while running. Build up an array of data points and values on the tunnel border of the while loop.

Find the maximum, minimum, and mean value of the temperature data and display them in numeric indicators ; **Functions » Analyze» Mathematics» Probability and Statistics**, and the **Array Max & Min** function can be found in **Functions » Array**.

Use the Write LabVIEW Measurements File Express VI, which can be found at **Functions » Output.** Once it is run, verify that the file was properly created by opening it in Notepad or by creating a VI that reads it back using the Read LabVIEW Measurements File.

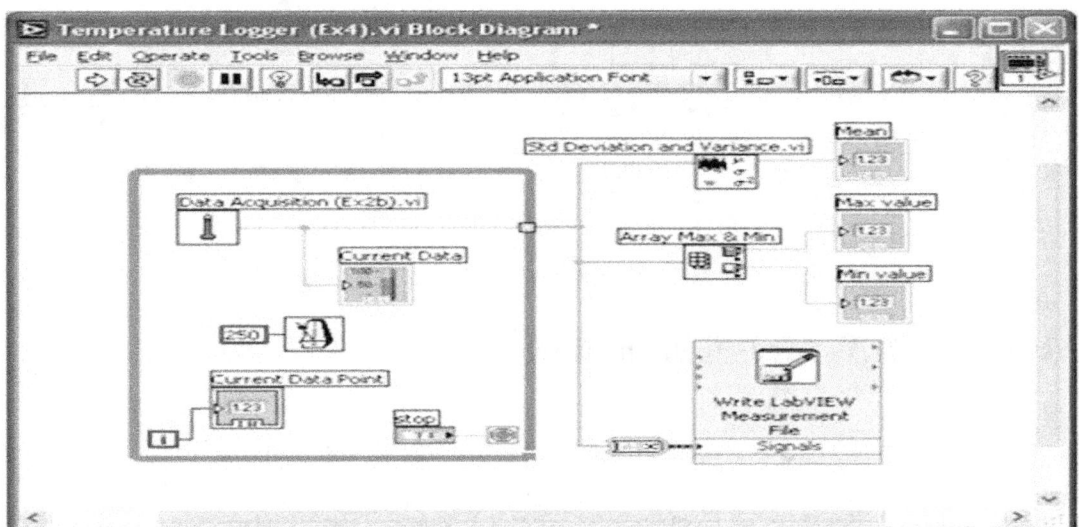

Temperature Logger.vi

2.7 Array Functions & Graphs:

Use the Array functions located on the **Functions» All Functions »Array** palette to create and manipulate arrays. Array functions include the following:

- **Array Size**—Returns the number of elements in each dimension of an array. If the array is *n*-dimensional, the **size** output is an array of *n* elements.
- **Initialize Array**—Creates an *n*-dimensional array in which every element is initialized to the value of **element**. Resize the function to increase the number of dimensions of the output array.
- **Build Array**—Concatenates multiple arrays or appends elements to an *n*-dimensional array. Resize the function to increase the number of elements in the output array.
- **Array Subset**—Returns a portion of an array starting at **index** and containing **length** elements.

26

- **Index Array**—Returns an element of an array at **index**. Also can use the Index Array function to extract a row or column of a 2D array to create a sub array of the original. To do so, wire a 2D array to the input of the function. Two **index** terminals are available. The top **index** terminal indicates the row, and the second terminal indicates the column. It can wire inputs to both **index** terminals to index a single element, or wire only one terminal to extract a row or column of data.

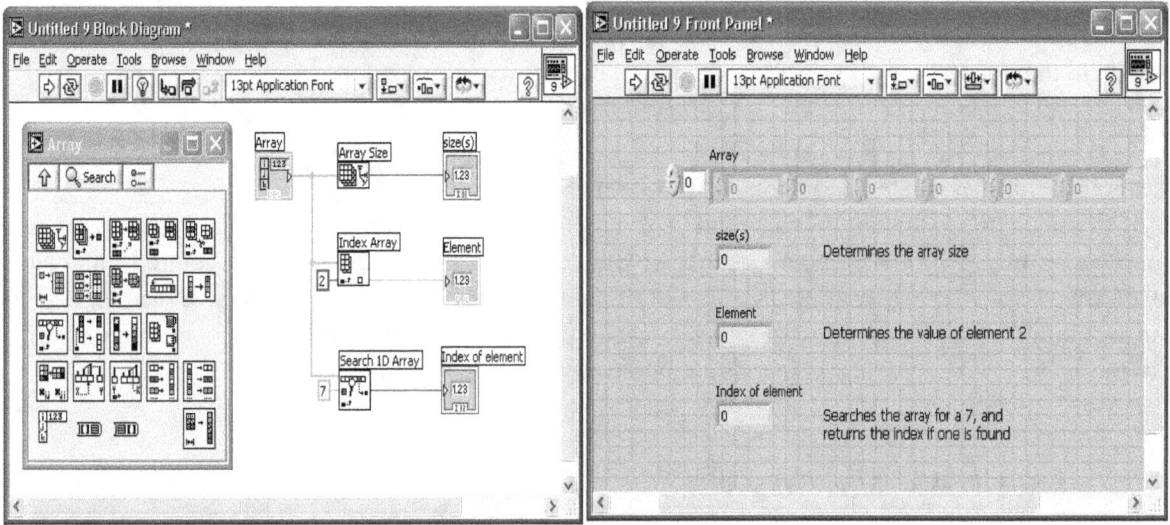

2.7.1 Array Functions – Build Array:

Build Array can perform two distinct functions. It Concatenates multiple arrays or appends elements to an *n*-dimensional array. Resize the function to increase the number of elements in the output array. To concatenate the inputs into a longer array of the same dimension as shown in the following array, right-click the function node and select **Concatenate Inputs** from the shortcut menu.

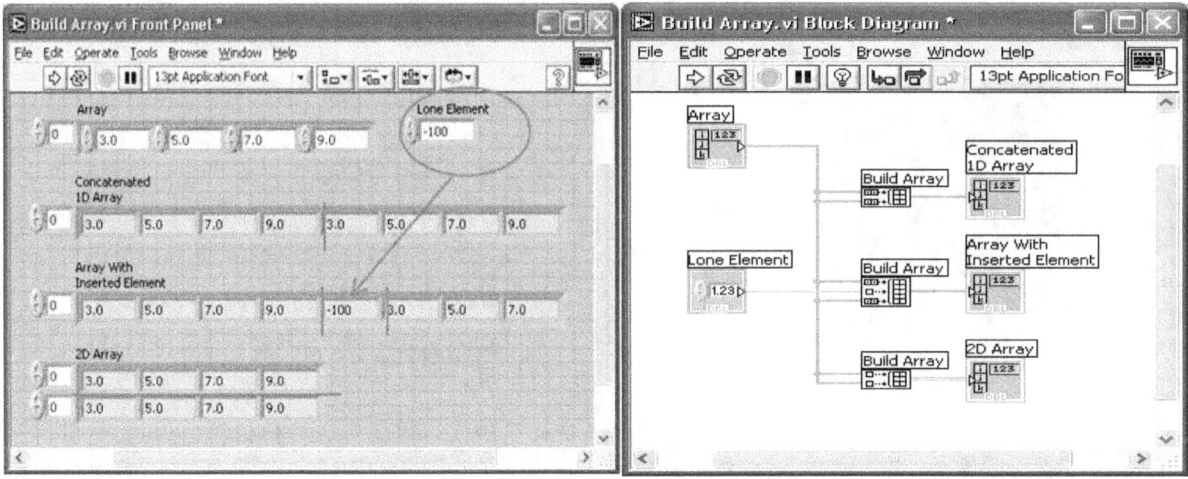

27

2.7.2 Graphs:

- Waveform Graph – Plot an array of numbers against their indices
- Express XY Graph – Plot one array against another

Digital Waveform Graph – Plot bits from binary data

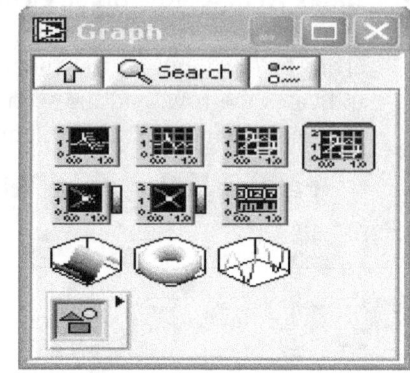

VIs with graphs usually collect the data in an array and then plot the data to the graph.

The graphs located on the **Controls» All Controls» Graph** palette include the waveform graph and XY graph. The waveform graph plots only single-valued functions, as in $y = f(x)$, with points evenly distributed along the x-axis, such as acquired time-varying waveforms. Express XY graphs display any set of points, evenly sampled or not. Resize the plot legend to display multiple plots. Use multiple plots to save space on the front panel and to make comparisons between plots. XY and waveform graphs automatically adapt to multiple plots.

> ➢ **Single-Plot Waveform Graphs**

The waveform graph accepts a single array of values and interprets the data as points on the graph and increments the x index by one starting at $x = 0$. The graph also accepts a cluster of an initial x value, a $.x$, and an array of y data.

Note: Right-Click on the Graph and choose Properties to Interactively Customize

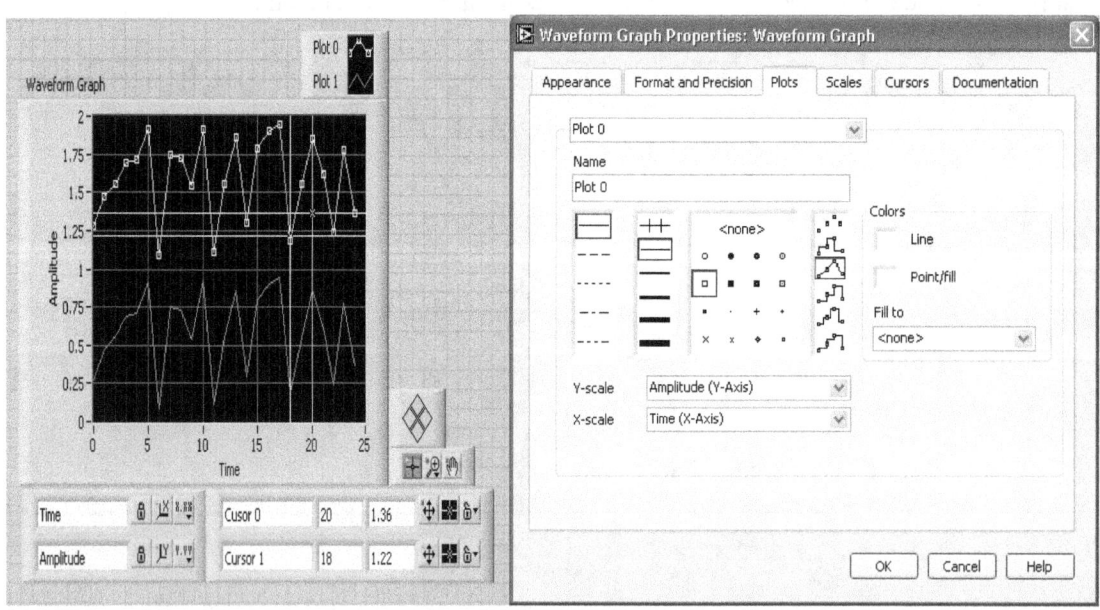

Graphs are very powerful indicators in LabVIEW. They are highly customizable, and can be used to concisely display a great deal of information. The properties page of the Graph allows you to display settings for plot types, scale and cursor options, and many other features of the graph.

Example 3 – Using Waveform Graphs:

The VI should use a while loop with a 100 millisecond delay to continuously generate sine and square waves and display them on a waveform graph. Use Simulate Signal Express VI from the **Functions »** **Input** palette to generate the signals. The frequency input to each function is chosen by the user.

- Change the colors, visible items, and plot styles of the graph.
- Experiment with some of the cursor and zooming options available.

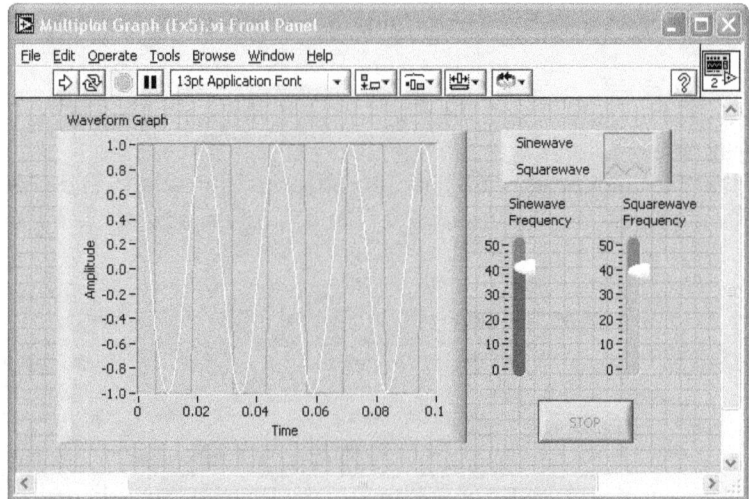

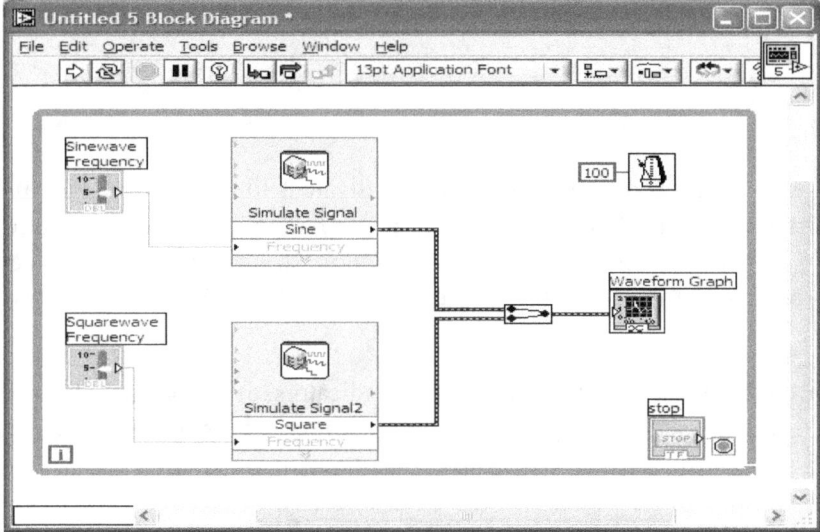

2.8 Strings, Clusters, & Error Handling:

A string is a sequence of displayable or non displayable (ASCII) characters. Strings often are used to send commands to instruments, to supply information about a test (such as operator name and date), or to display results to the user.

String controls and indicators are in the **Text Control** or **Text Indicator** sub palette of the **Controls** palette.

- Enter or change text by using the Operating or Text tool and clicking in the string control.
- Strings are resizable.
- String controls and indicators can have scrollbars: Right-click and select **Visible Items »**
  Scrollbar. The scroll bar will not be active if the control or indicator is not tall enough.

Strings

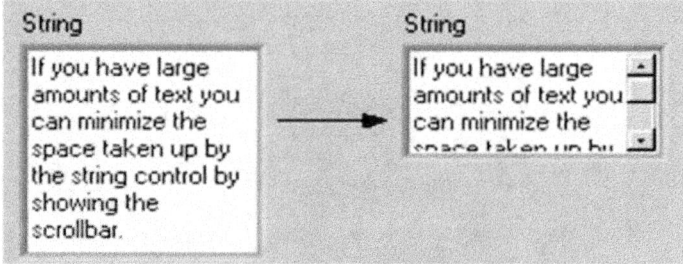

2.9 Clusters:

Clusters group like or unlike components together. Equivalent to a *record* in Pascal or a *struct* in C. Cluster components may be of different data types. Some Examples: Error information — Grouping a Boolean error flag, a numeric error code, and an error source string to specify the exact error. And, user information — Grouping a string indicating a user's name and an ID number specifying their security code.

All elements of a cluster must be either controls or indicators. It cannot have a string control and a Boolean indicator. Clusters can be thought of as grouping individual wires (data objects) together into a cable (cluster).

2.9.1 Creating a Cluster:

Demonstrate how to create a cluster front panel object by choosing **Cluster** from the **Controls » All Controls » Array & Cluster** palette.

- This option gives you a shell (similar to the array shell when creating arrays).
- It can be size the cluster shell when drop it.
- Right click inside the shell and add objects of any type.

Note: It can be even have a cluster inside of a cluster. The cluster becomes a control or an indicator cluster based on the first object you place inside the cluster.

Also, it can create a cluster constant on the block diagram by choosing **Cluster Constant** from the **Cluster** palette.

- This gives you an empty cluster shell.
- It could be size the cluster when you drop it.
- Put other constants inside the shell.

Note: It cannot place terminals for front panel objects in a cluster constant on the block diagram, nor can you place "special" constants like the Tab or Empty String constant within a block diagram cluster shell.

1. Select a **Cluster** shell

2. Place objects inside the shell

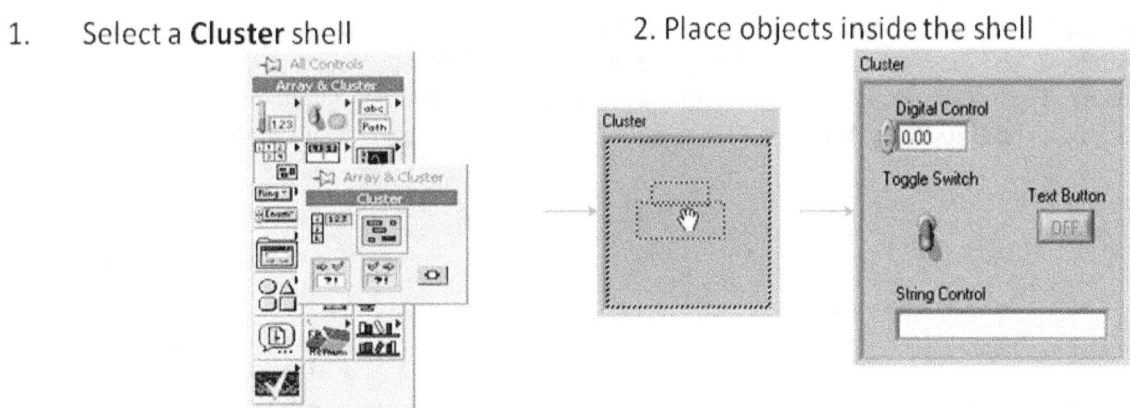

2.9.2 Cluster Functions:

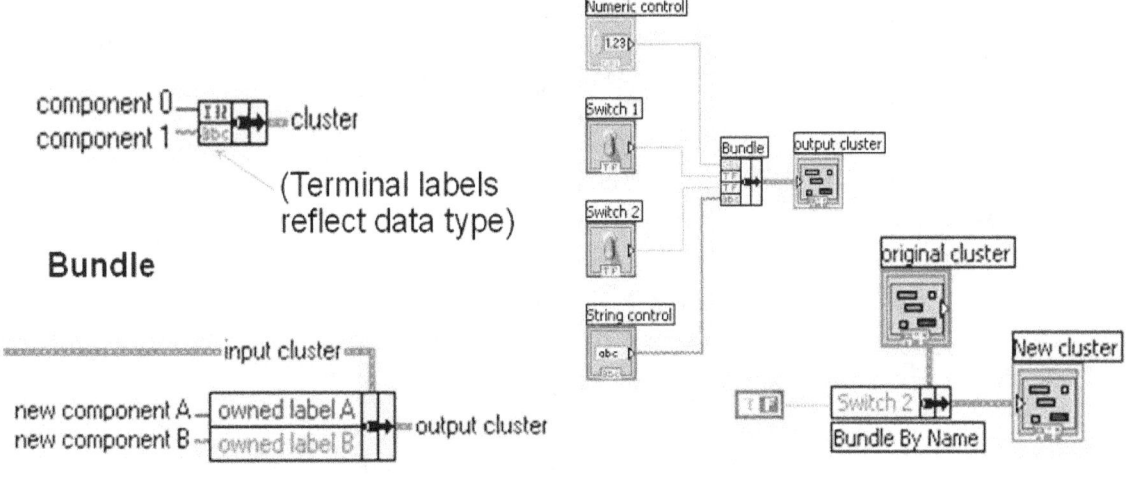

Bundle function—Forms a cluster containing the given objects.

Bundle by Name function—Updates specific cluster object values (the object must have an owned label).

Note: It must have an existing cluster wired into the middle terminal of the function to use Bundle By Name.

Unbundle function—Use to access *all* of the objects in the cluster.

Unbundle by Name function—Use to access *specific* objects (one or more) in the cluster.

Note: Only objects in the cluster having an owned label can be accessed. When unbundling by name, click on the terminal with the Operating tool to choose the element you want to access.

The **Unbundle** function must have exactly the same number of terminals as there are elements in the cluster. Adding or removing elements to the cluster breaks wires on the diagram. It can also obtain the **Bundle, Unbundle, Bundle by Name,** and **Unbundle by Name** functions by right clicking on the cluster terminal in the diagram and choosing **Cluster Tools** from the pop-up menu. When someone choose **Cluster Tools,** the **Bundle** and **Unbundle** functions automatically contain the correct number of terminals. The **Bundle by Name** and **Unbundle by Name** functions appear with the first cluster element.

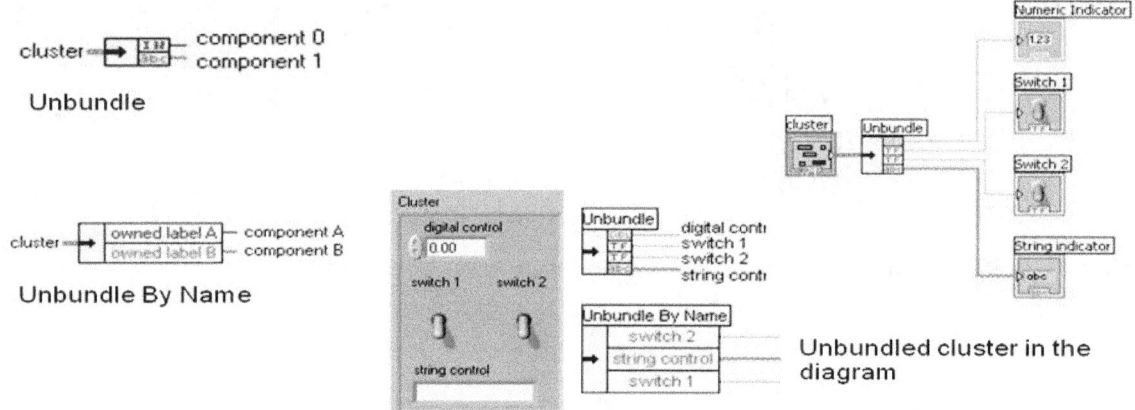

2.9.3 Error Clusters:

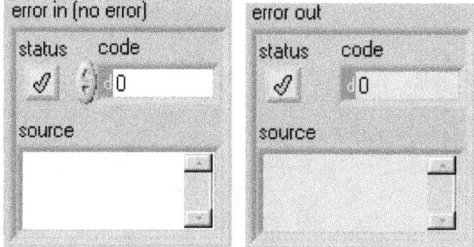

Figure 2-4 Error clusters are a powerful means of handling errors

32

The LabVIEW DAQ VIs, File I/O functions, networking VIs, and many other VI's use this method to pass error information between nodes. The error cluster contains the following elements:

- o **status**, a Boolean which is set to True if an error occurs.
- o **code**, a numeric which is set to a code number corresponding to the error that occurred.
- o **source**, a string which identifies the VI in which the error occurred.
- o Error information is passed from one sub VI to the next.
- o If an error occurs in one sub VI, all subsequent sub VIs are not executed in the usual manner.
- o Error Clusters contain all error conditions.
- o Automatic Error Handling.

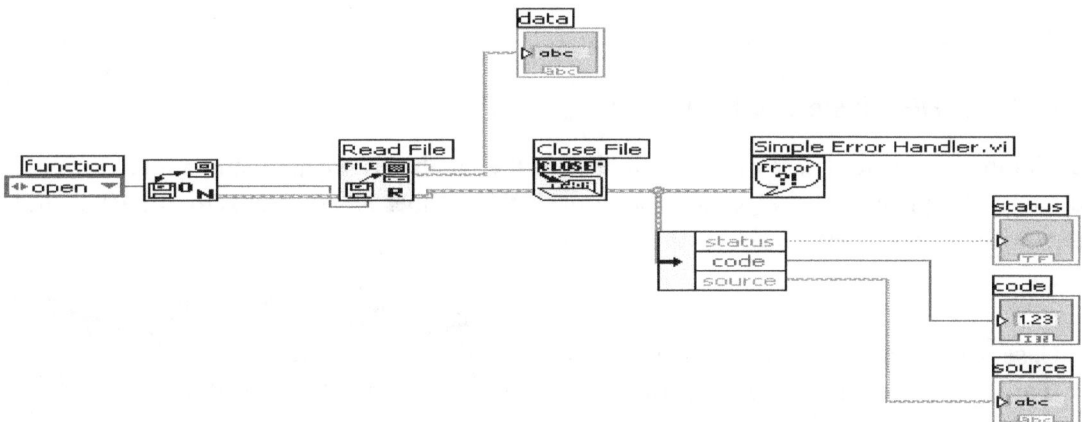

Error clusters are useful in determining the execution of sub VIs when an error is encountered. Note that, error clusters can be useful in determining program flow due to the dataflow programming paradigm. This can be especially useful when setting up sampling on more than one DAQ board simultaneously. The unbundle by name function shows the components of an error cluster.

2.10 Case Structures:

The Case structure allows to choose a course of action depending on an input value. In the **Execution Control** sub palette of the Functions palette. Analogous to an if-then-else statement in other languages. Like a deck of cards, it can seen only one case at a time. These are some important examples for the case structure activities as follows:

- • **Example - 1:** Boolean input: Simple if-then case. If the Boolean input is TRUE, the true case will execute; otherwise the FALSE case will execute.

- • **Example - 2:** Numeric input. The input value determines which box to execute. If out of range of the cases, LabVIEW will choose the default case.
- •

- **Example - 3:** String input. Like the numeric input case, the value of the string determines which box to execute. Stress that the value much match *exactly* or the structure will execute the default case.

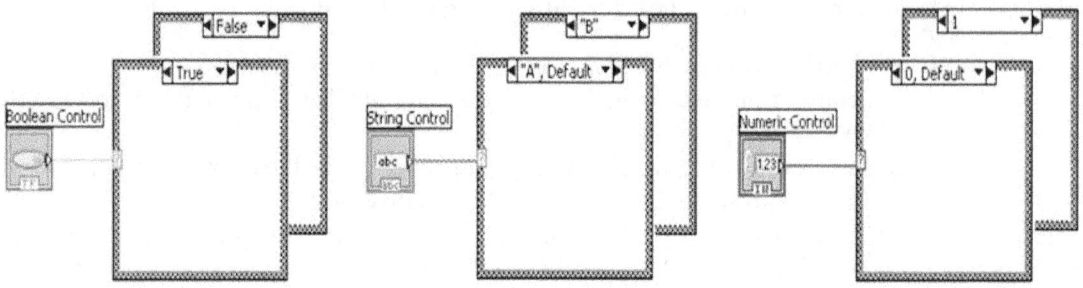

Figure 2-5 Case Structure

Example 4 – Error Clusters & Handling:

Create a VI that takes the square root of a number. If the number is greater than or equal to zero, the VI should return its square root, and generate no error. If the number is less than zero, the VI should return the value -9999.90, and insert an error into the error cluster.

Use a **Case structure** from the and the **Greater or Equal To 0.** function from the numeric palette to determine whether the VI will compute the square root or generate the error. The "False" case shown above is the error case. Use a **Bundle By Name** function from the cluster palette to insert a Boolean, Numeric, and String Constant into the Status, Code, and Source cluster items, respectively. The values of the constants should be True. Wire the new cluster into the Error Out indicator, and a constant value of -9999.90 to the Square Root indicator. The True case, not shown, should merely wire the error in control directly through the case to the error out indicator.

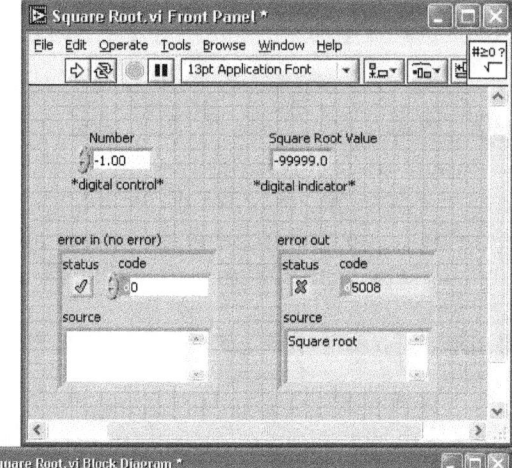

The Square Root Input should be wired to the **Square Root** function from the numeric palette, and the result should be wired out of the case into the Square Root indicator. Point out that this VI could be easily configured as a sub VI for a larger piece of code, and troubleshooting and debugging is much easier when errors clusters are properly used.

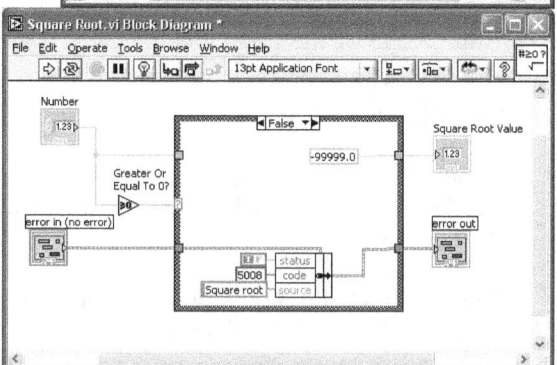

34

2.11 Sequence Structures:

In a text-based language, program statements execute in the order in which they appear. In data flow, a node executes when data is available at all its input terminals. Sometimes it is hard to tell the exact order of execution. Often, certain events must take place before other events. When you need to control the order of execution of code in your block diagram, you can use a sequence structure.

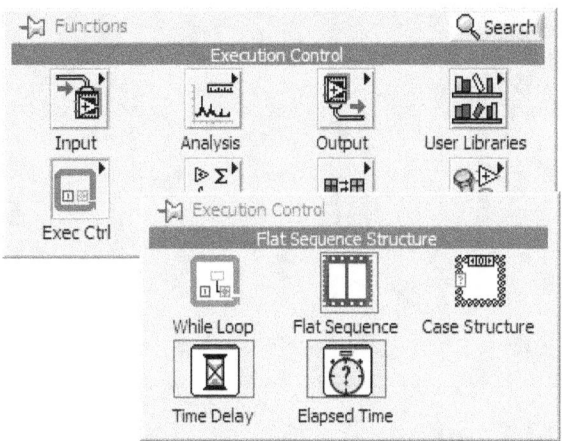

Sequence structure: Used to control the order in which nodes in a diagram will execute.

In the **Execution Control** sub palette.

- Looks like a frame of film.
- Used to execute diagrams sequentially.
- Right-click on the border to create a new frame.

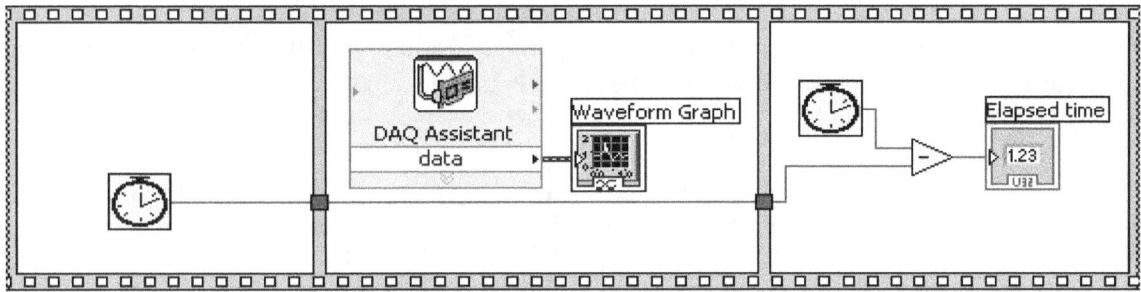

Formula Nodes:

Sometimes it is preferable to program mathematical expressions with text-based function calls, rather than icons .

Formula Node: Allows you to implement complicated equations using text-based instructions.

- Located in the **Structures** sub palette.
- Resizable box for entering algebraic formulas directly into block diagrams.
- To add variables, right-click and choose **Add Input** or **Add Output**. Name variables as they are used in formula. (Names are case sensitive.)
- Statements must be terminated with a semicolon.
- When using several formulas in a single formula node, every assigned variable (those appearing on the left hand side of each formula) must have an output terminal on the formula node. These output terminals do not need to be wired, however.

35

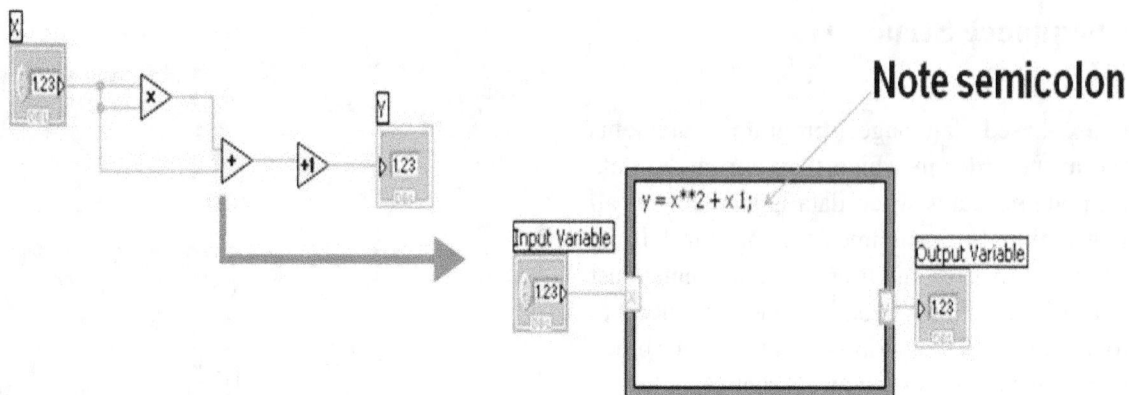

2.12 Printing:

LabVIEW offers many options for printing VIs. From the standard File » Print... menu, the user can print a hard copy of his or her VI, or print the VI to a file for storage or publishing. Using the Print Panel VI in LabVIEW allows the user to programmatically print the results of a test. VIs can also be configured to print automatically after execution. " This option is set in VI Properties » Print Options"

For more advanced applications, LabVIEW has report generation tools that allow the user to create custom reports for individual applications. The VI returns, and report properties, such as the author, company, and number (‾

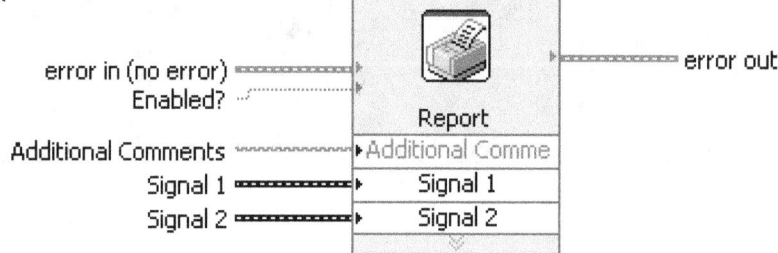

2.13 Documenting VIs:

By opening a VI's properties a developer can add documentation to his or her VI. The documentation placed in the Description field of the VI Documentation window shows up in Context Help, and prints with the VI. If a user has an application that is continually developing, he or she can track changes in the development with Revision History, also found in VI Properties. Any control or indicator on the Front Panel can be documented by right-clicking and choosing Description and Tip.

The Description information appears in the Context Help menu when a user hovers the mouse over the object, and the Tip information pops up in a small strip next to the mouse when the user pauses above the object. Much like comments in a text-based language, the developer may want to explain a portion of his code, or provide directions on a front panel. Either of these needs can be met by using the labeling tool to create a free text box with as many instructions or explanations as necessary.

2.13.1 Simple VI Architecture:

When making quick lab measurements, you do not need a complicated architecture. Your program may consist of a single VI that takes a measurement, performs calculations, and either displays the results or records them to disk. The measurement can be initiated when the user clicks on the run arrow. In addition to being commonly used for simple applications, this architecture is used for "functional" components within larger applications. You can convert these simple VIs into sub VIs that are used as building blocks for larger applications.

Functional VI that produces results when run
- No "start" or "stop" options
- Suitable for lab tests, calculations

Example: Convert C to F.vi

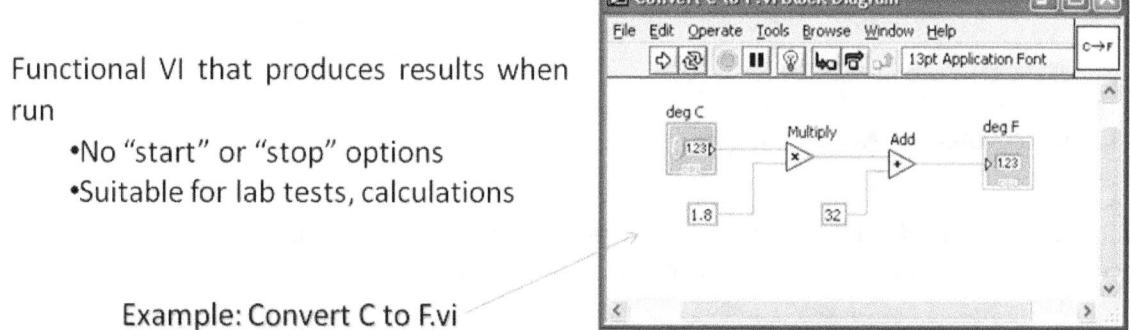

2.13.2 General VI Architecture:

Three Main Steps

- Startup
- Main Application
- Shutdown

In designing an application, generally have up to three main phases:

- Startup—Use this area to initialize hardware, read configuration information from files, or prompt the user for data file locations.
- Main Application—Generally consists of at least one loop that repeats until the user decides to exit the program, or the program terminates for other reasons such as I/O completion.
- Shutdown—This section usually takes care of closing files, writing configuration information to disk, or resetting I/O to its default state.

For simple applications, the main application loop can be fairly straightforward. When you have complicated user interfaces or multiple events (user action, I/O triggers, and so on), this section can get more complicated.

2.13.3 State Machine Architecture:

In this model, it scan the list of possible events, or states, and then map that to a case. For the VI shown above, the possible states are *startup*, *idle*, *event 1*, *event 2*, and *shutdown*.

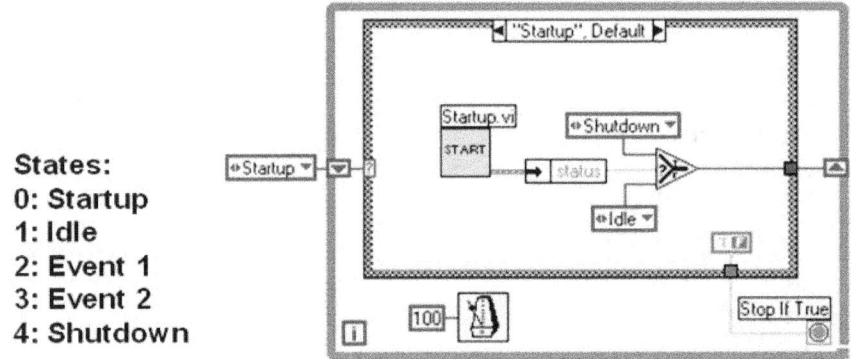

States:
0: Startup
1: Idle
2: Event 1
3: Event 2
4: Shutdown

These states are stored in an enumerated constant. Each state has its own case where placed the appropriate nodes. While a case is running, the next state is determined based on the current outcome. The next state to run is stored in the shift register. If an error occurs an any of the states, the shutdown case is called.

The advantage of this model is that a diagram can become significantly smaller (left to right), making it easier to read and debug. One drawback of the Sequence structure is that it cannot skip or break out of a frame. This method solves that problem because each case determines what the next state will be as it runs.

A disadvantage to this approach is that with the simple approach shown above, you can lose events. If two events occur at the same time, this model handles only the first one, and the second one is lost. This can lead to errors that are difficult to debug because they may occur only occasionally. More complex versions of the state machine VI architecture contain extra code that builds a queue of events (states) so that you do not miss any events.

Example 5 – Simple State Machine:

In this example it will create a VI using state machine architecture. The VI will have an idle state, where it waits for input. When the user presses a button, the VI will go to State 1. State 1 will generate a pop-up dialog box that allows the sure to proceed to State 2 or start over. From State 2 the user can choose to exit the program or start over.

From the File menu, instead of selecting **New VI**, select **New…**, and click on Start from template. Browse to …\Program Files\National Instruments\LabVIEW 2012\templates\State Machine.vit, and open it. This opens a template for a state machine using strings to control the state. Examine the template, and then save it in another directory before you begin working on it.

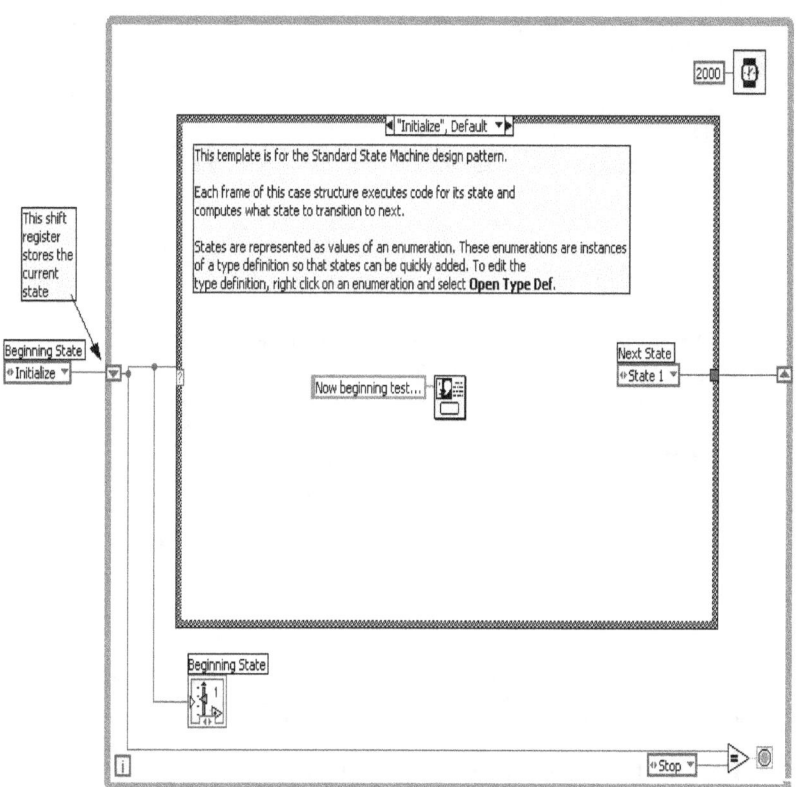

The first state it will make is the Initialize state. Notice this state has already been created for us, and is the default state. In the Initialize case, place a **One Button Dialog** (Functions » Time & Dialog), and wire "Initialized. Click OK to proceed" into the message terminal. Replace the text "Next State" with "Idle".

2.14 Remote Panel Web Publishing Tool:

To embed your VI into a web page, simply open the Web Publishing Tool from the Tools pull-down menu. The Web Publishing Tool creates a simple html file with the LabVIEW front panel embedded inside it. From the tool,it can create a title, and supply text before and after the front panel object. Once it has created the html file, one can preview it in a web browser, and save the file to disk. Saving the file places the html file into LabVIEW www directory, which is the default directory for the LabVIEW web server.

It can start the web server from the Web Publishing Tool, or from Tools » Options… » Web Server: Configuration. (Note: in the Tools » Options… menu you can also configure access rights and exposed VI's for the web server.

- Click Save to Disk and VI is embedded into an HTML file
- After file is saved, it can be reopened and customized in any HTML editor

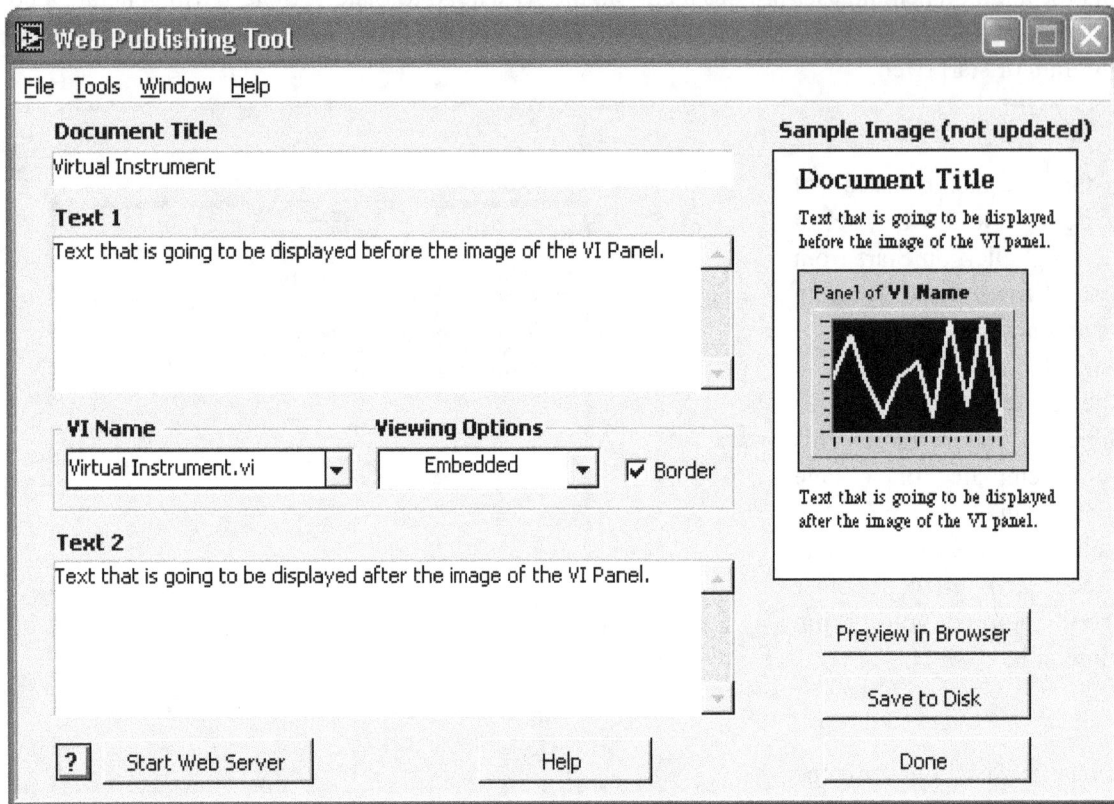

2.14.1 Remote Front Panels - Resources:

NI Developer Zone (zone.ni.com)

- – Search for Remote Front Panel
- – Tutorials & Instructions Are Available for Download
- – Information on Incorporating Web Cameras into Remote Panel Applications

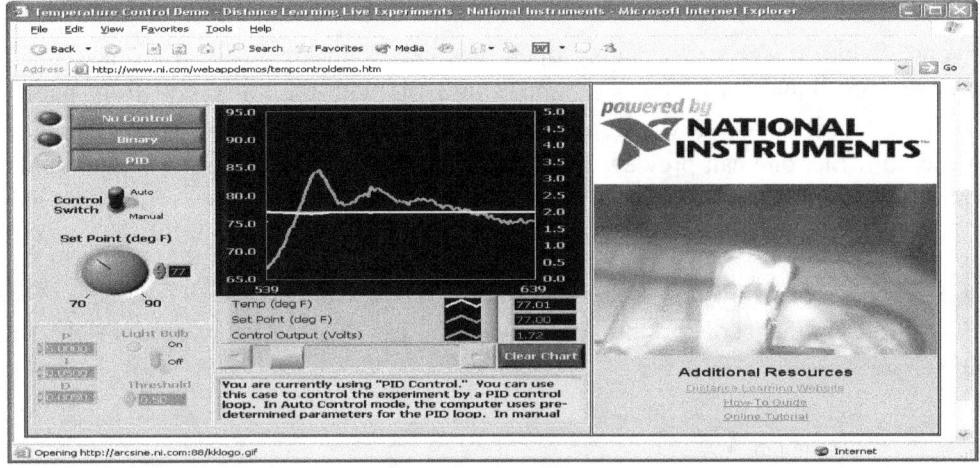

LabVIEW has many other tools available for your development. While time constraints don't permit us to go into great detail, it is a good idea to highlight some of the advanced capabilities offered by LabVIEW. Students who are interested in a specific topic can search the LabVIEW Example Finder or visit NI's Developer Zone for more information.

- Property Nodes – Properties are all the qualities of a front panel object. With *property nodes*, you can set or read such characteristics as foreground and background color, data formatting and precision, visibility, descriptive text, size and location on the front panel, and so on. Search for property nodes in the Find Examples window, there are 24 example programs that ship with LabVIEW.

- Local Variables – Local Variables break the dataflow programming paradigm, allowing data to be passed without wires.

- Global Variables – A special kind of VI, used to store data in front panel objects for the purpose of data exchange between VI's.

- Data Socket – A platform independent means of exchanging data between computers and applications, based on TCP-IP, but without the complexity of low level programming.

- Binary File I/O—allows a user to stream data to a disk in high speed applications.

Chapter 3

Samples & Examples

3.1 Vertical Pointer Slide

Create a user interface (Front Panel) with the following controls (right-click to open the Controls palette then select "Modern" → "Numeric"):

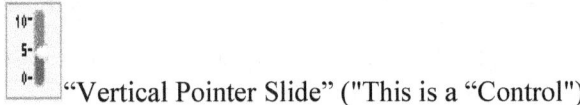 "Vertical Pointer Slide" ("This is a "Control")

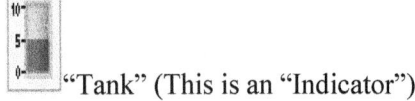 "Tank" (This is an "Indicator")

The Front Panel should look like in the figure 3-1:

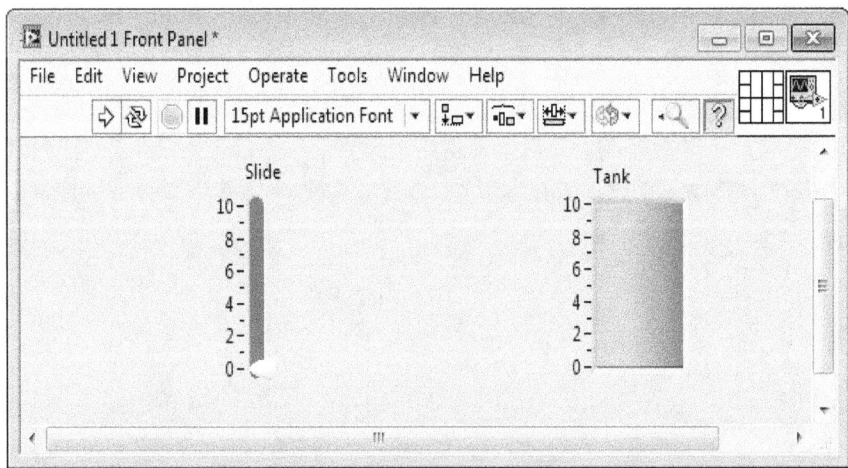

Figure 3-1

The Block Diagram should look like in the figure 3-2:

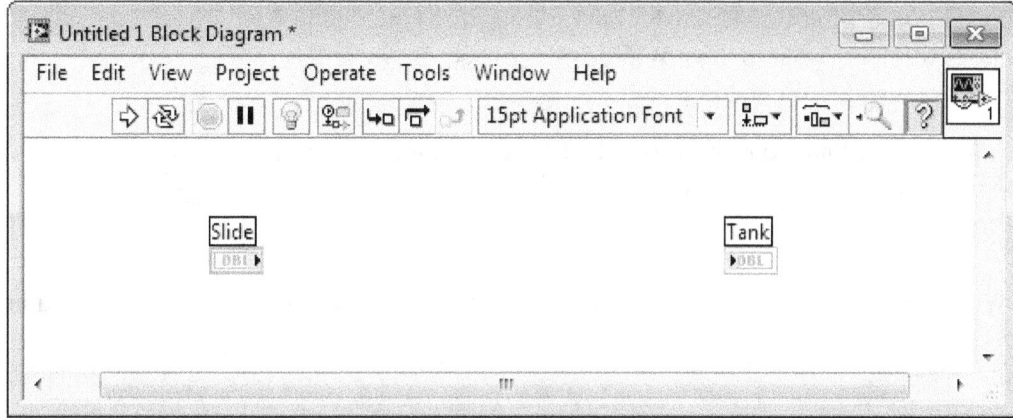

Figure 3-2

Wire them together like this:

Then, set the slider on a specific value:

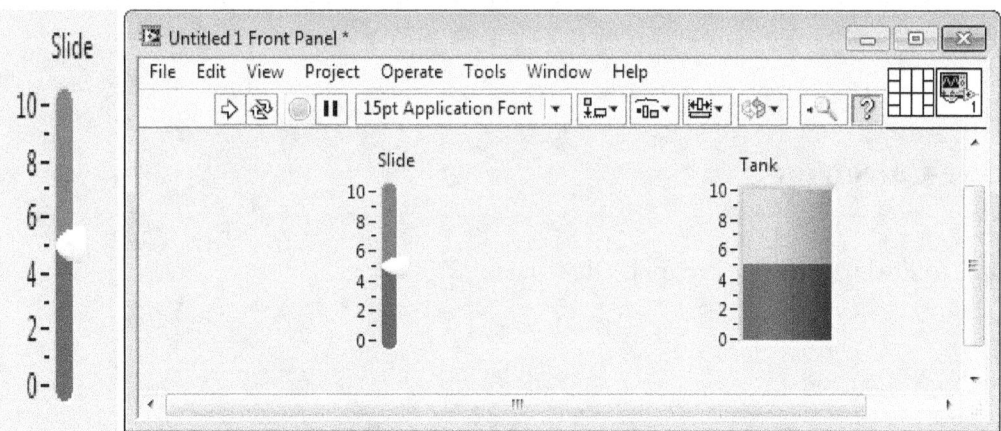

Run the program by clicking the "**Run**" button:

 Then click the "**Run Continuously**" button and change the value for the "Slide" and observe what's happens.

In some situations the "Run Continuously" button is handy to use, but in most situations we rather want to use a While Loop instead. So we will extend the program with a While Loop.

Right-click to open the Functions palette on the Block Diagram and select a While Loop ("Structures" → "While Loop").

Drag the While Loop over the blocks we already have on the Block Diagram so it looks something like below:

⇨ Run the program by clicking the "**Run**" button and change value on the Slider and observe what's happening.

3.2 Simple Calculator

Create the following simple program that adds 2 numbers.

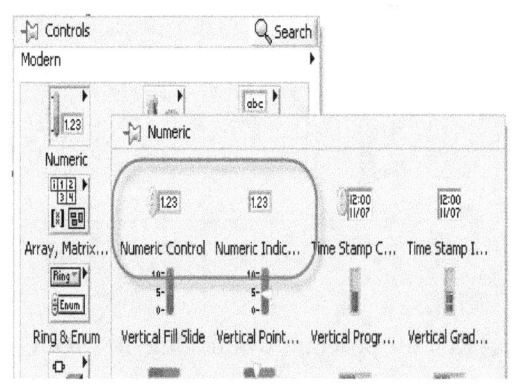

Add 2 Numeric Controls and name them"Number1" and "Number2"

Add a Numeric Indicator and name it "Answer"

Insert an "Add" function from the Functions palette:

Front Panel:

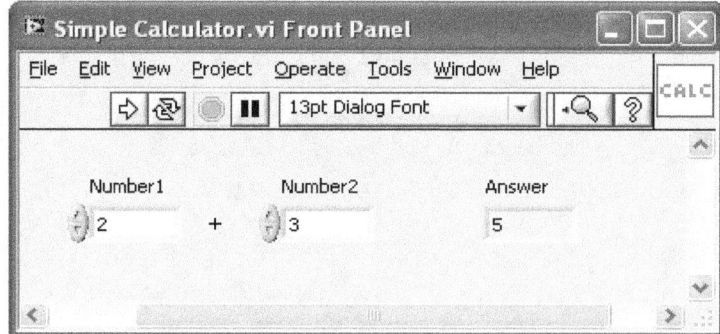

Push the Run button 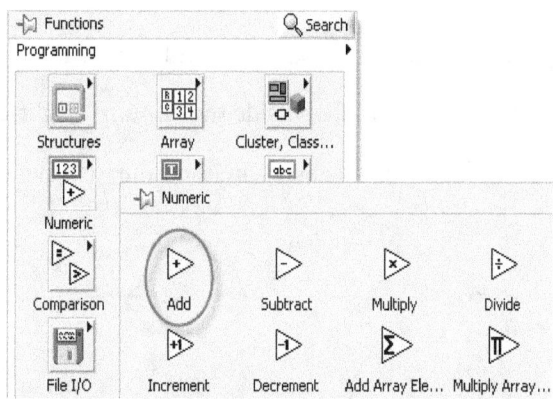 in the Toolbar to do the calculations.

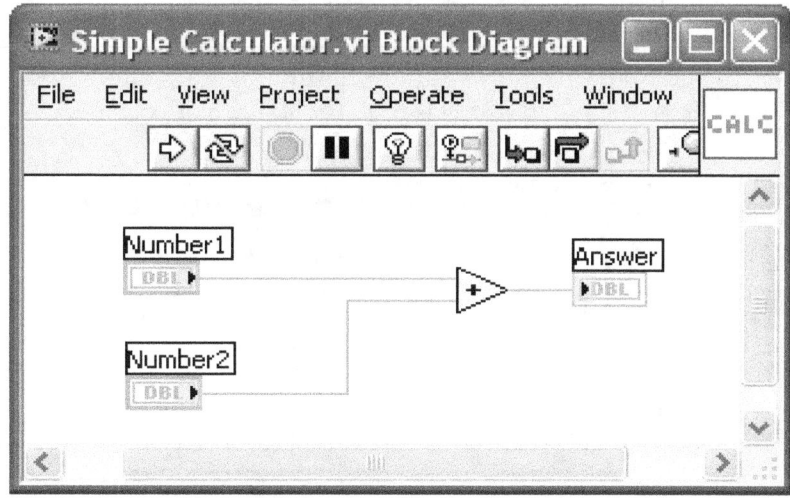

Extend your calculator (make sure to rename your file) so it can handle the 4 basic calculations (Add, Subtract, Multiply, Divide).

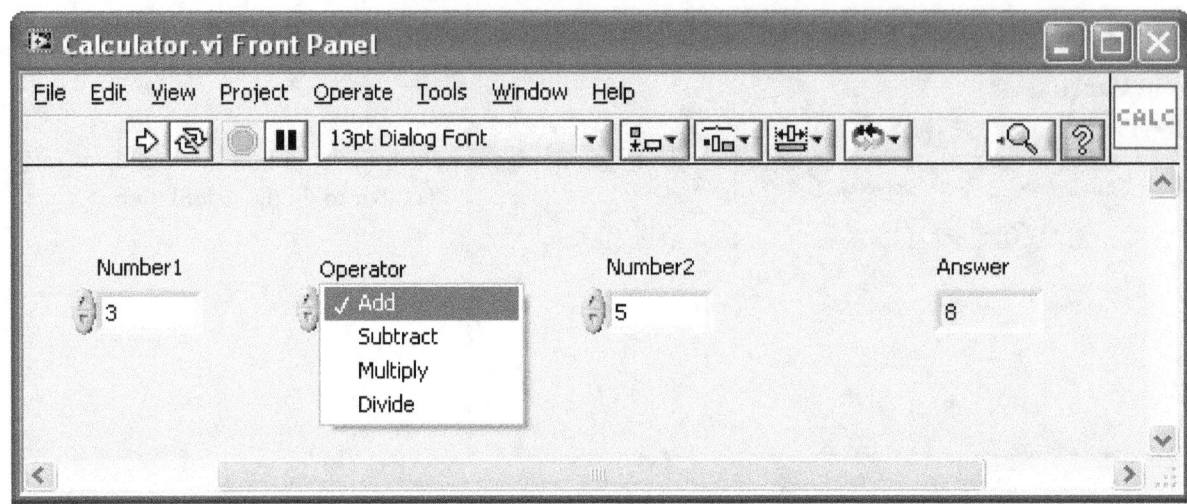

For selecting the different Operators ("Add", "Subtract", "Multiply", "Divide"), by using "Enum":

→ Enter "Add", "Subtract", "Multiply", "Divide" inside the "Enum" (right-click and select "Add Item After"). By using the "**Case Structure**" for this and wire the Enum ("Operator") to the ⍰ sign on the "case structure".

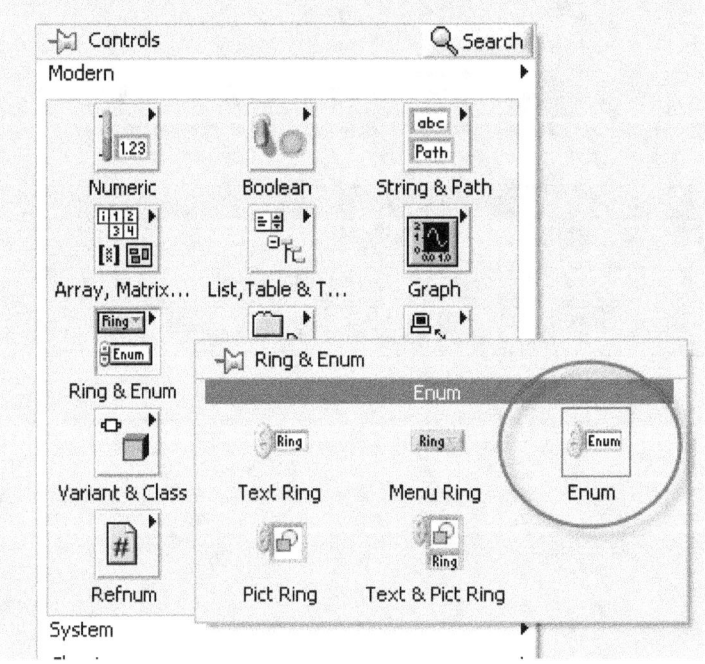

The Block Diagram should look like the figure 3-3 when you are finished:

46

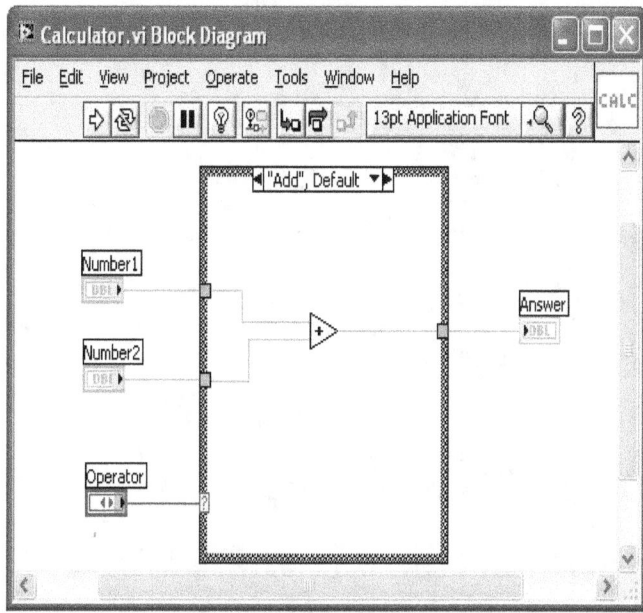

Figure 3-3

➢ Task : While Loop

Create the following file & program: **While Loop.vi**

This program will update the date and time each second using a While loop until you stop the program using the Stop button.

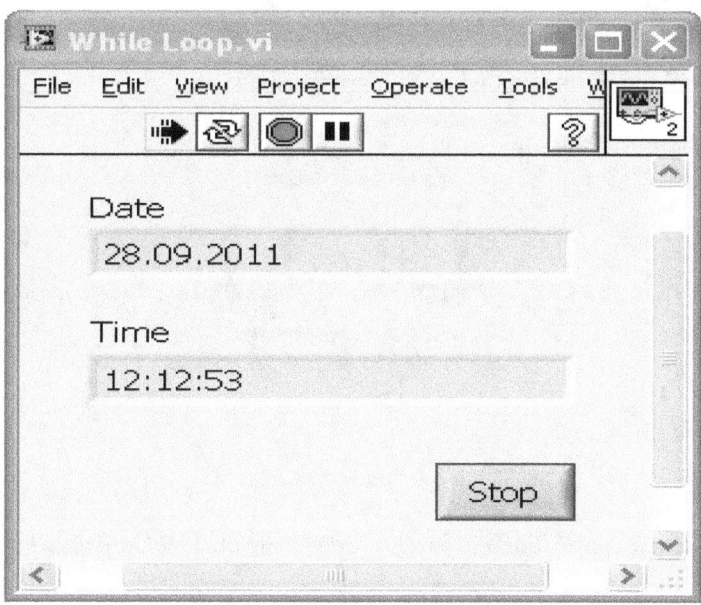

→ Add 2 String Indicators and name them "Date" and "Time".

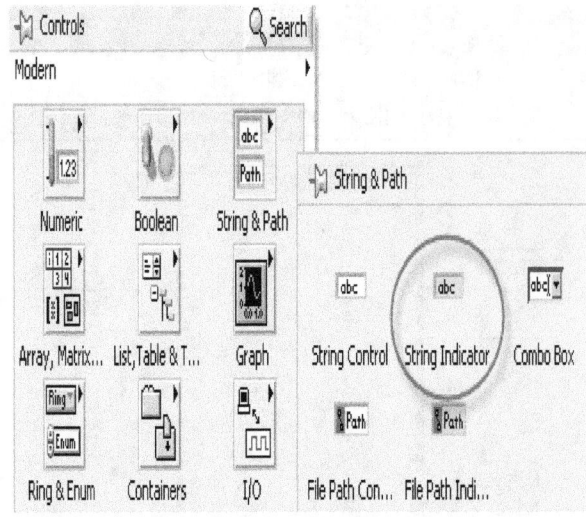

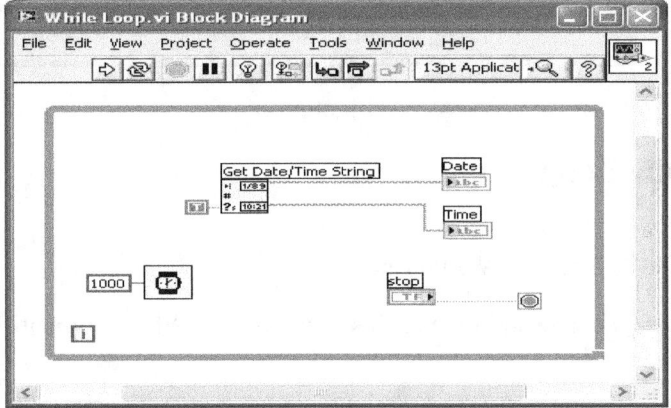

The Block Diagram should look like this when you are finished:

→ Add a **While Loop** to this program.

Get Date/Time String:

→ Use the "Get Date/Time String" function in order to get the current Date and Time from the computer:

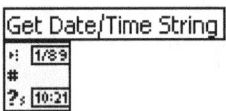

Wait (ms):

→ The Date and Time strings should be updated every second. You can use the "Wait (ms)" for that:

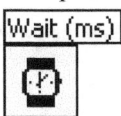

3.3 Advanced Calculator

Create the following program: **Advanced Calculator.vi**

→ Add the User Interface objects as shown in the Figure above. The calculations will be performed when you click the "Calculate" button.

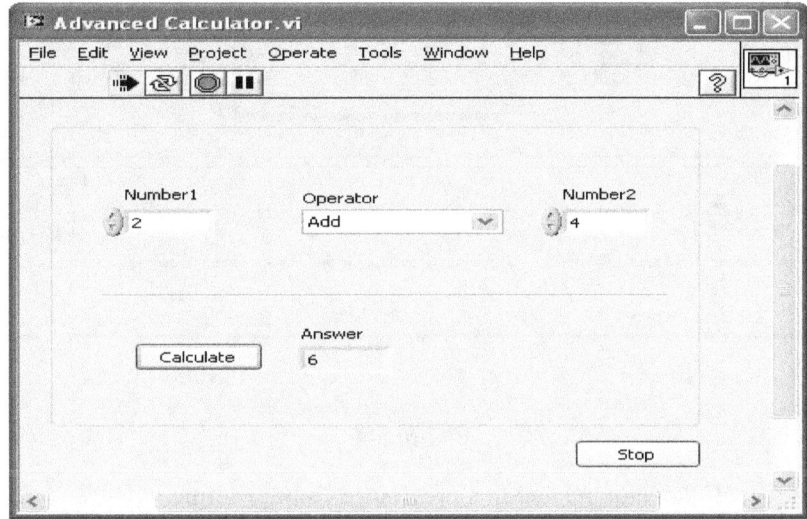

Add an "Event Structure" inside the While Loop.. The Block Diagram should look like the figure 3-4 when it's finished:

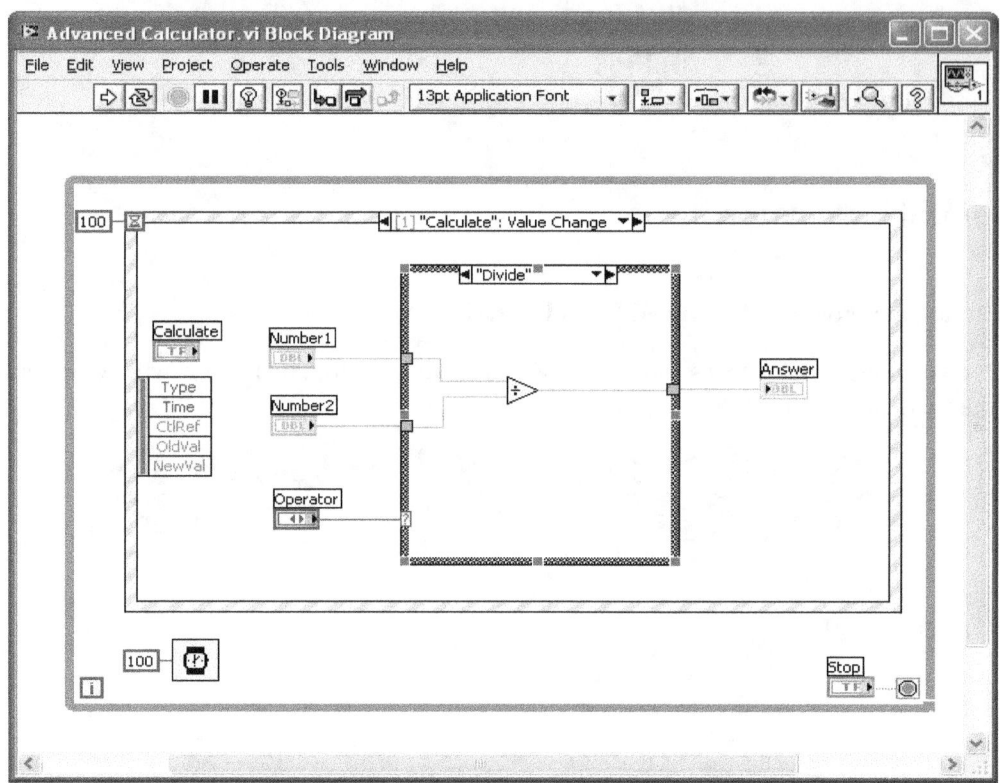

Figure 3-4

Create a Sub VI from the Calculator created in a previous example, define inputs and outputs as in the illustration above.

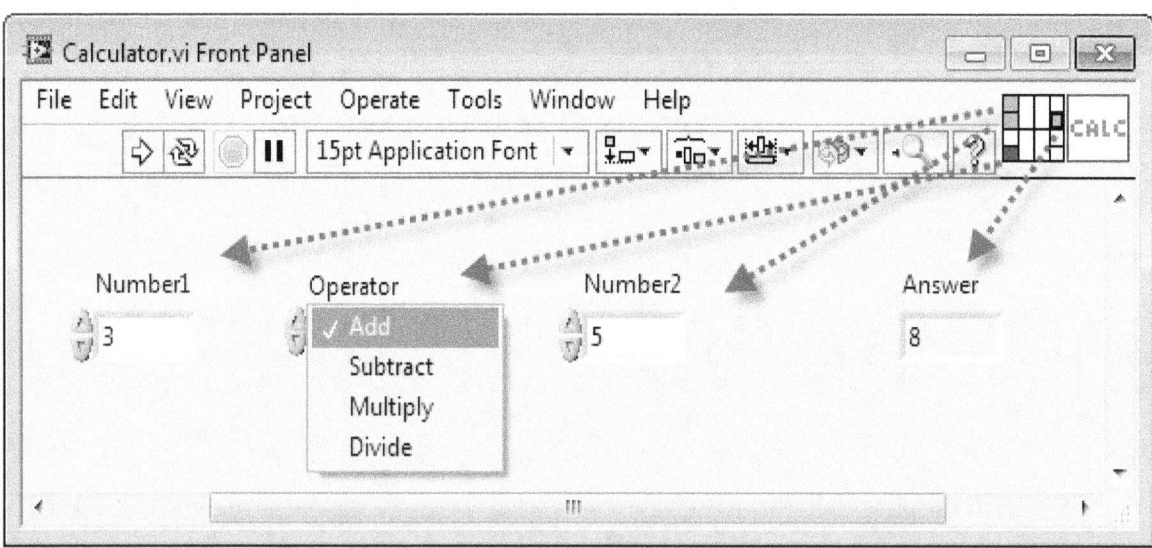

This includes creating a nice icon using the "Icon Editor": → Right-click in the upper right window and select "Edit Icon" in order to open the "Icon Editor".

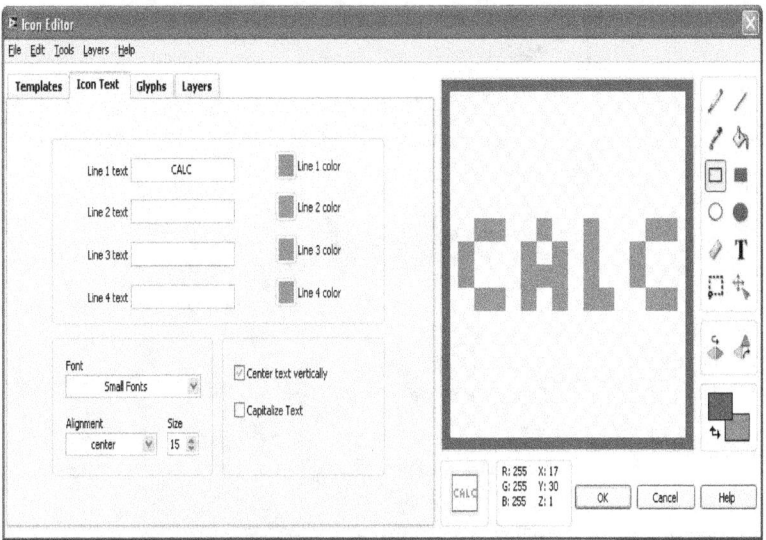

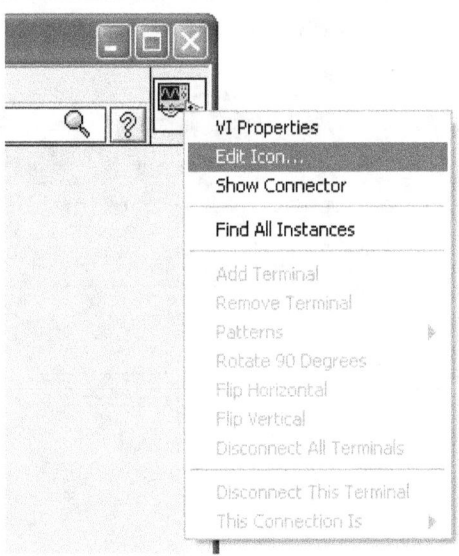

→ Create a new blank VI and open the Block Diagram.

→ Insert the Sub VI by right-click and select "Select a VI" in the Functions palette.

→ Right-click on the Sub VI inputs and select "Create → Control"

The following sketches are shown these steps:

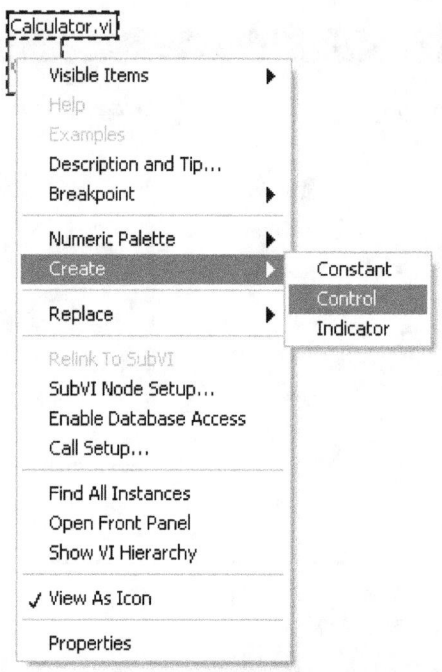

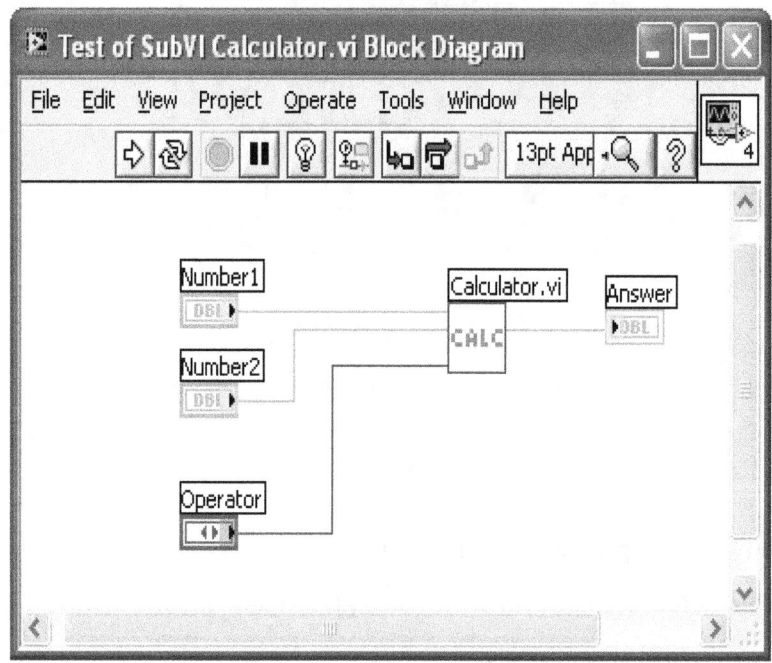

The Front Panel could look like this: → Test your program by using the Run button 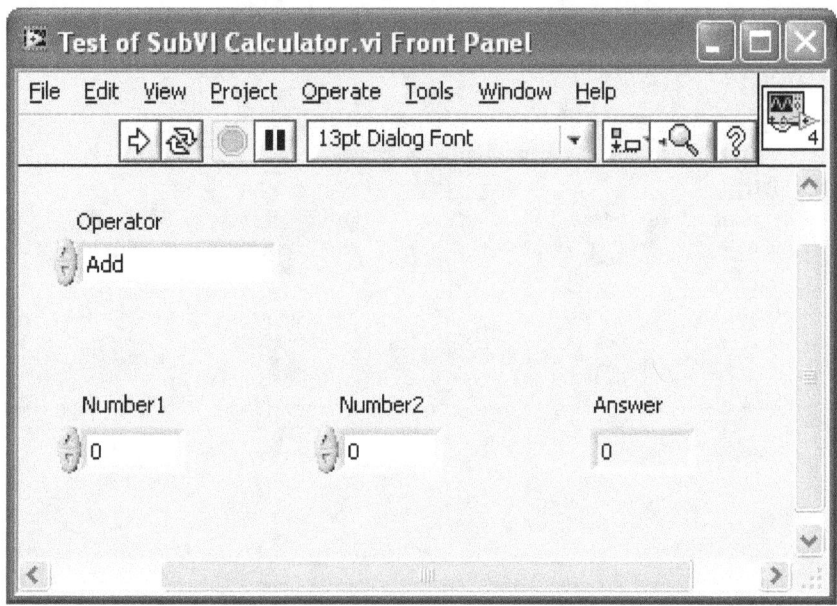 in the Toolbar.

3.4 Plotting

Front Panel:

Use the "Waveform Chart" Control:

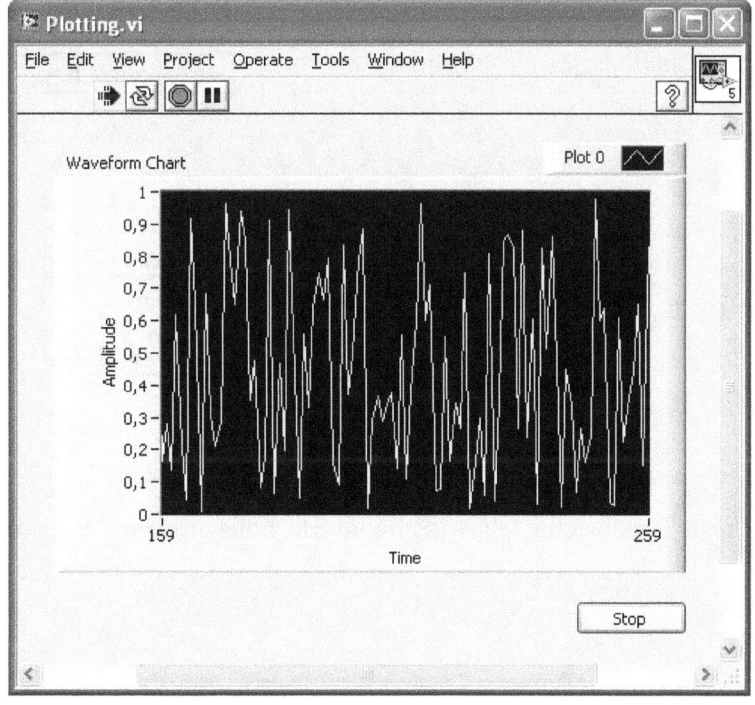

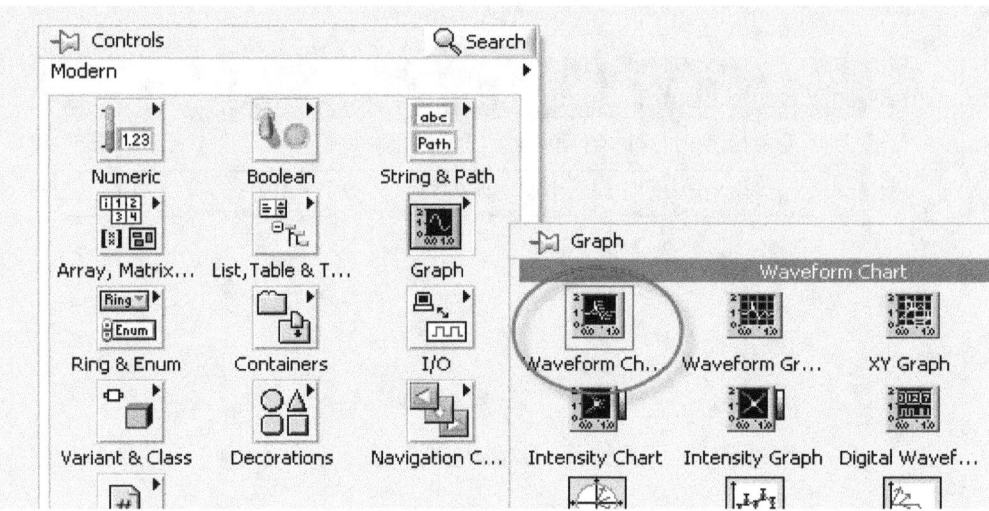

Block Diagram:

Random Number (0-1):

→ Use the "Random Number (0-1)" function in order to create some random data in your plot:

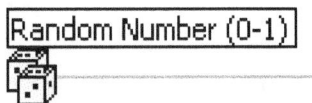

Waveform Chart:

→ Insert the "Waveform Chart" inside the "While Loop".

Wit (ms):

→ Update the chart every second by using the "Wait (ms)":

The Block Diagram should look like the following figure when this program finished:

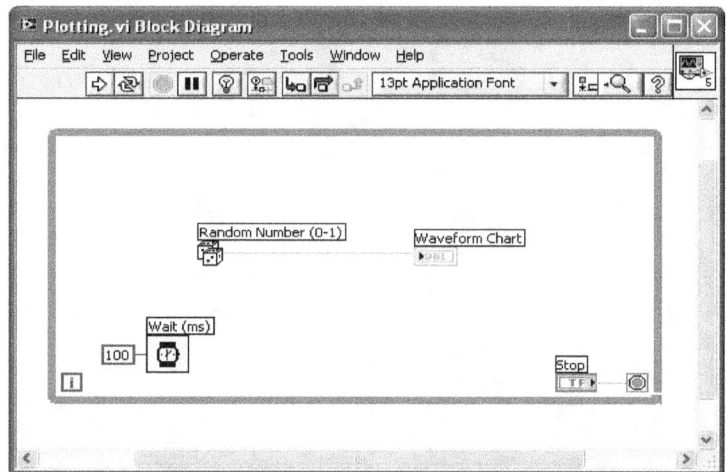

Home Work : *Convert the temperature from degrees Celsius to degrees Fahrenheit ?*

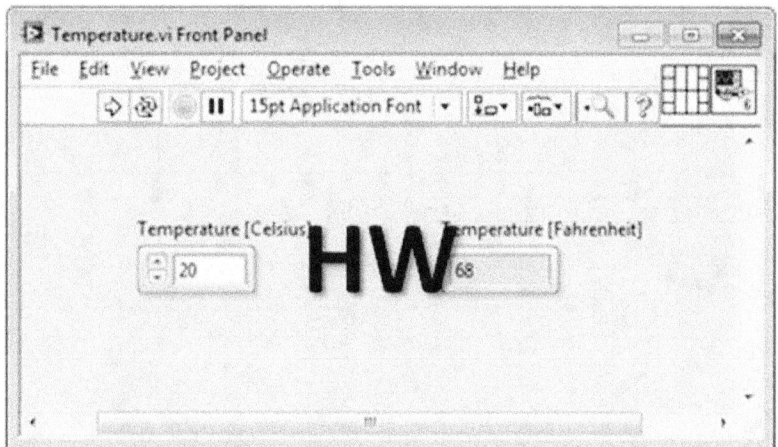

3.5 Projects for Students

3.5.1 Home Alarm System

In this project , the main program has a loop. This loop will start with initial time ; array zero and false constant,..The inputs will go to arm / disarm Sub VI. The arm/disarm Sub VI will check to see if the code entered is correct and arm/disarm the alarm. See the front panel of this system shown in figure 3-5.

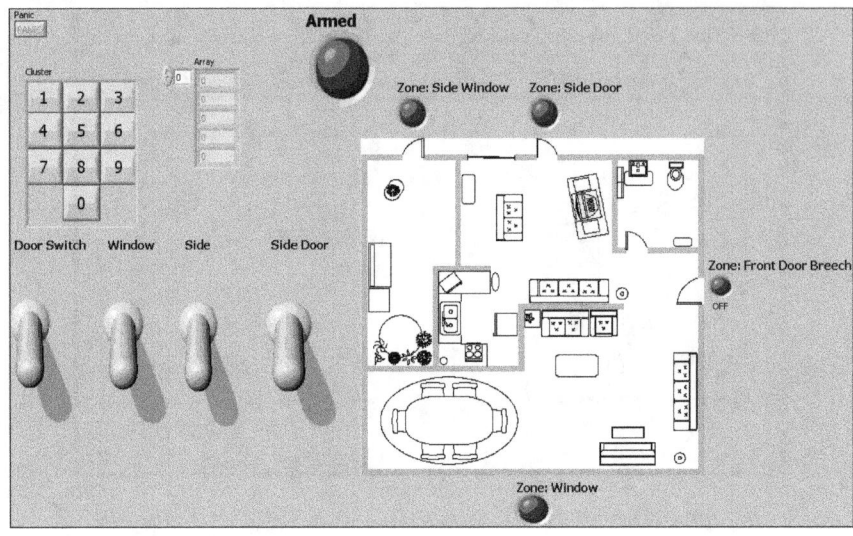

Figure 3-5

The block diagram of the Alarm system is shown in the figure 3-6:

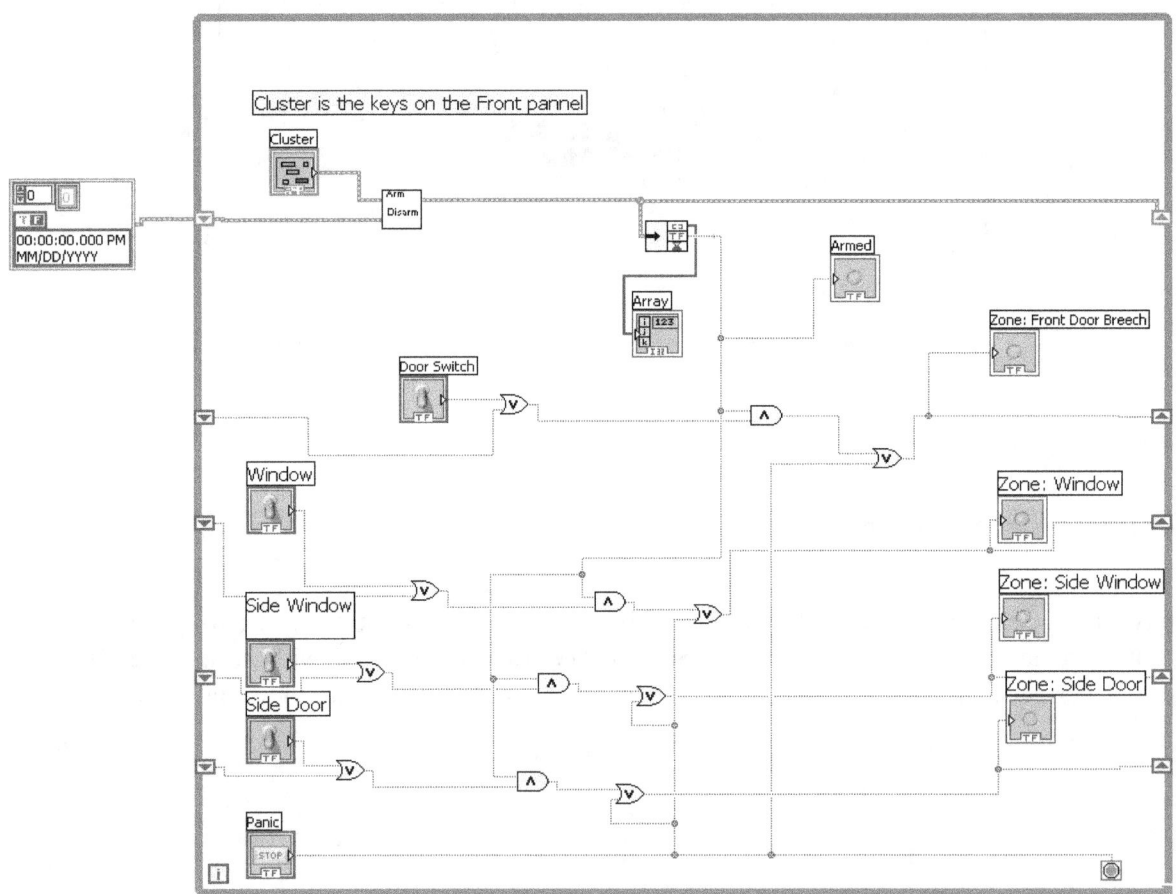

Figure 3-6

56

3.5.2 GPA Calculator

This calculator has been designed to calculate the credits you need to improve your current Grade Point Average. see figure3-7

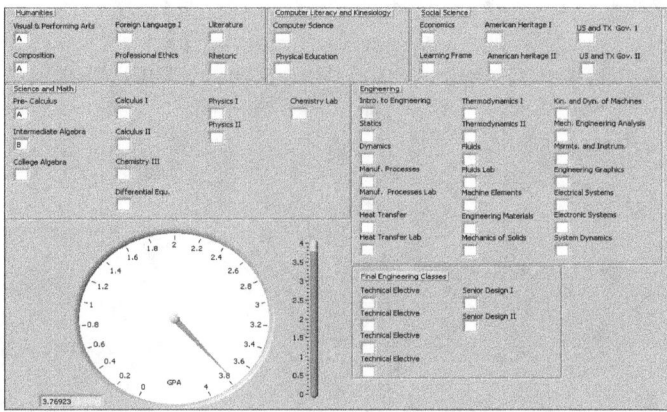

Figure 3-7

The block diagram of the GPA calculator is shown in figure 3-8:

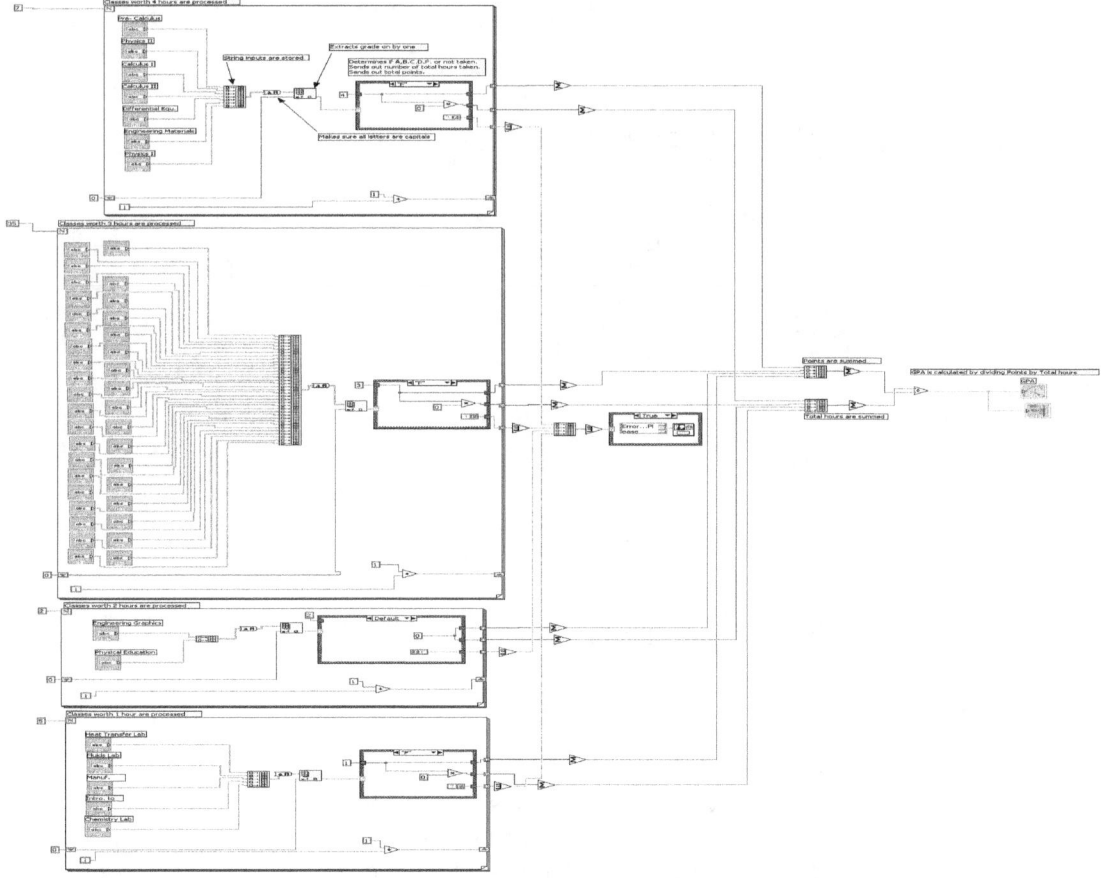

Figure 3-8

3.5.3 Ohm- meter

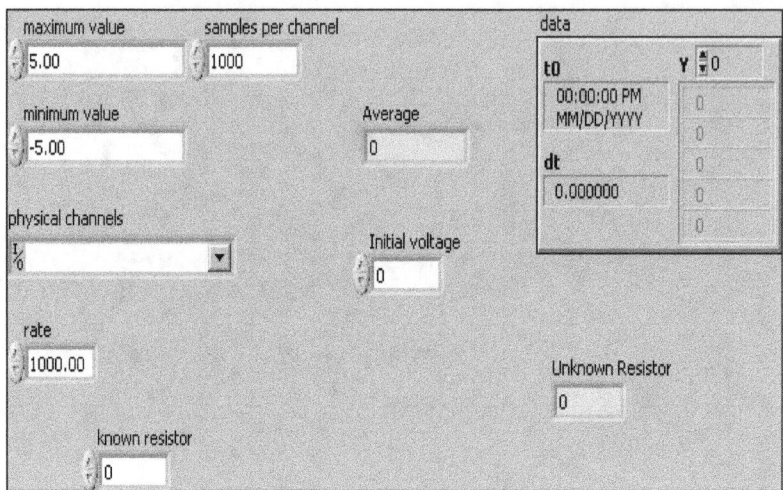

Figure 3-9 shows the Ohm meter front panel

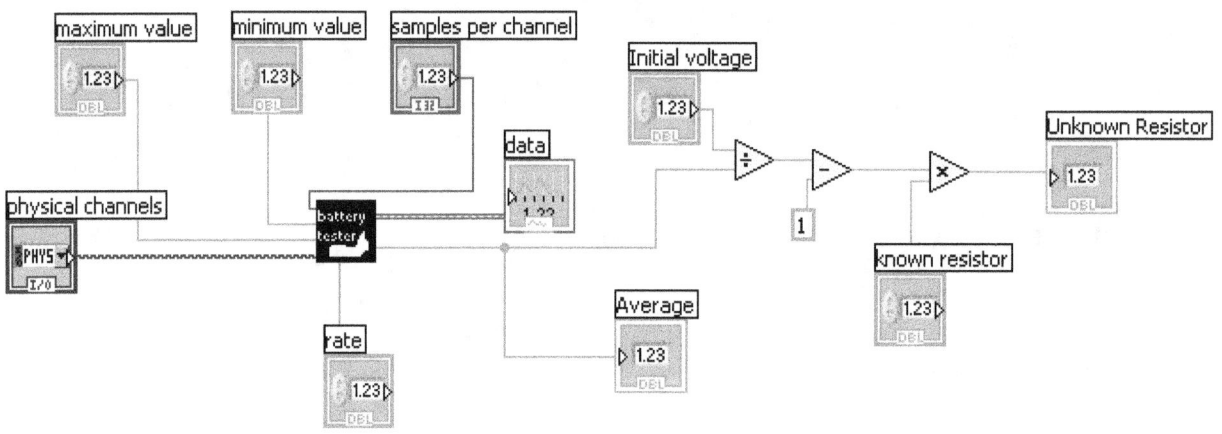

Figure 3-10 shows the block diagram of the Ohm meter

3.5.4 Hangman Game

The word to guess is represented by a row of dashes, giving the number of letters, numbers and category. If the guessing player suggests a letter or number which occurs in the word, the other player writes it in all its correct positions. If the suggested letter or number does not occur in the word, the other player draws one element of the hanged man **stick figure** as a **tally mark**.

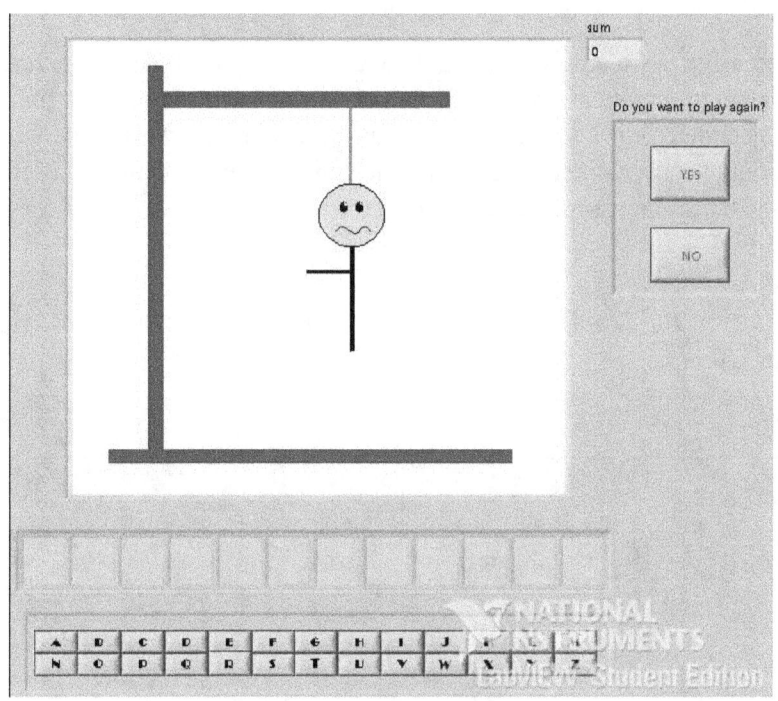

Figure 3-11 shows the Hangman Game front panel

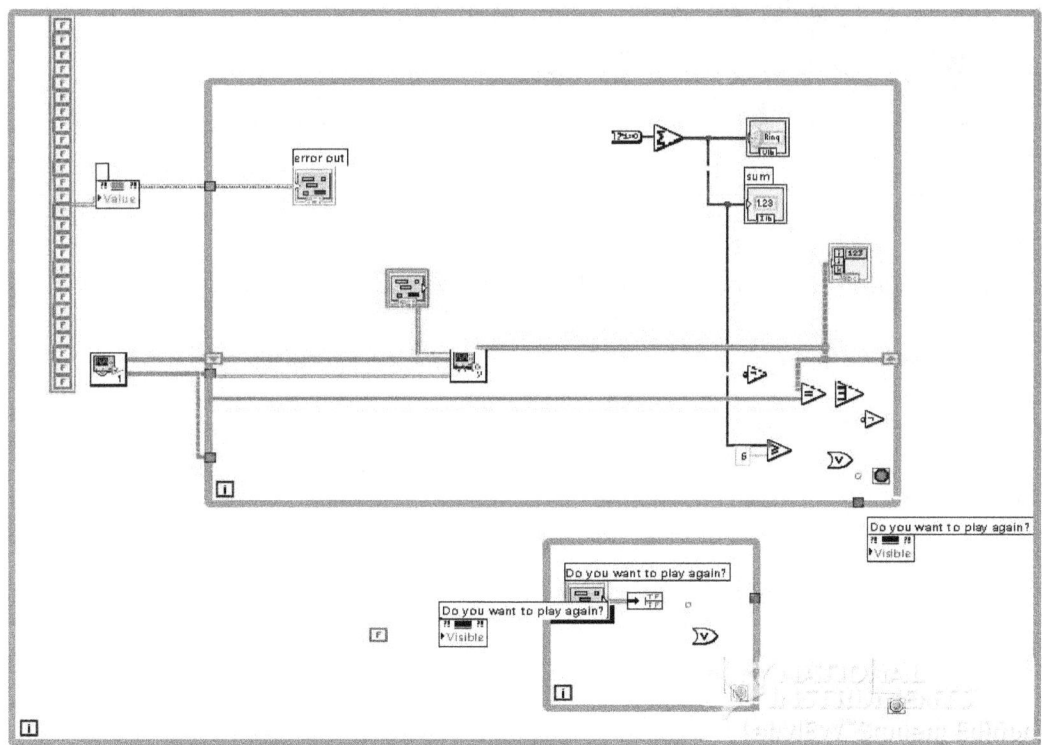

Figure 3-12 shows the block diagram of the Hangman Game

3.5.5 Heat Transfer Fin Solver

The study of heat transfer, a **fin** is a surface that extends from an object to increase the rate of heat transfer to or from the environment by increasing convection. The amount of conduction, convection, or radiation of an object determines the amount of heat it transfers. Increasing the temperature difference between the object and the environment, increasing the convection heat transfer coefficient, or increasing the surface area of the object increases the heat transfer.

This program is a solver for fins in Heat Transfer. It has used the MatLab Script application with LabVIEW, see figure 3-13.

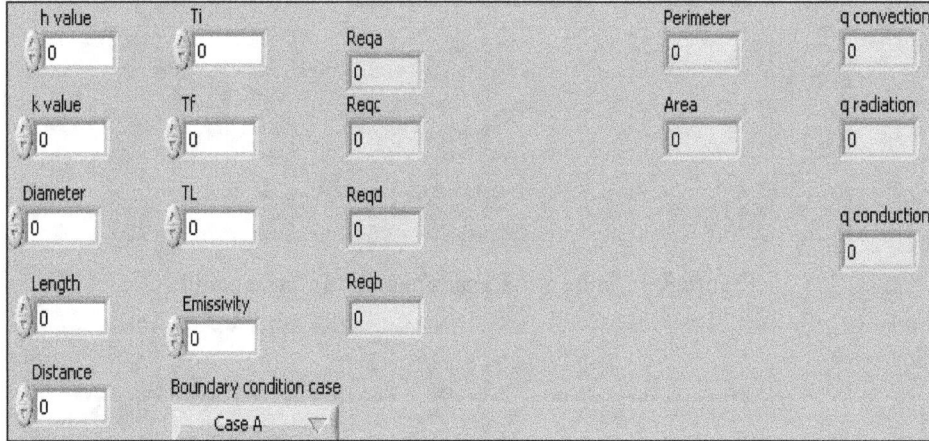

Figure 3-13

It has a block diagram gets Mat Lab scripts inside its application, figure 3-14 shows this solver.

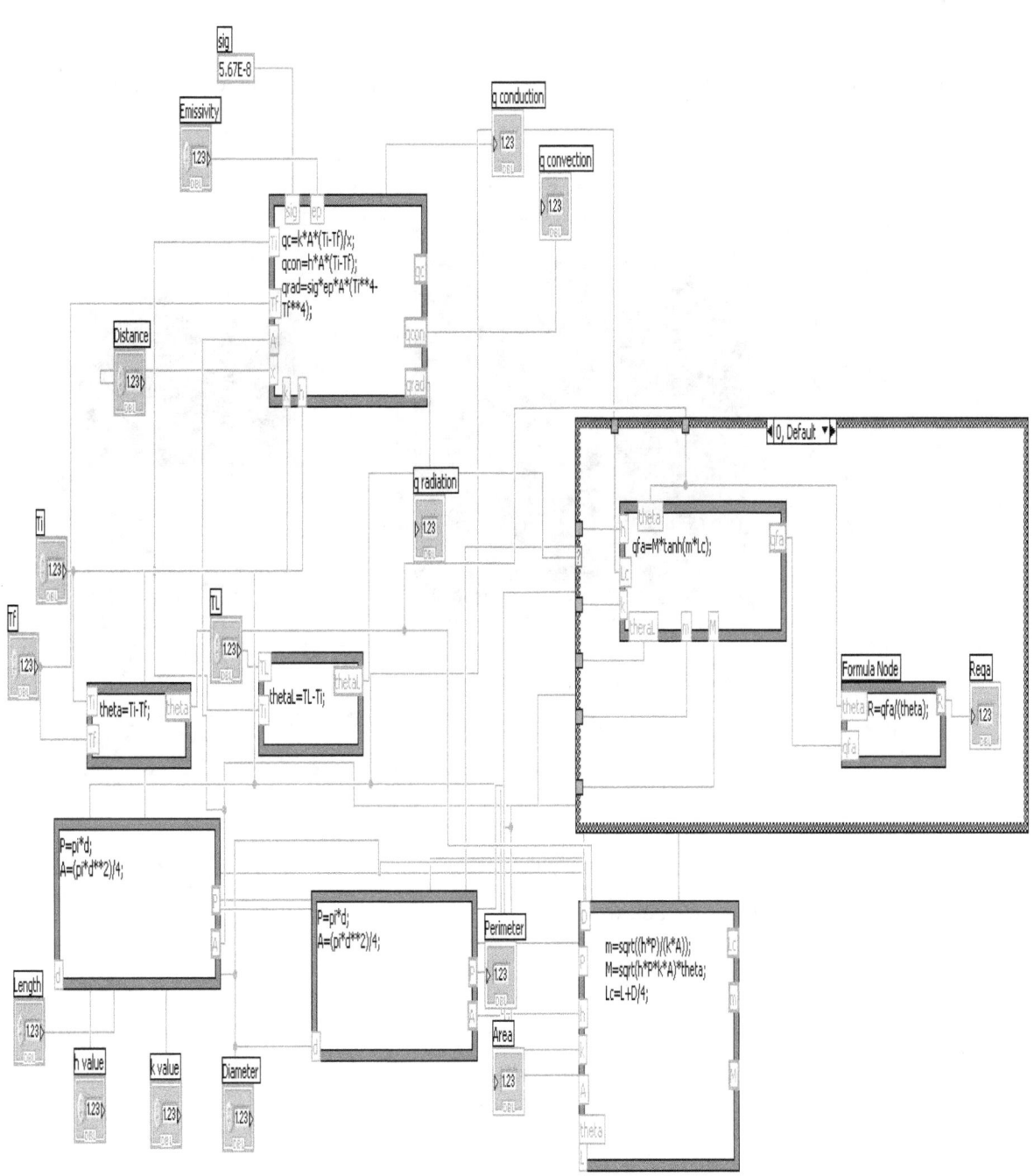

Figure 3-14

61

3.5.6 Battle Tanks

Select appropriate angle to match the angle of target, the front panel is shown in the figure 3-15.

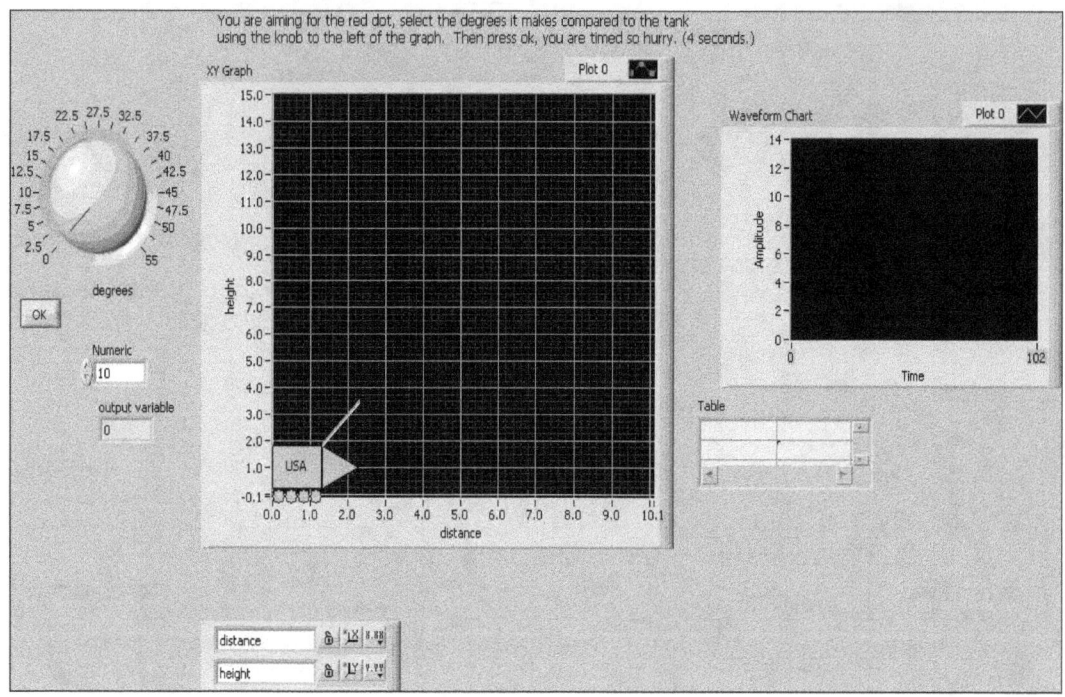

Figure 3-15

Battle Tanks Block Diagram is shown below:

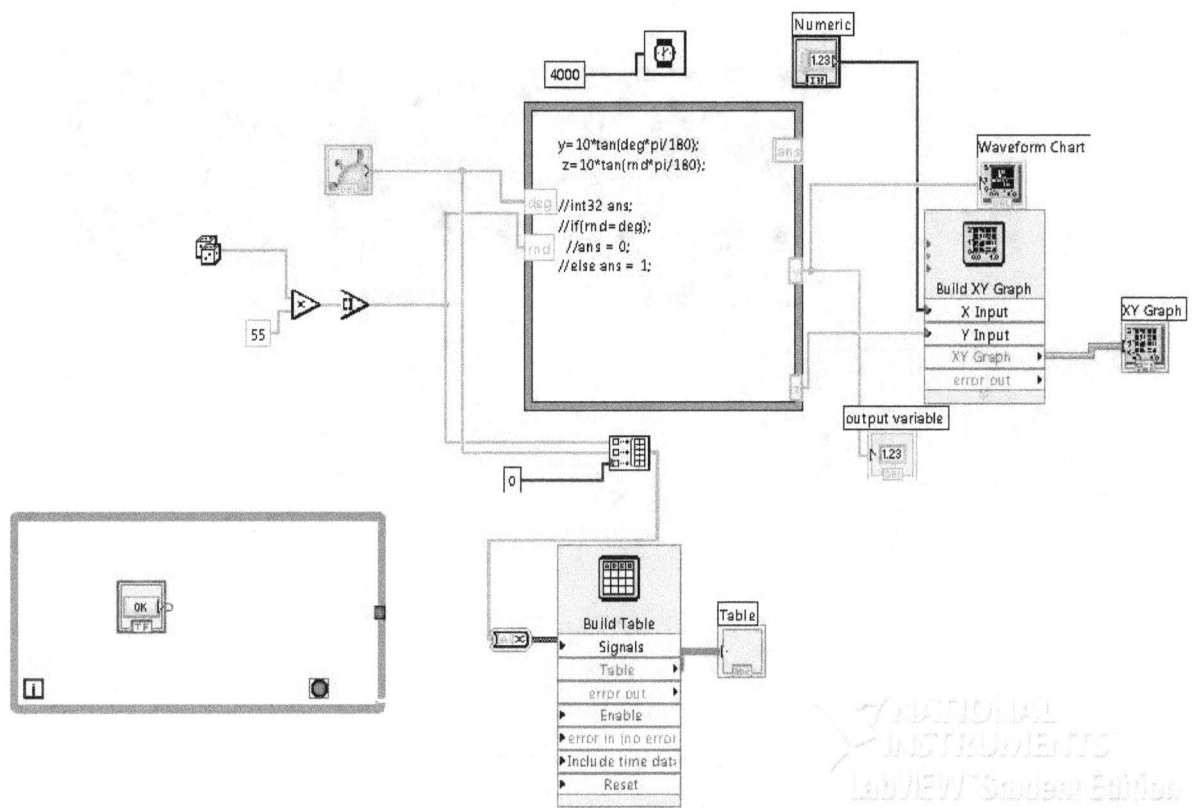

3.5.7 Oscilloscope

An **oscilloscope**, is a type of electronic test instrument that allows observation of constantly varying signal voltages, usually as a two-dimensional graph of one or more electrical potential differences using the vertical or *y*-axis, plotted as a function of time (horizontal or *x*-axis), see figure 3-16 and 3-17 to show its front panel and block diagram respectively.

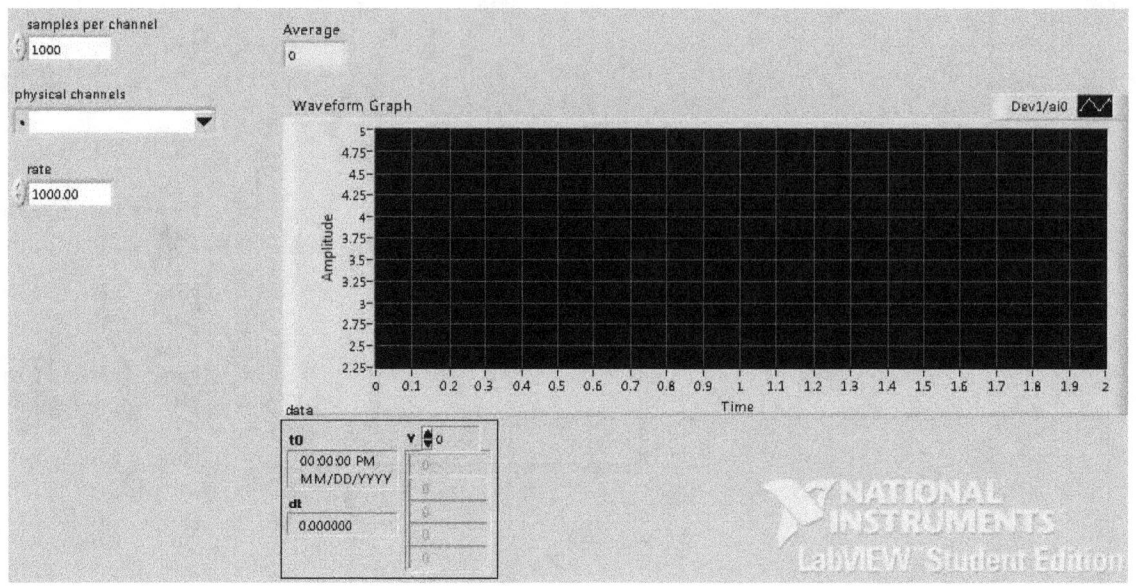

Figure 3-16

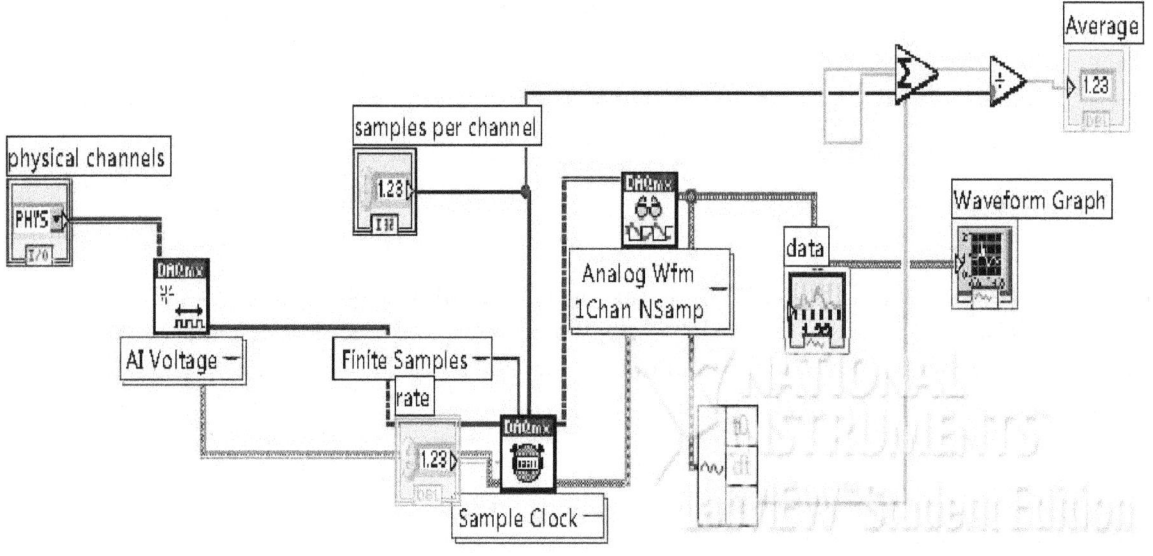

Figure 3-17

3.5.8 Signal Filter Experiment

The figure 3-18 is shown both signal filter front panel and block diagram.

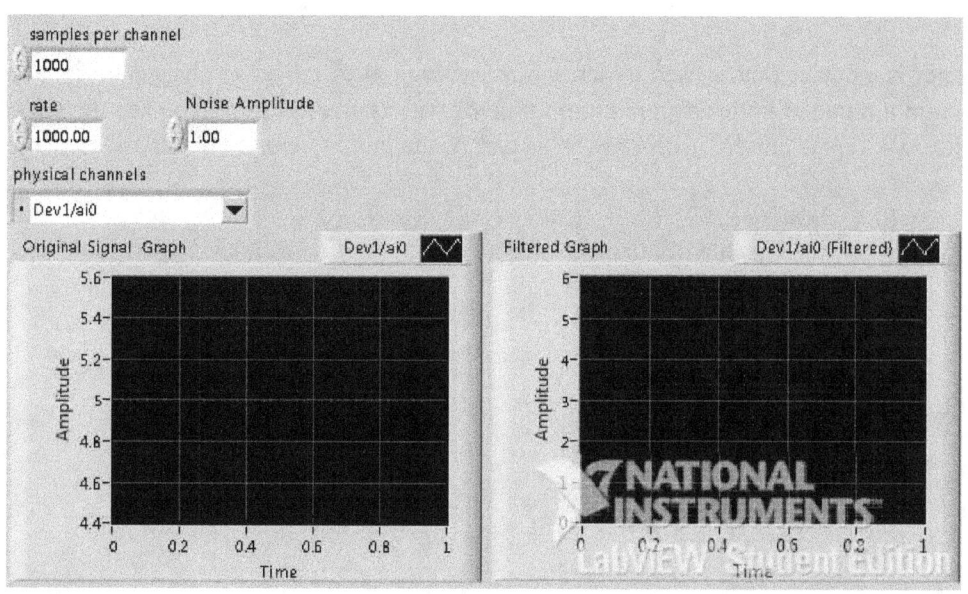

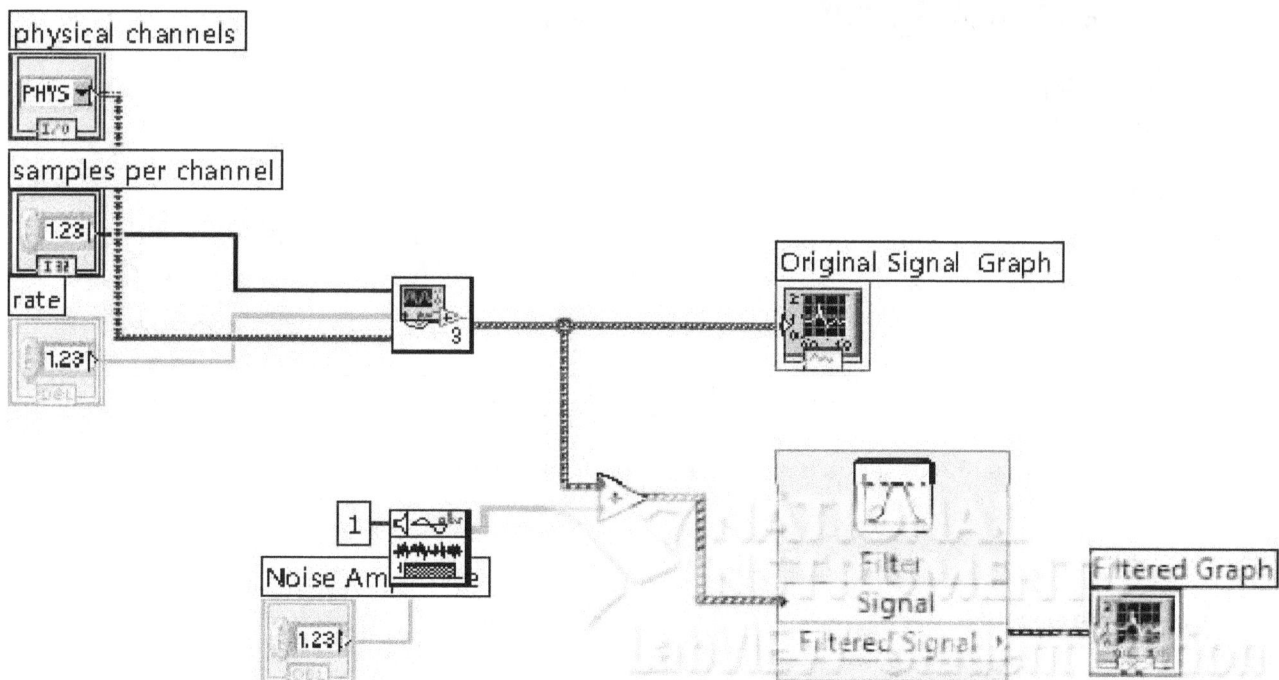

Figure 3-18

65

3.5.9 Battery Tester

A battery tester is a device that is used to determine the amount of power or charge left in a battery. Such devices come in a range of styles depending on type of battery being examined. The device can come as an item on its own, on the battery itself or even as part of the item where the battery is used.

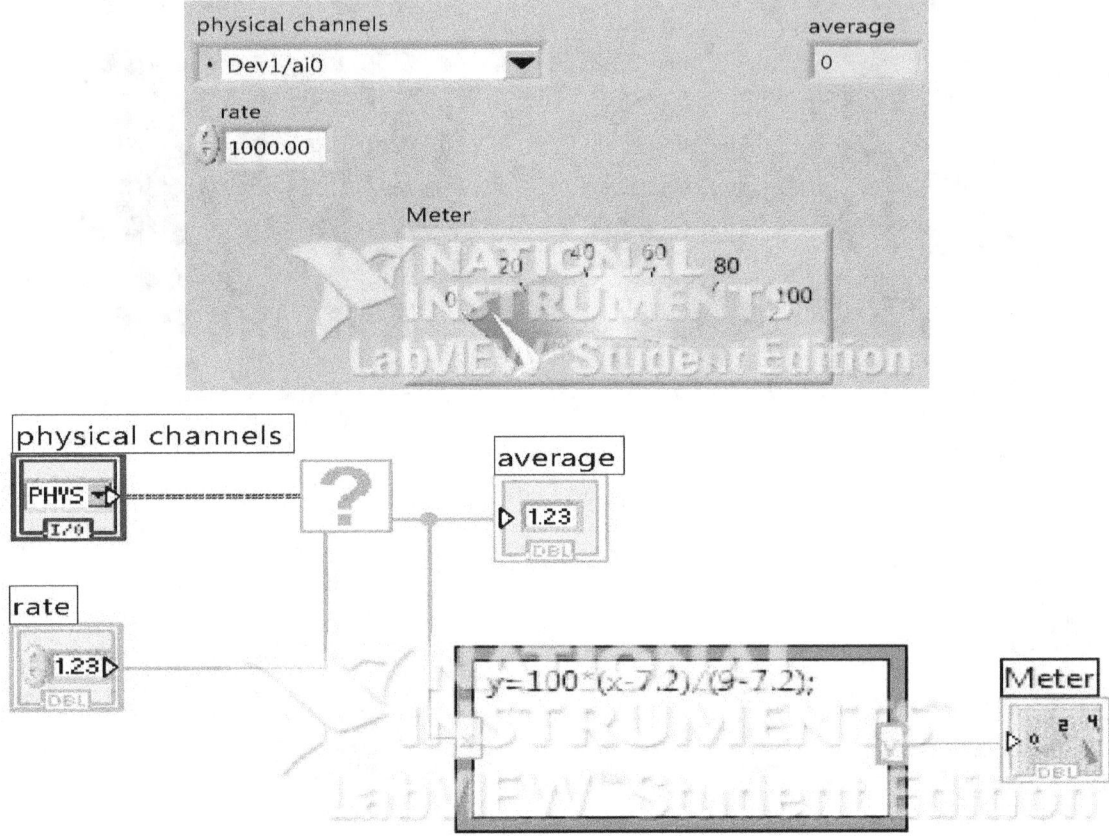

Figure 3-19 shows both front panel & block diagram of the Battery Tester

3.5.10 Guitar Tuner

It is an electronic device used by musicians to detect and display the pitch of notes played on musical instruments. The simplest tuners use LED lights to indicate approximately whether the pitch of the note played is lower, higher, or approximately equal to the desired pitch. The figure 3-20 is shown the block diagram of the Guitar Tuner.

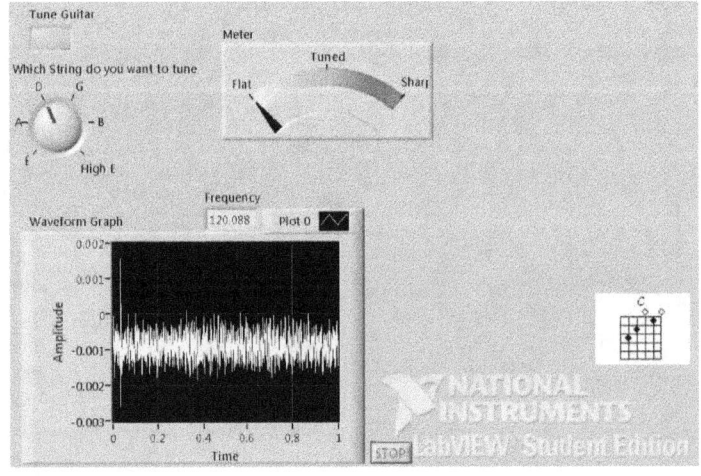

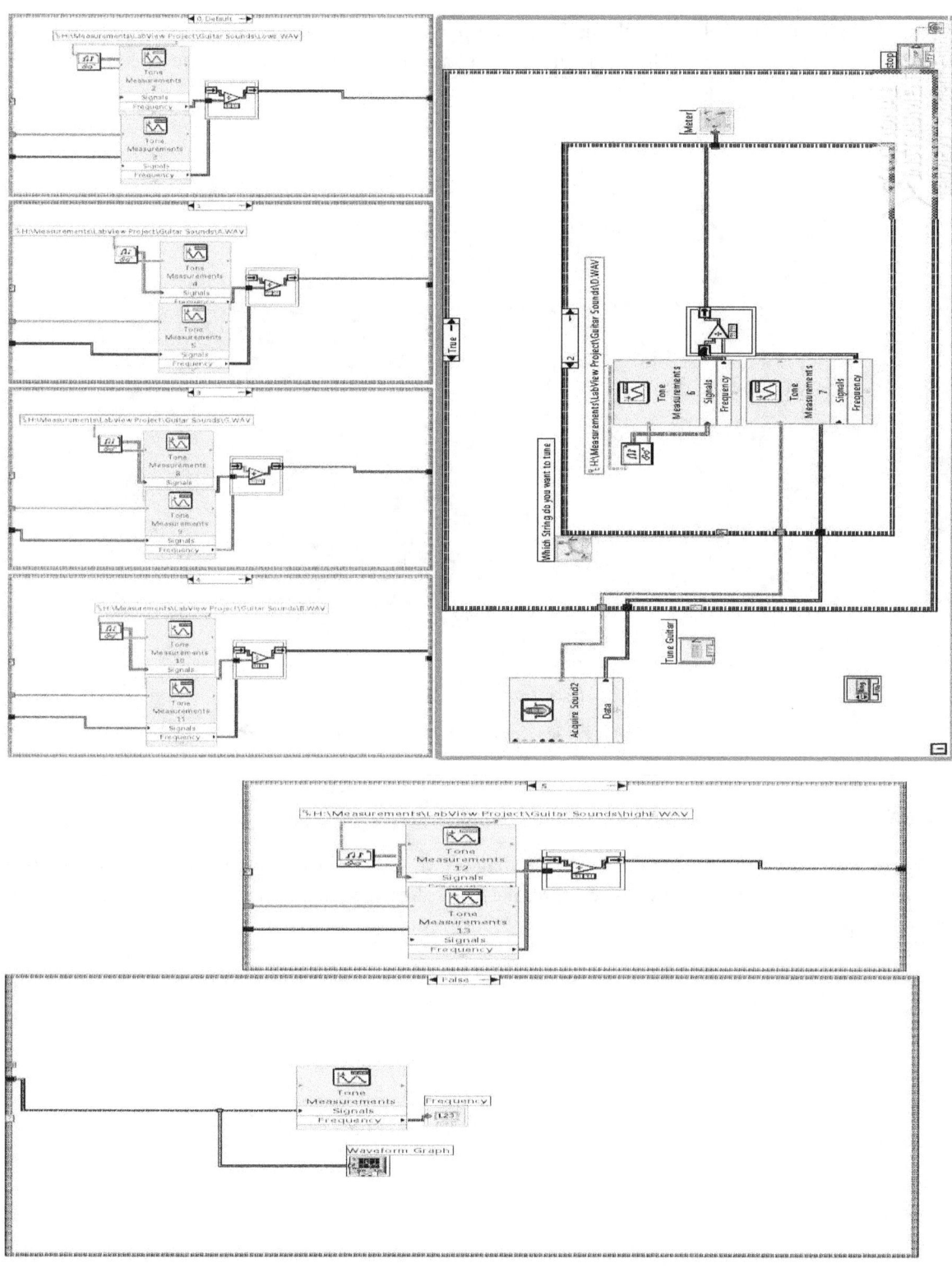

Figure 3-20

3.5.11 Heliocentric Model of the Solar System

It is the astronomical model in which the Earth and planets revolve around a relatively stationary Sun at the center of the Solar System. This program plots out accurate heliocentric distance of each planet in our solar system, see its front panel below.

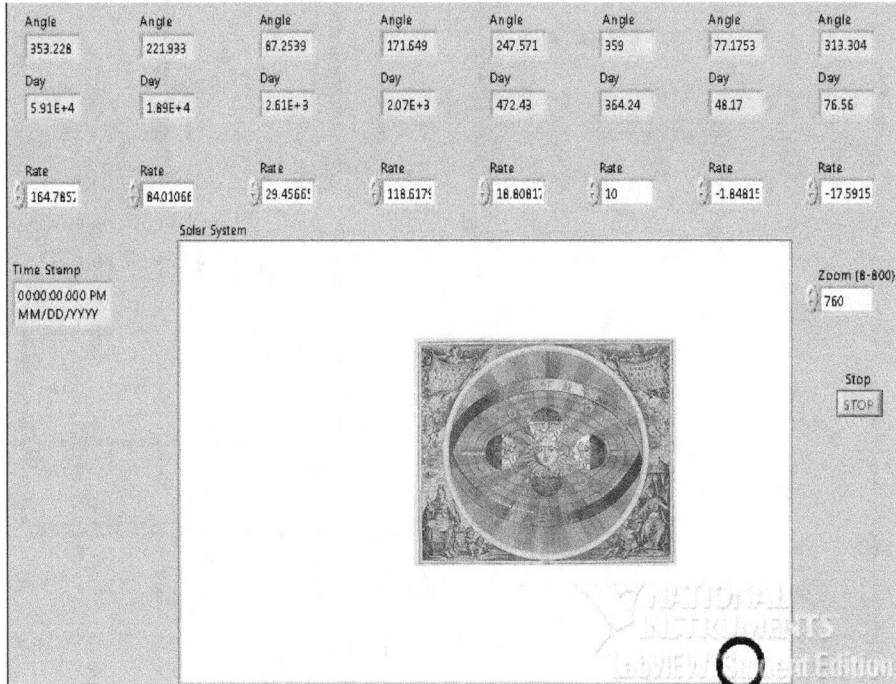

Home Work:

It is preferred to draw and sketch out the block diagram of the Heliocentric model of the solar system, as seen in the front panel figure above. Therefore, the block diagram should match the same functions shown in the front panel.

Chapter 4

LabVIEW Programming

4.1 Customizing the LabVIEW Environment:

Description: LabVIEW has lots of possibilities for customizing the appearance and the use of the LabVIEW environment. Select "Options…" from the Tools menu like in figure 4-1.

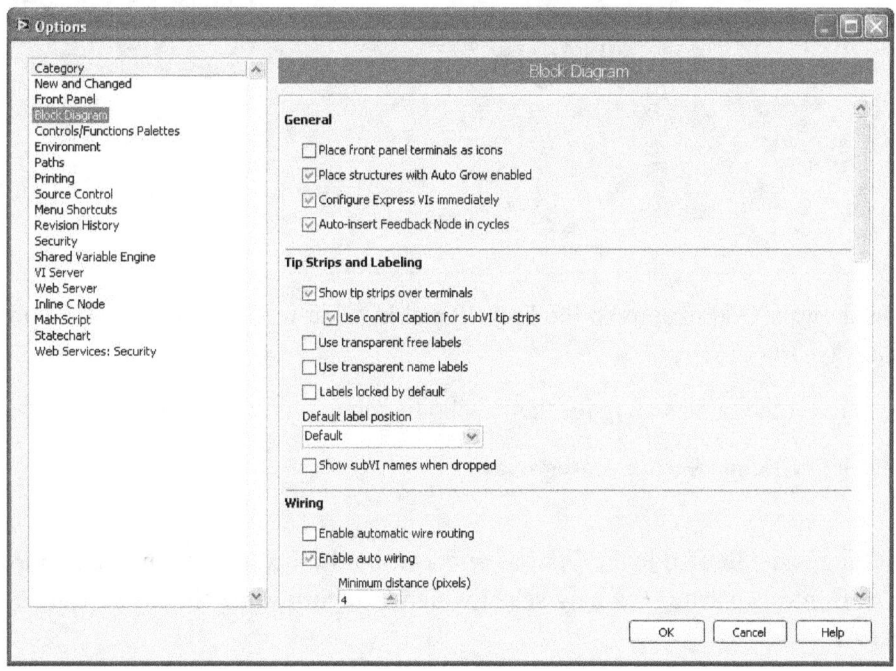

Figure 4-1

In this example you will customize the LabVIEW Environment so it bests fits your demands. The example will go through the most important settings in the Options window (Select "Options…" from the Tools menu). The default settings is not necessary the best, here are some recommendations for setting up the LabVIEW environment.

➢ Try the different settings above and see the difference and make your own personal choice. Setting: "**Place front panel terminals as icons**" (Category: Block Diagram – General).

➢ Setting: "**Enable automatic wire routing**" (Category: Block Diagram – Wiring) as shown in the figure below.

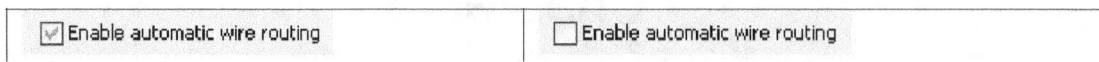

This prevents LabVIEW from automatically connecting adjacent blocks. The setting to the right is my personal (and recommended) favorite. When you use the setting to the right you have more control and you may easy switch between the tools using the **Tab** key. Although it seems useful to have auto wiring enables, it is my experience that the auto wiring is a little annoying since it tends to draw wires between blocks when you do not want any wire.

➢ Setting: "**Show front panel grid**" (Category: Front Panel – Front Panel Grid)

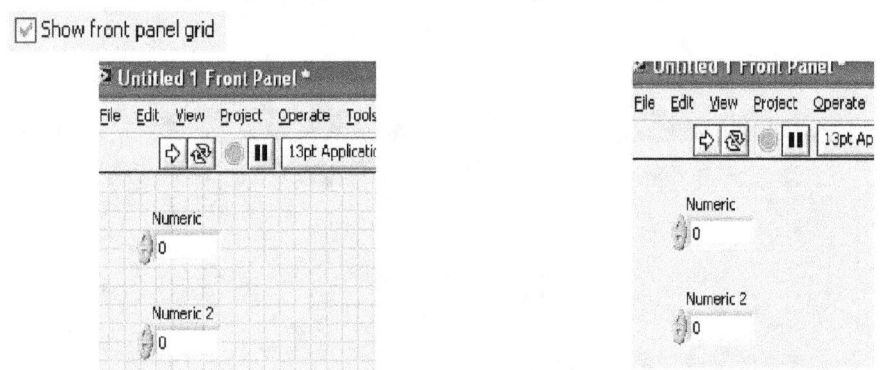

This setting shows a Grid pattern on the Front Panel in "Edit mode", this setting could be distracting, but that is not always true.

Note: One may set the same setting for the Block Diagram.

➢ Setting: "**Change Visible Categories**"

The next setting is not located in the Options window, but I think it is worth mention is this context. In the Functions or Controls palette, you may select which Categories that should be visible. I recommend that you "Select All".

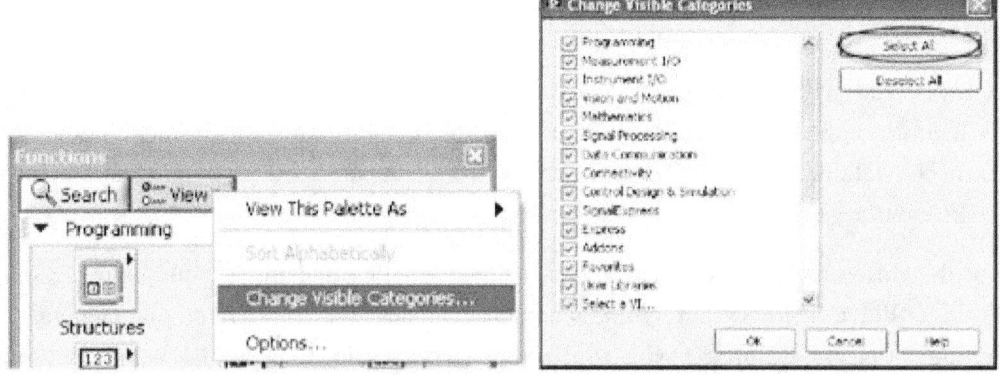

4.2 LabVIEW Wires and Variables:

Description: In text-based programming languages, you store and access data with functions through the use of variables. In the LabVIEW graphical programming language, wires implicitly handle all of the data storage and access that are associated with variables in text-based languages. Think of wires as a path for data to flow. Data comes into block diagram objects through a wire and can leave only through a wire. Local (or Global) Variables are used to pass data when a wire in some situations cannot be used.

- **Wires**: Create a program where you use **Wires** and **Shift Registers** to update data as shown in figure 4-2.

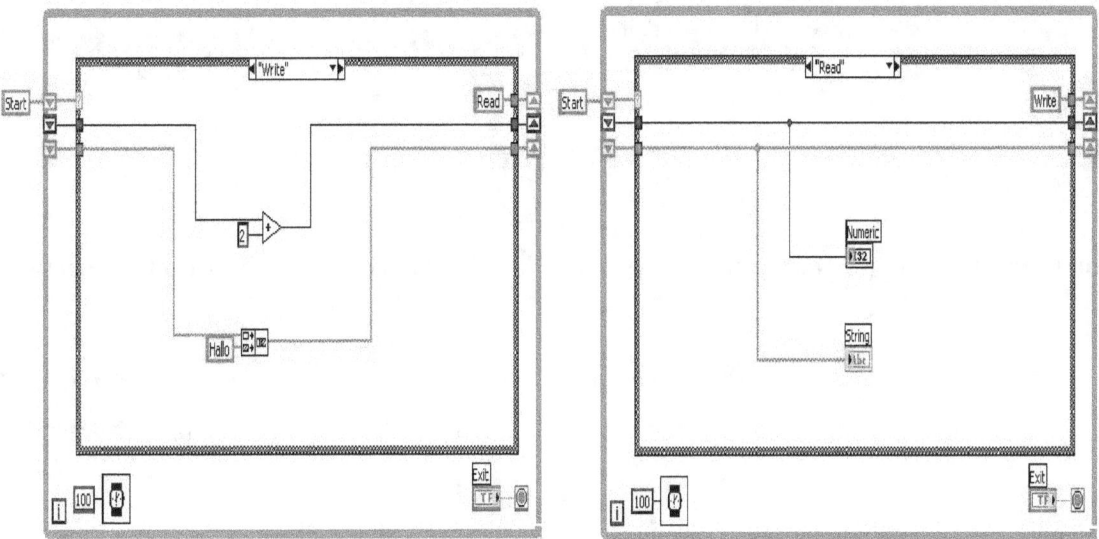

Figure 4-2

- **Local Variables:** The Local Variable item is located on the Structures palette on the Block Diagram, see the figure 4-3.

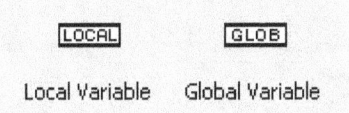

Figure 4-3

When someone places a Local variable on the Block Diagram, it looks like a Question mark as seen in figure 4-4. Then you right-click on the Local variable and choose "Select Item" and select which Control/Indicator you want to connect it to. Another way to create a local variable is to right-click on a Control/Indicator either on the Front Panel or the Block Diagram and select "Create → Local Variable".

71

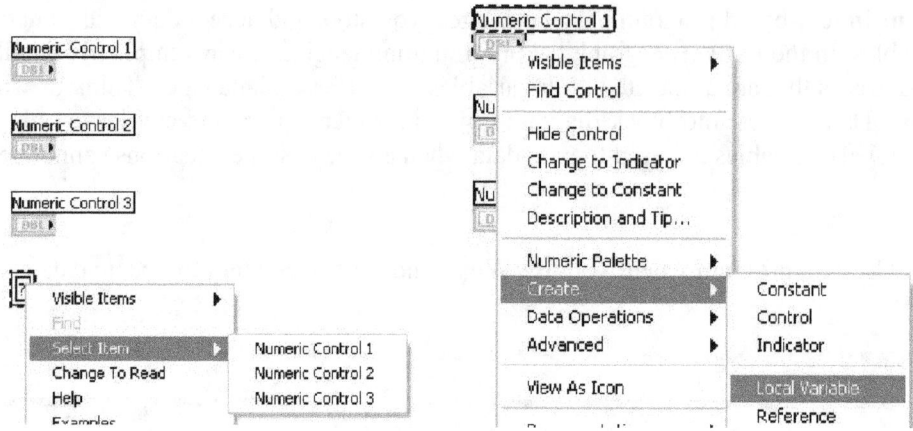

Figure 4-4

Demo: Create the same program as in the previous task and use Local Variables instead as seen the figure below.

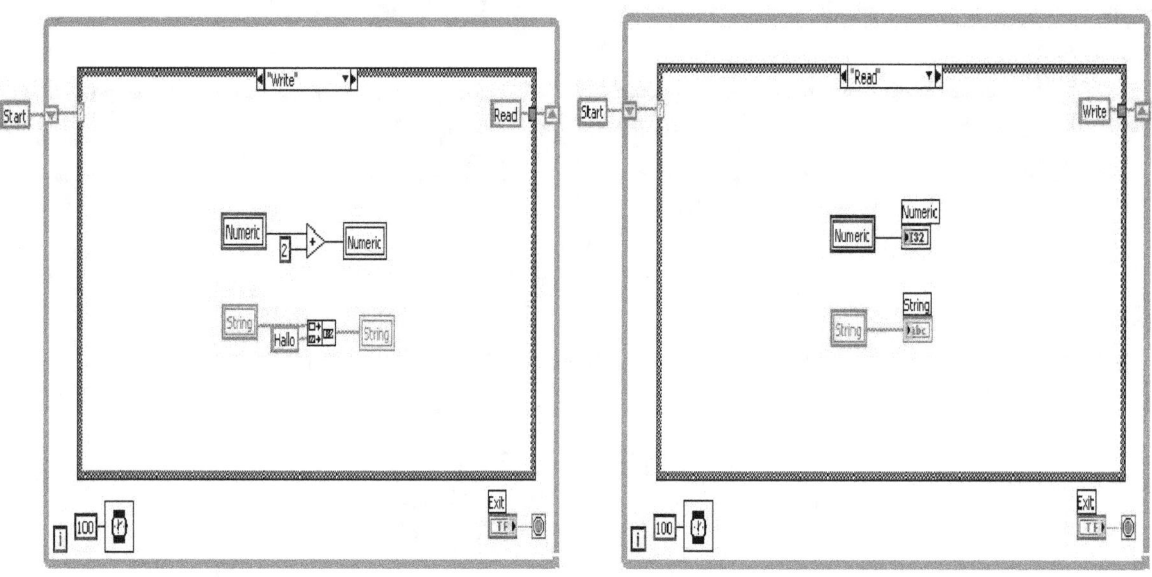

4.2.1 Global Variables:

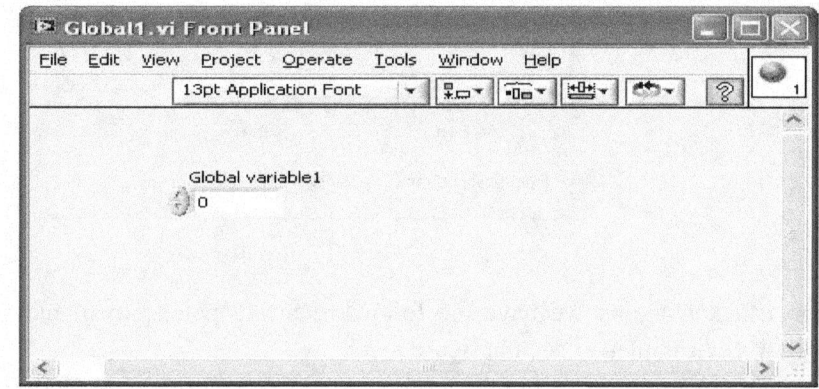 Use global variables to access and pass data among several VIs. When you create a global variable, LabVIEW automatically creates a special global VI, which has a front panel but no block diagram. The Global Variable item is located on the Structures palette on the Block Diagram. When you place a Local variable on the Block Diagram, it looks like a Question mark with a globe, as seen in figure 4-5. Double-clink on the item in order to create the Global Variable.

Figure 4-5

Demo: Create 2 VIs that uses a Global variable to exchange data between them as seen in the figure below.

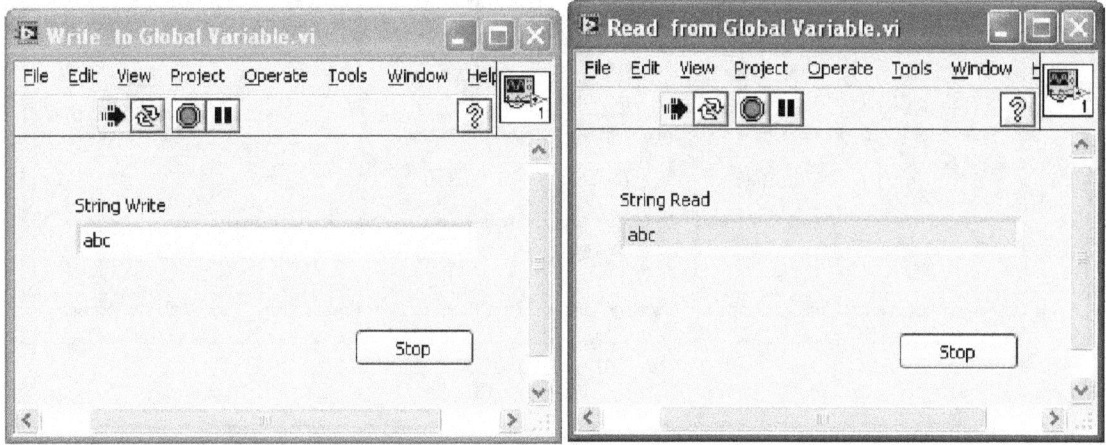

4.3 LabVIEW Strings:

Description: Working and manipulating with strings is an important part in LabVIEW development. On the Front panel we have the following String controls and indicators available from the Control palette shown in figure 4-6.

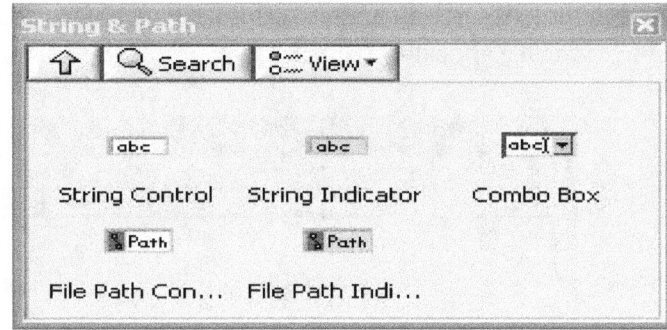

Figure 4-6

On the Block Diagram we have the following String palette available from the Functions palette in LabVIEW environment as seen in figure 4-7.

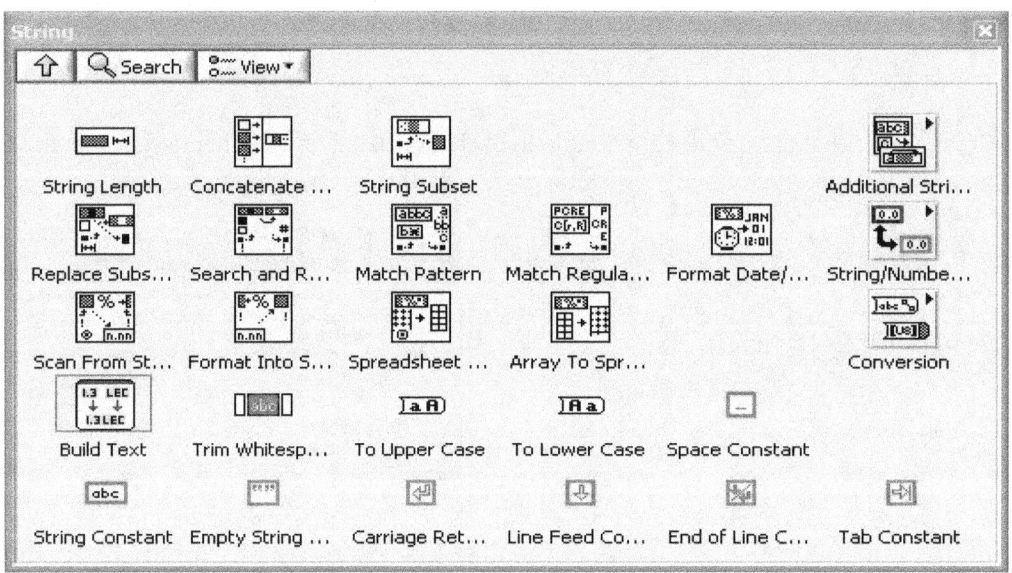

Figure 4-7

In the following Example you will learn how to use strings and string manipulation in LabVIEW:

Demo 1: Concatenate Strings. Create the following:

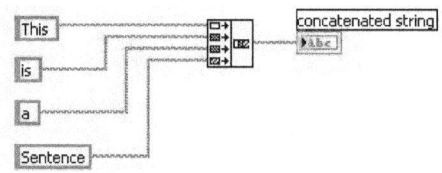

Demo 2: Search and Replace String. Create the following:

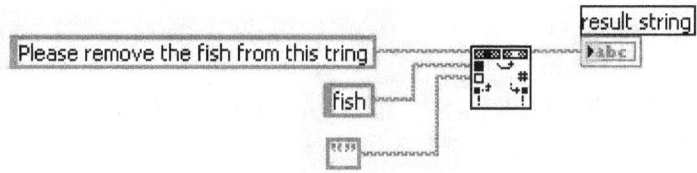

Demo 3: Match Pattern. Create the following:

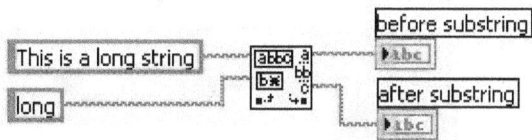

Demo 4: Format into String. Create the following:

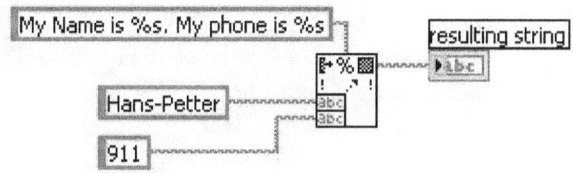

Demo 5: String to Number and **Number to String**. Create the following:

Demo 6: Create the following Example. You will need all you have learned above and more.

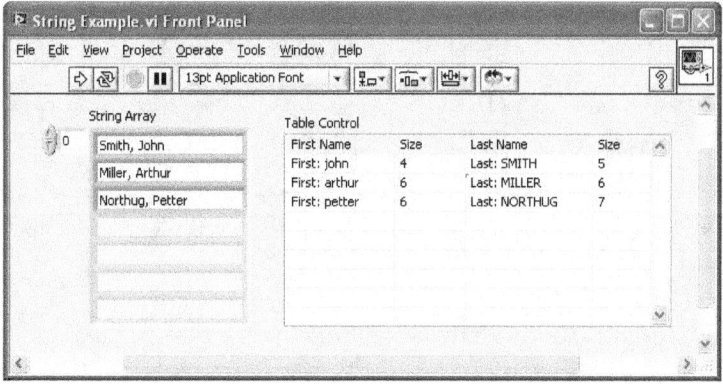

4.4 LabVIEW Arrays:

Description: Working and manipulating with Arrays is an important part in LabVIEW development. Arrays are very powerful to use in LabVIEW. In all your applications you would probably use both One-Dimensional Arrays and Two-Dimensional Arrays. On the Front Panel using the Control palette we can create an array as follows (Array, Matrix & Cluster sub palette):

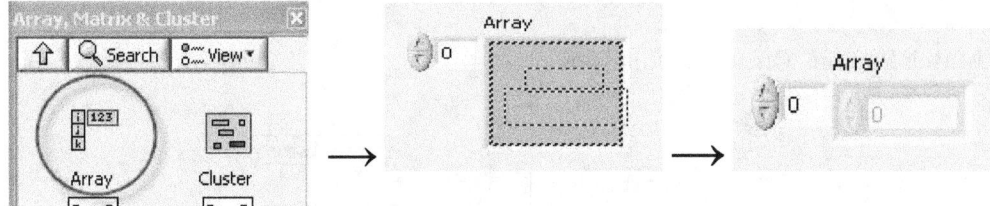

Drag and drop the empty Array on the Front Panel, next you find a Control or Indicator (Numeric, String, Boolean) and drag it into the empty Array. You can create an Array of (almost) any kind of Control or Indicator. 2D or multidimensional Array, just drag the mouse in the Index display to the left and increase the dimension, as seen in figure 4-8.

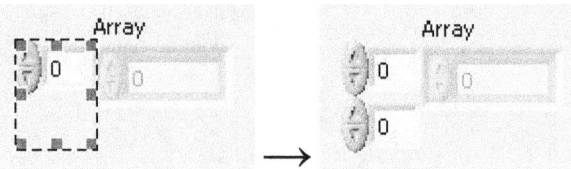

Figure 4-8

On the Block Diagram the following Array palette is available from the Functions palette in LabVIEW:

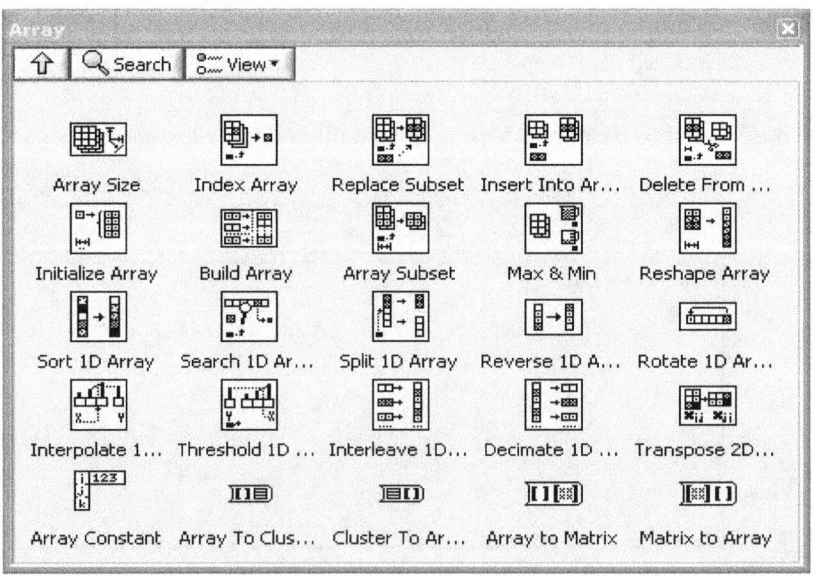

76

All these functions are basic (but very useful) array functions you will probably be using in all your applications and VIs.

4.4.1 Build Array:

This function concatenates multiple arrays or appends elements to an n-dimensional array. Try the simple example shown in figure 4-9. This example using the Build Array function inside a For loop in order build an array with 10 elements.

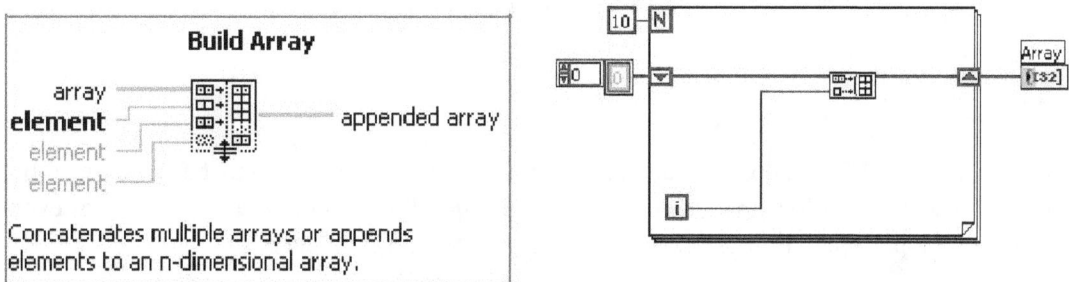

Figure 4-9

4.4.2 Index Array:

This function returns the element or sub array of n-dimension array at index. It is always useful to find a specific value in an array. The Index Array is extendible, so you can drag it out to find more than one elements. Try the simple example as seen in the figure 4-10.

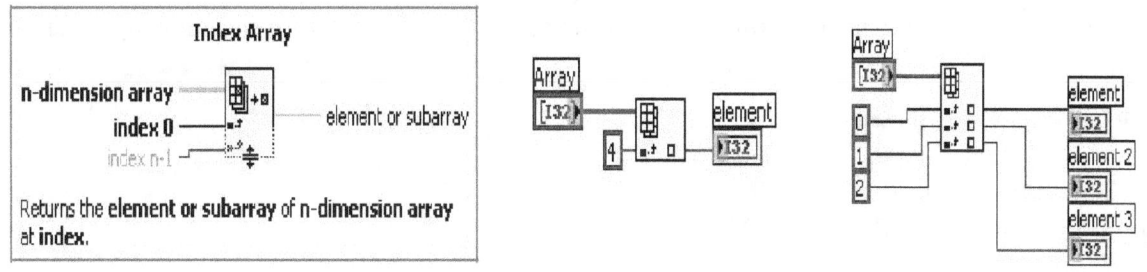

Figure 4-10

4.4.3 Array Size:

This function returns the number of elements in each dimension of array. Try the simple example below.

The example finds the size of an arbitrary array, see figure 4-11.

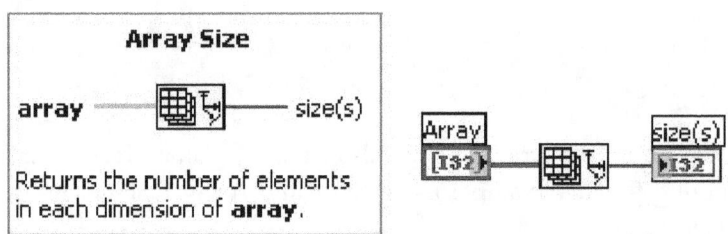

Array Size

array ———— ⊞ —— size(s)

Returns the number of elements
in each dimension of **array**.

Figure 4-11

4.4.4 Auto-indexing:

If you wire an array to a For Loop, you can read and process every element in that array by enabling auto-indexing. You also can enable auto-indexing by configuring a For Loop to return an array of every value generated by the loop. Create a simple example in order to see the difference.

4.5 LabVIEW Sub VI:

Description: Sub VIs are very useful in LabVIEW. Using Sub VI helps you manage changes and debug the Block Diagram quickly. You can also easily reuse your code. Sub VIs are equal to **functions** in text based languages.

Demo 1: Create a Sub VI that performs a linear scaling $y = ax + b$. Where, a , b and x are inputs, and y is an output.

The Procedure is as follows:

Step 1: Create a New VI (File→ New VI) (Blank VI)

Step 2: Give the VI a Name (**Linear Scaling.vi**)

Step 3: Create your Front Panel with your necessary Controls and Indicators

Step 4: Create your Block Diagram. The Block Diagram could look something like the figure below:

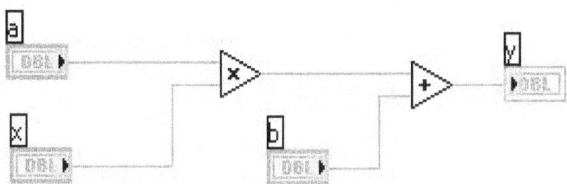

Step 5: Create the Input and Output Connectors. Right-click on the little icon in the upper right corner and select "Show Connector".

 Select the Wire tool and click on the wanted connector, then click on the Control or Indicator on the Front Panel you want to connect to this connector.

Step 6: Create an Icon using the Icon Editor. Right-click on the little icon in the upper right corner and select "Edit Icon…".

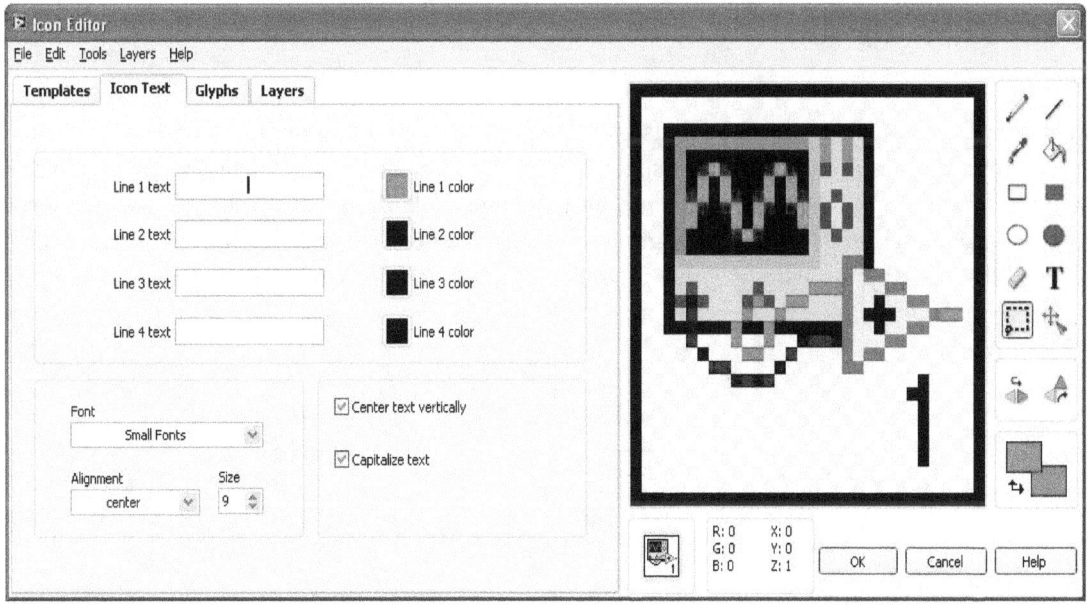

Step 7: Create a new VI that you use to test your Sub VI as seen below;

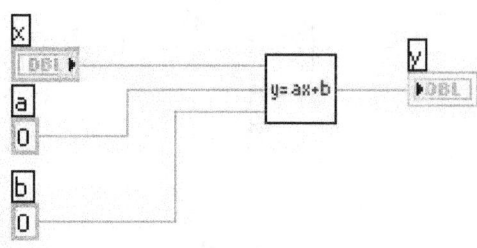

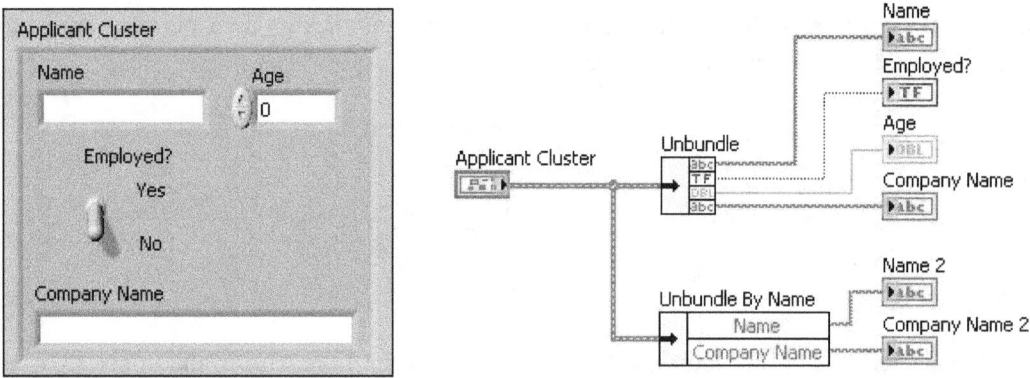

4.6 LabVIEW Clusters:

Description: Clusters group data elements of mixed types, such as a bundle of wires, as in a telephone cable, where each wire in the cable represents a different element of the cluster. A cluster is similar to a record or a structure in text-based programming languages. Bundling several data elements into clusters eliminates wire clutter on the block diagram and reduces the number of connector pane terminals that sub VIs need. Like an array, a cluster is either a control or an indicator. A cluster <u>cannot</u> contain a mixture of controls and indicators.

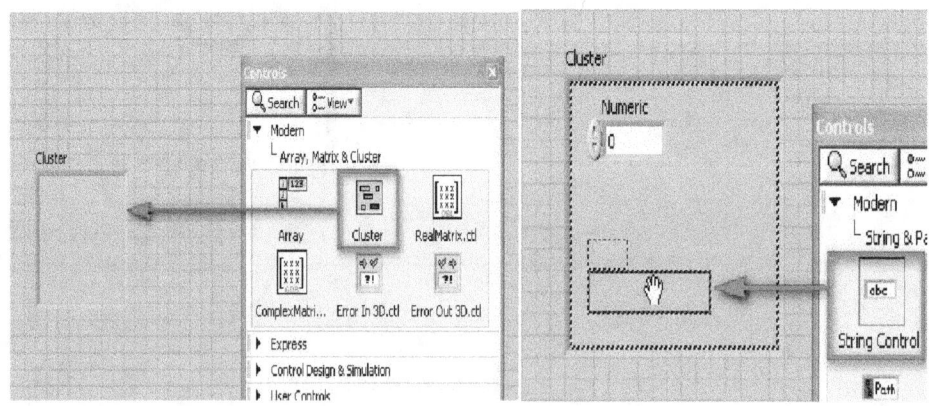

4.6.1 Cluster functions:

In the Cluster, Class & Variant sub palette on the Block Diagram we have the following Cluster functions we may use to manipulate and get data in or out of a cluster. In this example we will create clusters and use these functions.

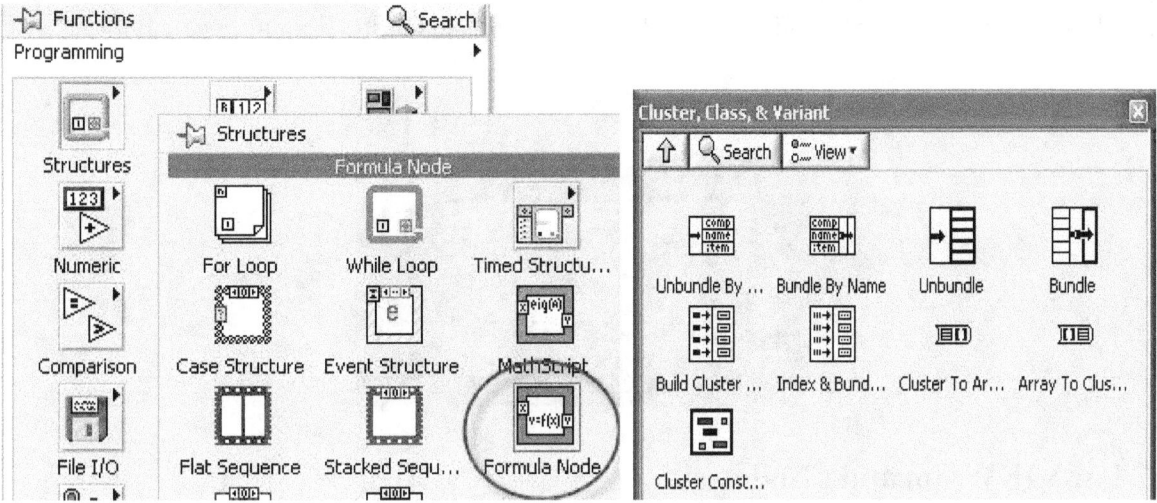

4.6.2 Unbundle/Unbundle By Name:

Use the **Unbundle** functions to disassemble a cluster into its individual elements. Use the **Unbundle by Name** function to return specific cluster elements you specify by name. You can also resize these functions for multiple elements using the mouse. Create the following code:

4.6.3 Bundle/Bundle By Name:

Use the **Bundle** function to assemble a cluster from individual elements. To wire elements into the Bundle function, use your mouse to resize the function. Create the following diagram:

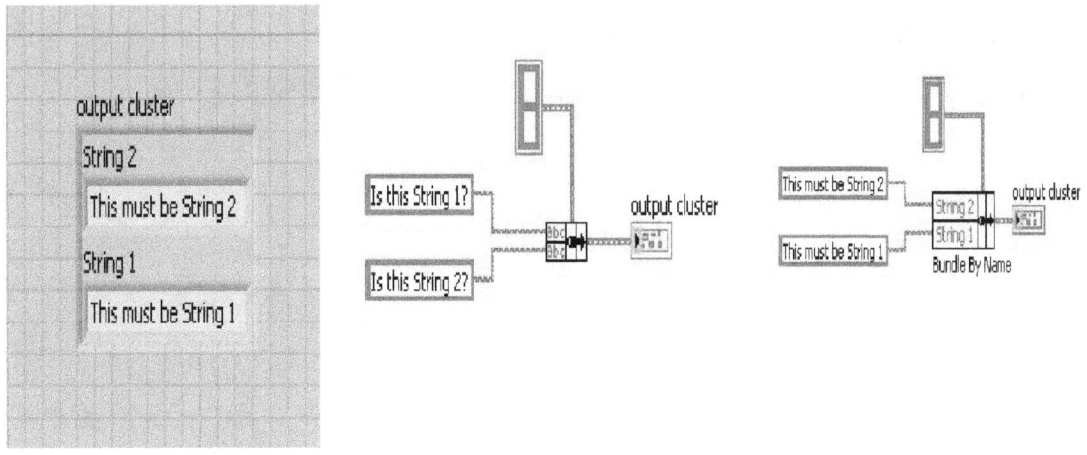

4.6.4 Cluster Order:

Cluster elements have a logical order unrelated to their position in the shell. The first object you place in the cluster is element 0, the second is element 1, and so on. If you delete an element, the order adjusts automatically. The cluster order determines the order in which the elements appear as terminals on the Bundle and Unbundle functions on the block diagram. You can view and modify the cluster order by right-clicking the cluster border and selecting Reorder Controls in Cluster from the shortcut menu.

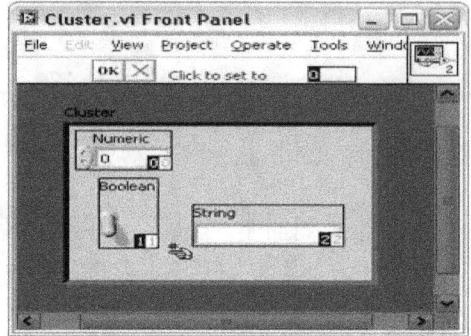

4.7 LabVIEW Formula Node:

Description: A Formula Node in LabVIEW evaluates mathematical formulas and expressions similar to C on the block diagram. In this way you may use existing C code directly inside your LabVIEW code. It is also useful when you have "complex" mathematical expressions.

> **Create a simple Sub VI**: where you use the Formula Node to calculate *a (slope)* and *b (intercept)* in the equation $y = ax + b$ when you have two points (x_1, y_1) and (x_2, y_2).

The Procedure is as follows:

<u>Step 1:</u> Create a New VI (File →New VI) (Blank VI)

<u>Step 2:</u> Give the VI a Name (**Linear Scaling.vi**)

<u>Step 3:</u> Create your Front Panel with your necessary

Controls and Indicators.

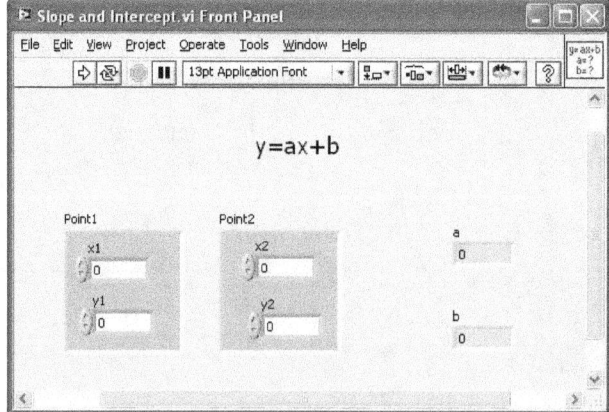

Step 4: Switch to your Block Diagram (**Ctrl+E**).

Step 5: Add the **Formula Node** to you Block Diagram.

Step 6: Add Inputs and Outputs.

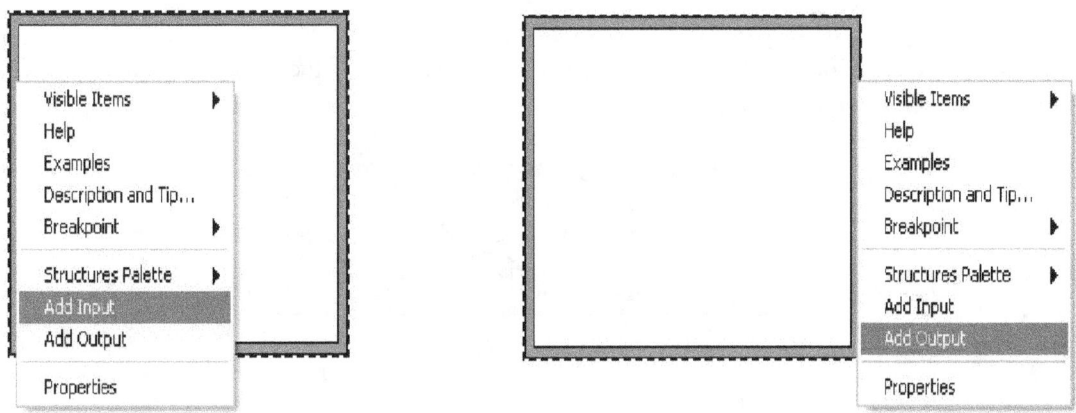

Step 7: Create your C-code inside your Formula Node.

The formula for finding the slope (a) and intercept (b) is as follows:

$$y - y_1 = \frac{y_2 - y_1}{x_2 - x_1}(x - x_1), where\ a = \frac{y_2 - y_1}{x_2 - x_1}$$

The Block Diagram could look something like the following figure:

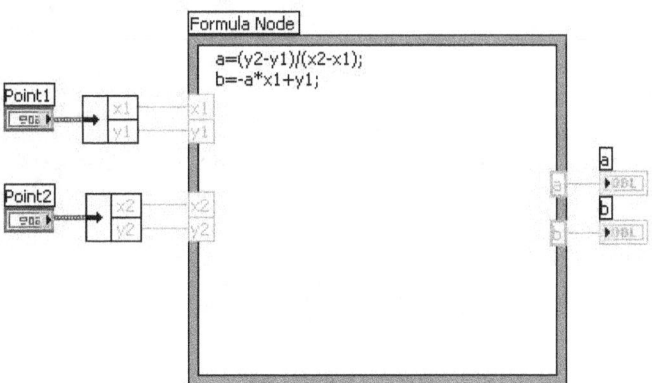

Step 8: Create the Input and Output Connectors. Right-click on the little icon in the upper right corner and select "Show Connector".

Step 9: Create an Icon using the Icon Editor. Right-click on the little icon in the upper right corner and select "Edit Icon…".

Step 10: Create a new VI that you use to test your Sub VI.

4.8 LabVIEW Debugging:

Description: LabVIEW offers different debugging techniques, such as **Highlight Execution**, **Probes**, **Breakpoints** and **Single Stepping**. In this example we will use these debugging techniques on an existing VI.

> **Highlight Execution:** View an animation of the execution of the block diagram by clicking the **Highlight Execution** button.

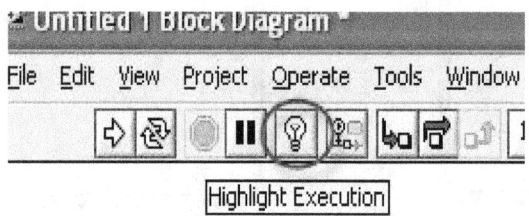

Execution highlighting shows the movement of data on the block diagram from one node to another using bubbles that move along the wires. Use execution highlighting to see how data values move from node to node through a VI.

> **Probes:** Use the Probe tool to check intermediate values on a wire as a VI runs. Use the Probe tool if you have a complicated block diagram with a series of operations, any one of which might return incorrect data. If data is available, the probe immediately updates and displays the data during execution. Add a Probe by right clicking a wire and select "Probe". Probes are also available from the Tools Palette.

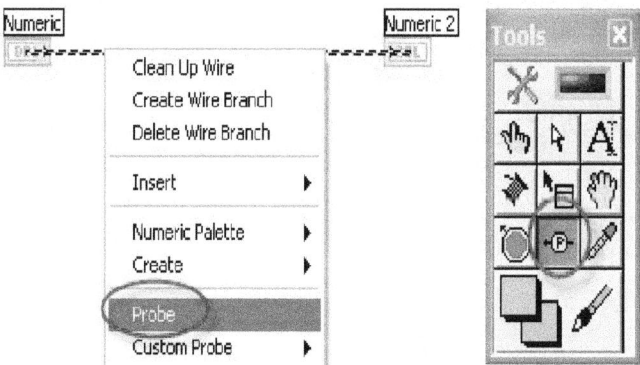

Use the **Probe Watch window** to manage and watch values in all your probes:

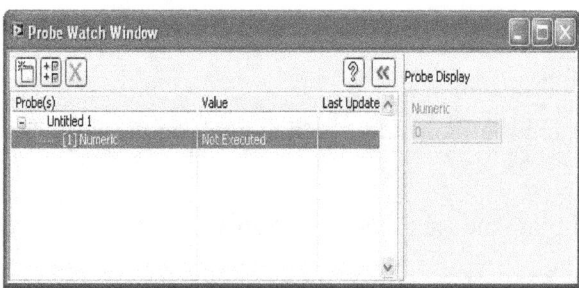

➢ **Breakpoints**: Use the Breakpoint tool to place a breakpoint on a VI, node, or wire and pause execution at that location. When you set a breakpoint on a wire, execution pauses after data passes through the wire and the Pause button appears red. Right-click on a wire to set a breakpoint or use the Tools Palette as shown below.

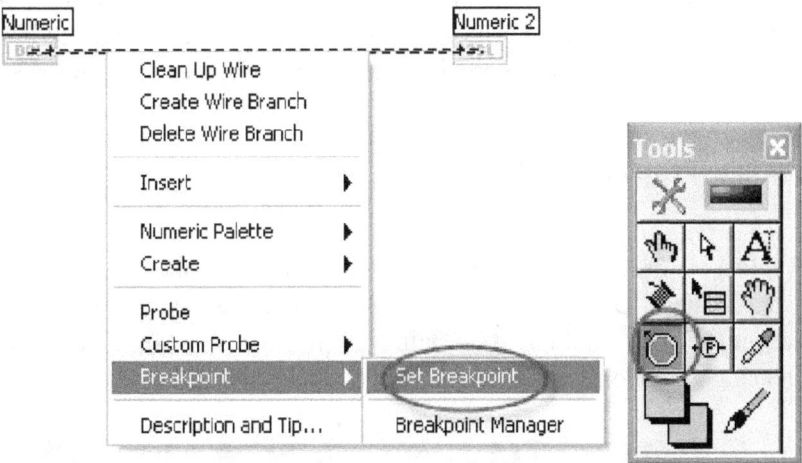

The **Breakpoint Manager** (available from the View menu) lists all your breakpoints.

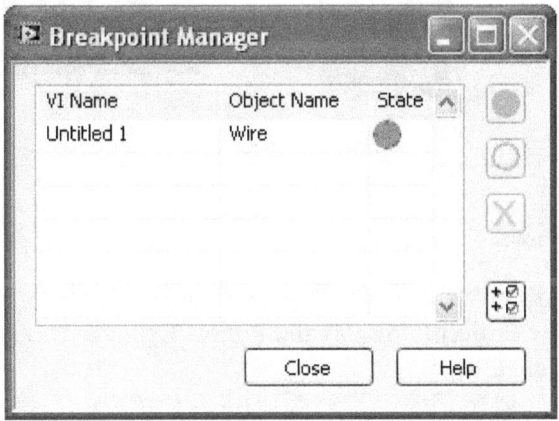

➢ **Single Stepping**: Single-step through a VI to view each action of the VI on the block diagram as the VI runs. The single-stepping buttons, shown as follows, affect execution only in a VI or sub VI in single-step mode.

Note: When you single-step through a VI, nodes blink to indicate they are ready to execute. If you single-step through a VI with execution highlighting on, an execution glyph appears on the icons of the sub VIs that are currently running.

4.9 LabVIEW Project Explorer:

Description: Projects in LabVIEW consist of VIs, files necessary for those VIs to run properly, and supplemental files such as documentation or related links. Use the Project Explorer window to manage projects in LabVIEW. In the Project Explorer window, you can use folders and libraries to group together items, and you can use a list of VI hierarchies called Dependencies to keep track of items a VI depends on.

4.9.1 Create a new Project in LabVIEW:

Create several folders in order to organize your project. Create a folder for your Main VI and a folder for your Sub VI, etc. Create a "Dummy" Application with some VIs that you insert into these folders. At the end create an **executable** file (.exe) of your main VI, e.g., "MyApp.exe". The Example below simulates a process system using a Random generator inside a While Loop.

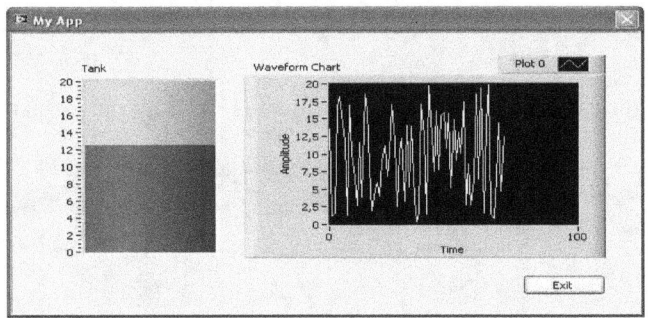

The Procedure is as follows:

Step 1: Create a **New Project**. Use the Project Explorer window to create and edit LabVIEW projects. Select File» New Project to display the Project Explorer window, or select "Empty Project" in the "Getting Started" window.

Step 2: Create Folders and Files (VIs) and test your application.

Step 3: Create an **Executable** file (.exe) of your Application.

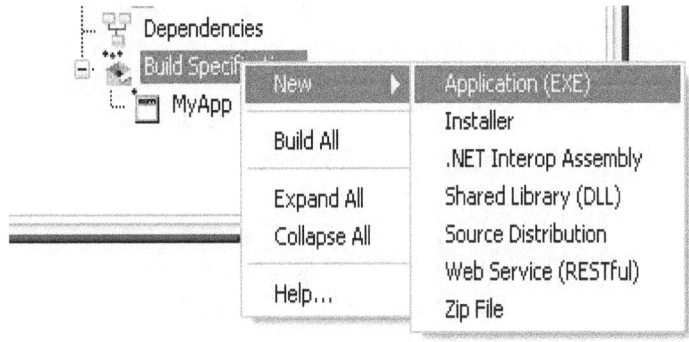

Step 4: ConFigure the **Properties** for your Executable Application:

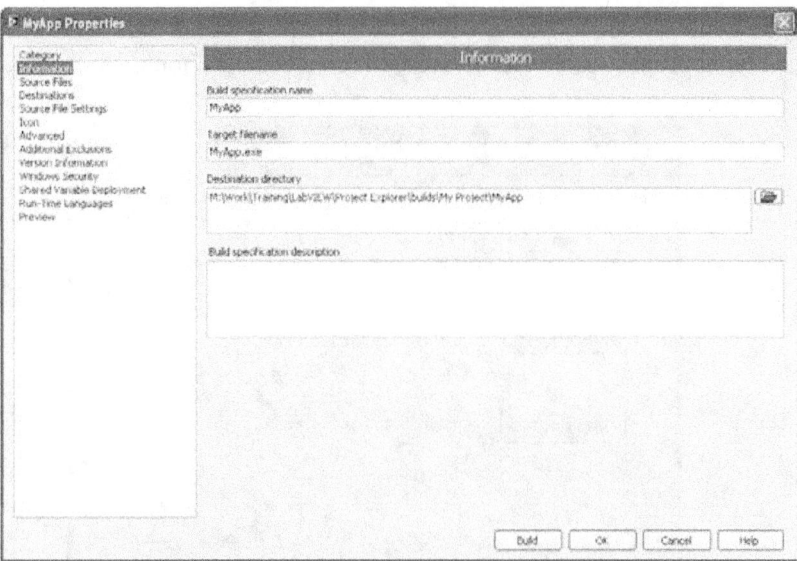

Step 5: Finished. The Project Explorer could look something like the figure below:

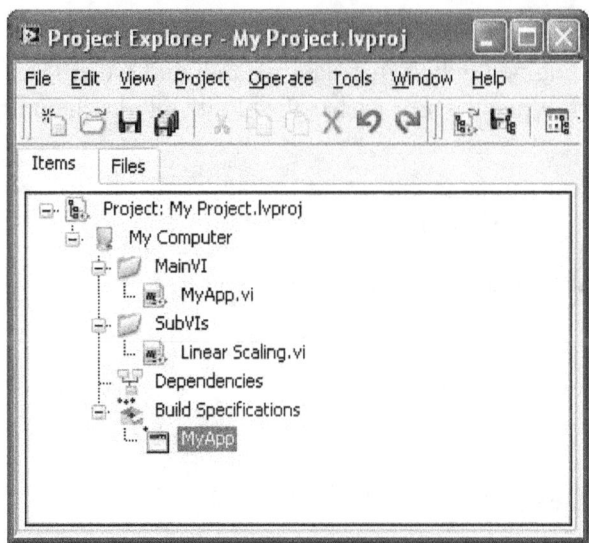

4.10 LabVIEW State Machine:

Description: The **state machine** is one of the fundamental architectures LabVIEW developers frequently use to build applications quickly. The State Machine approach in LabVIEW uses a **Case structure** inside a **While loop** to handle the different states in the program, and the transitions between them. A **Shift Register** is used to save data from and between the different states.

4.10.1 Create a simple program in LabVIEW

Using the State Machine approach. The User Interface (Front Panel) should look something like figure 4-12:

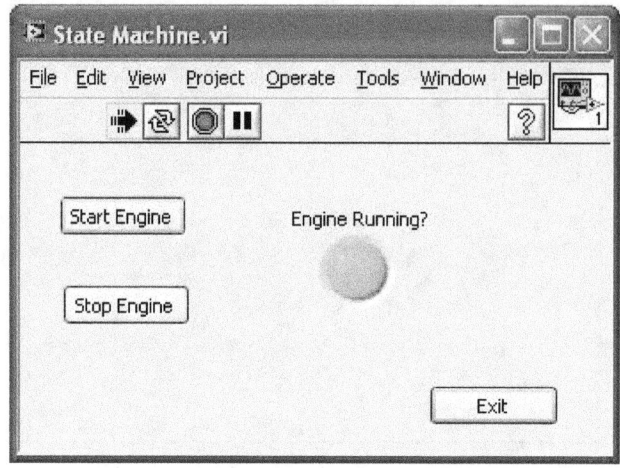

Figure 4-12

When the user push the "Start Engine" button, the light gets on and when the user push the "Stop Engine" button, the light gets off. When the user pushes the "Exit" button, the program will stop.

The Procedure is as follows:

Step 1: Create a **New VI** (File →New VI) (Blank VI)

Step 2: Create your **Front Panel** as in the Example above.

Step 3: Add a **While Loop** to your VI

Step 4: Add a **Case Structure** to your VI <u>inside</u> the While Loop

Step 5: Add a **Shift Register** to your While Loop by right-click on the While Loop and select "Add Shift Register".

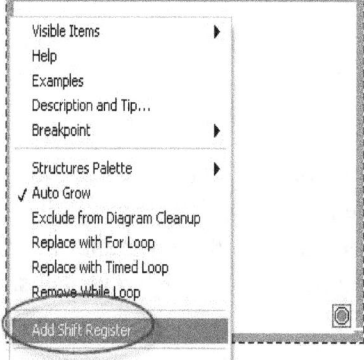

Step 6: Add a Case called Wait in your Case Structure. Create an **Event Structure** inside this case. Add Event for your 3 different buttons; "Start Engine", "Stop Engine" and "Exit".

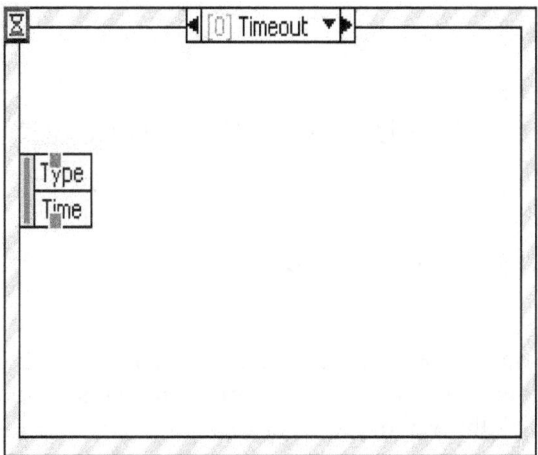

Step 7: Finish the Code.

Here is an extract of the program:

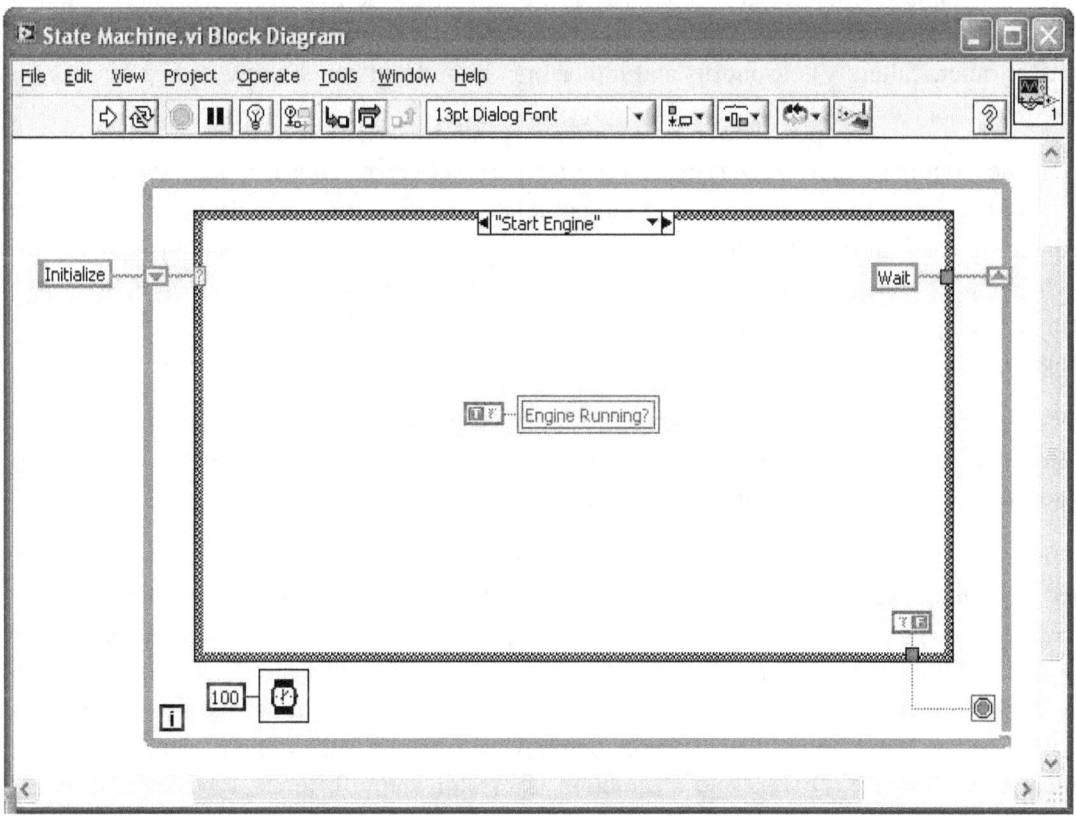

Chapter 5

Database Communication

5.1 Introduction to Dataflow programming

As we've seen in the previous chapter, the programming language used in LabVIEW has referred to dataflow programming language. Execution is determined by the structure of a graphical block diagram (the LV-source code) on which the programmer connects different function-nodes by drawing wires.

These wires propagate variables and any node can execute as soon as all its input data become available. Since this might be the case for multiple nodes simultaneously, dataflow is inherently capable of parallel execution. Multi-processing and multi-threading hardware is automatically exploited by the built-in scheduler, which multiplexes multiple OS threads over the nodes ready for execution.

Also, LabVIEW ties the creation of user interfaces (called front panels) into the development cycle. LabVIEW programs/subroutines are called virtual instruments (VIs). Each VI has three components: a block diagram, a front panel, and a connector panel. The last is used to represent the VI in the block diagrams of other, calling VIs. Controls and indicators on the front panel allow an operator to input data into or extract data from a running virtual instrument.

However, the front panel can also serve as a programmatic interface. Thus a virtual instrument can either be run as a program, with the front panel serving as a user interface, or, when dropped as a node onto the block diagram, the front panel defines the inputs and outputs for the given node through the connector pane. This implies each VI can be easily tested before being embedded as a subroutine into a larger program. The graphical approach also allows non-programmers to build programs simply by dragging and dropping virtual representations of lab equipment with which they are already familiar.

LabVIEW programming environment, with the included examples and the documentation, makes it simple to create small applications. This is a benefit on one side, but there is also a certain danger of underestimating the expertise needed for good quality programming.

For complex algorithms or large-scale code, it is important that the programmer possess an extensive knowledge of the special LabVIEW syntax and the topology of its memory management. The most advanced LabVIEW development systems offer the possibility of building stand-alone applications. Furthermore, it is possible to create distributed applications, which communicate by a client/server scheme, and are therefore easier to implement due to the inherently parallel nature of dataflow-code. One benefit of LabVIEW over other development environments is the extensive support for accessing instrumentation hardware. Drivers and abstraction layers for many different types of instruments and buses are included or are available for inclusion. The abstraction layers offer standard software interfaces to communicate with hardware devices. Besides, the provided driver interfaces save program development time.

5.2 Database Systems

A database is an integrated collection of logically related records or files consolidated into a common pool that provides data for one or more multiple uses.

One way of classifying databases involves the type of content, for example: bibliographic, full-text, numeric, and image. Other classification methods start from examining database models or database architectures. The data in a database is organized according to a database model, and the relational model is the most common one used.

A Database Management System (**DBMS**) consists of software that organizes the storage of data. A DBMS controls the creation, maintenance, and use of the database storage structures of organizations and of their end users. It allows organizations to place control of organization-wide database development in the hands of Database Administrators (DBAs) and other specialists. In large systems, a DBMS allows users and other software to store and retrieve data in a structured way.

Database management systems are usually categorized according to the database model that they support, such as the network, relational or object model. The model tends to determine the query languages that are available to access the database. One commonly used query language for the relational database is SQL, although SQL syntax and function can vary from one DBMS to another. A great deal of the internal engineering of a DBMS is independent of the data model, and is concerned with managing factors such as performance, concurrency, integrity, and recovery from hardware failures. In these areas there are large differences between products, figure 5-1 is shown the block diagram of the SQL VI in LabVIEW.

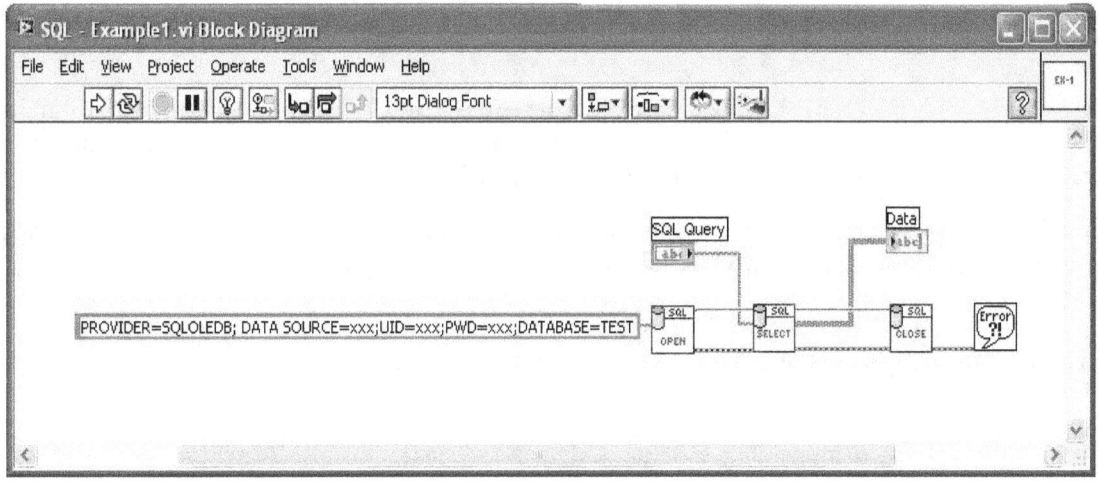

Figure 5-1

5.2.1 RDBMS Components

A Relational Database Management System (DBMS) consists of the following components:

91

- ☐ **Interface drivers** - A user or application program initiates either schema modification or content modification. These drivers are built on top of SQL. They provide methods to prepare statements, execute statements, fetch results, etc. An important example is the ODBC driver.

- ☐ **SQL engine** - This component interprets and executes the SQL query. It comprises three major components (compiler, optimizer, and execution engine).

- ☐ **Transaction engine** - Transactions are sequences of operations that read or write database elements, which are grouped together.

- ☐ **Relational engine** - Relational objects such as Table, Index, and Referential integrity constraints are implemented in this component.
- ☐ **Storage engine** - This component stores and retrieves data records. It also provides a mechanism to store metadata and control information such as undo logs, redo logs, lock tables, etc.

5.2.2 Data warehouse and Relational Database

A data warehouse stores data from current and previous years — data extracted from the various operational databases of an organization. It becomes the central source of data that has been screened, edited, standardized and integrated so that it can be used by managers and other end-user professionals throughout an organization.

A relational database matches data using common characteristics found within the data set. The resulting groups of data are organized and are much easier for people to understand. For example, a data set containing all the real-estate transactions in a town can be grouped by the year the transaction occurred; or it can be grouped by the sale price of the transaction; or it can be grouped by the buyer's last name; and so on.

Such a grouping uses the relational model (a technical term for this is schema). Hence, such a database is called a "relational database." The software used to do this grouping is called a relational database management system. The term "relational database" often refers to this type of software.

Relational databases are currently the predominant choice in storing financial records, manufacturing and logistical information, personnel data and much more. Strictly, a relational database is a collection of relations (frequently called tables).

5.2.3 Real-Time Databases and Database Management Systems

A real-time database is a processing system designed to handle workloads whose state may change constantly. This differs from traditional databases containing persistent data, mostly unaffected by time. For example, a stock market changes rapidly and dynamically.

Real-time processing means that a transaction is processed fast enough for the result to come back and be acted on right away. Real-time databases are useful for accounting, banking, law, medical records, multi-media, process control, reservation systems, and scientific data analysis. As computers increase in power and can store more data, real-time databases become integrated into society and are employed in many applications. There are Database Management Systems (DBMS), such as:

- ☐ Microsoft SQL Server

- ☐ Oracle

- ☐ Sybase

- ☐ dBase

- ☐ Microsoft Access

- ☐ MySQL from Sun Microsystems (Oracle)

- ☐ DB2 from IBM

Note: This chapter will focus on Microsoft Access and Microsoft SQL Server.

5.2.4 MDAC

The Microsoft Data Access Components (**MDAC**) is the framework that makes it possible to connect and communicate with the database. MDAC includes the following components:

- **ODBC** (Open Database Connectivity)

- **OLE DB**

- **ADO** (ActiveX Data Objects)

MDAC also installs several data providers you can use to open a connection to a specific data source, such as an MS Access database.

1. ODBC:

Open Database Connectivity (ODBC) is a native interface that is accessed through a programming language that can make calls into a native library. In MDAC this interface is defined as a DLL. A separate module or driver is needed for each database that must be accessed.

2. OLE DB:

OLE allows MDAC applications access to different types of data stores in a uniform manner. Microsoft has used this technology to separate the application from the data store that it needs to access. This was done because different applications need access to different types and sources of data, and do not necessarily need to know how to access technology-specific functionality. The technology is conceptually divided into consumers and providers. The consumers are the applications that need access to the data, and the provider is the software component that exposes an OLE DB interface through the use of the Component Object Model (or COM).

3. ADO (ActiveX Data Objects)

ActiveX Data Objects (ADO) is a high level programming interface to OLE DB. It uses a hierarchical object model to allow applications to programmatically create, retrieve, update and delete data from sources supported by OLE DB. ADO consists of a series of hierarchical COM-based objects and collections, an object that acts as a container of many other objects. A programmer can directly access ADO objects to manipulate data, or can send an SQL query to the database via several ADO mechanisms.

5.3 Relational Databases

A relational database matches data using common characteristics found within the data set. The resulting groups of data are organized and are much easier for people to understand. For example, a data set containing all the real-estate transactions in a town can be grouped by the year the transaction occurred; or it can be grouped by the sale price of the transaction; or it can be grouped by the buyer's last name; and so on.

Such a grouping uses the relational model (a technical term for this is schema). Hence, such a database is called a "relational database." The software used to do this grouping is called a relational database management system. The term "relational database" often refers to this type of software. Relational databases are currently the predominant choice in storing financial records, manufacturing and logistical information, personnel data and much more.

5.3.1 Tables

The basic units in a database are tables and the relationship between them. Strictly, a relational database is a collection of relations (frequently called tables).

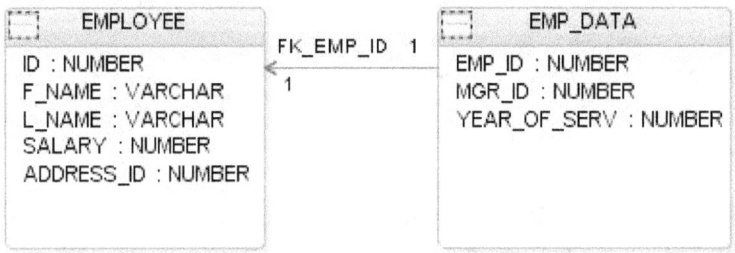

5.3.2 Unique Keys and Primary Key

In relational database design, a **unique key** or primary key is a candidate key to uniquely identify each row in a table. A unique key or primary key comprises a single column or set of columns. No two distinct rows in a table can have the same value (or combination of values) in those columns. Depending on its design, a table may have arbitrarily many unique keys but at most one primary key. A unique key must uniquely identify all possible rows that exist in a table and not only the currently existing rows. Examples of unique keys are Social Security numbers or ISBNs.

A **primary key** is a special case of unique keys. The major difference is that for unique keys the implicit NOT NULL constraint is not automatically enforced, while for primary keys it is enforced. Thus, the values in unique key columns may or may not be NULL. Another difference is that primary keys must be defined using another syntax.

Primary keys are defined with the following syntax:

```
CREATE TABLE   table_name (

    id_col INT,

    CONSTRAINT tab_pk PRIMARY KEY(id_col),

    ...)
```

If the primary key consists only of a single column, the column can be marked as such using the following syntax:

```
CREATE TABLE     table_name (

    id_col  INT PRIMARY KEY,

    col2    CHARACTER VARYING(20),

    ...

)
CREATE TABLE   table_name (

    id_col    INT,

            CHARACTE

    col2    R           VARYING(20)

key_col     SMALLINT,

...

CONSTRAINT key_unique UNIQUE(key_col),


...)
```

The definition of unique keys is syntactically very similar to primary keys. Likewise, unique keys can be defined as part of the CREATE TABLE SQL statement.

```
    CONSTRAINT key_unique UNIQUE(key_col),

    ...)
```

Or if the unique key consists only of a single column, the column can be marked as such using the following syntax:

```
CREATE TABLE   table_name (
```

5.3.3 Foreign Key

In the context of relational databases, a foreign key is a referential constraint between two tables. The foreign key identifies a column or a set of columns in one table that refers to a column or set of columns in another table.

The columns in the referencing table must be the primary key or other candidate key in the referenced table. The values in one row of the referencing columns must occur in a single row in the referenced table. Thus, a row in the referencing table cannot contain values that don't exist in the referenced table.

This way references can be made to link information together and it is an essential part of database normalization. Multiple rows in the referencing table may refer to the same row in the referenced table. Most of the time, it reflects the one (master table, or referenced table) to many (child table, or referencing table) relationship.

The referencing and referenced table may be the same table, i.e. the foreign key refers back to the same table. Such a foreign key is known as self-referencing or recursive foreign key. A table may have multiple foreign keys, and each foreign key can have a different referenced table. Each foreign key is enforced independently by the database system.

Therefore, cascading relationships between tables can be established using foreign keys. Improper foreign key/primary key relationships or not enforcing those relationships are often the source of many database and data modeling problems.

Foreign keys can be defined as part of the CREATE TABLE SQL statement.

```
CREATE TABLE table_name (

    id      INTEGER  PRIMARY KEY,

    col2    CHARACTER VARYING(20),

    col3    INTEGER,

    ...

    CONSTRAINT col3_fk FOREIGN KEY(col3)
        REFERENCES other_table(key_col),

    ... )
```

If the foreign key is a single column only, the column can be marked as such using the following syntax:

```
CREATE TABLE table_name (

id      INTEGER       PRIMARY KEY,
```

col2   CHARACTER VARYING(20),

col3   INTEGER REFERENCES other_table (column_name),

...)

5.3.4 Views

In database theory, a view consists of a stored query accessible as a virtual table composed of the result set of a query. Unlike ordinary tables in a relational database, a view does not form part of the physical schema: it is a dynamic, virtual table computed or collated from data in the database. Changing the data in a table alters the data shown in subsequent invocations of the view. Views can provide advantages over tables:

- Views can represent a subset of the data contained in a table

- Views can join and simplify multiple tables into a single virtual table

- Views can act as aggregated tables, where the database engine aggregates data (sum, average etc) and presents the calculated results as part of the data

- Views can hide the complexity of data; for example a view could appear as Sales2000 or Sales2001, transparently partitioning the actual underlying table

- Views take very little space to store; the database contains only the definition of a view, not a copy of all the data it presents

- Views can limit the degree of exposure of a table or tables to the outer world

Syntax:

```
CREATE VIEW <ViewName>
AS

…
```

5.3.5 Functions

In SQL databases, a user-defined function provides a mechanism for extending the functionality of the database server by adding a function that can be evaluated in SQL statements. The SQL standard distinguishes between scalar and table functions. A scalar function returns only a single value (or NULL), whereas a table function returns a (relational) table comprising zero or more rows, each row with one or more columns. User-defined functions in SQL are declared using the CREATE FUNCTION statement.

Syntax:

```
CREATE  FUNCTION  <FunctionName>
      (@Parameter1     <datatype>,     @
      Parameter2 <datatype>,


      …)



RETURNS <datatype> AS

…
```

5.3.6 Stored procedures

A stored procedure is executable code that is associated with, and generally stored in, the database. Stored procedures usually collect and customize common operations, like inserting a table into a relation, gathering statistical information about usage patterns, or encapsulating complex business logic and calculations. Frequently they are used as an application programming interface (API) for security or simplicity. Stored procedures are not part of the relational database model, but all commercial implementations include them. Stored procedures are called or used with the following syntax:

```
CALL procedure(…)
```

or

```
EXECUTE procedure(…)
```

Stored procedures can return result sets, i.e. the results of a SELECT statement. Such result sets can be processed using cursors by other stored procedures by associating a result set locator, or by applications. Stored procedures may also contain declared variables for processing data and cursors that allow it to loop through multiple rows in a table. The standard Structured Query Language provides IF, WHILE, LOOP, REPEAT, CASE statements, and more. Stored procedures can receive variables, return results or modify variables and return them, depending on how and where the variable is declared.

5.3.7 Triggers

A database trigger is procedural code that is automatically executed in response to certain events on a particular table or view in a database. The trigger is mostly used for keeping the integrity of the information on the database. For example, when a new record (representing a new worker) added to the employees table, new records should be created also in the tables of the taxes, vacations, and salaries.

The syntax is as follows:

CREATE TRIGGER <TriggerName> ON <TableName>
FOR INSERT, UPDATE, DELETE

AS

…

5.4 Structured Query Language (SQL)

SQL (Structured Query Language) is a database computer language designed for managing data in relational database management systems (RDBMS).

5.4.1 Queries

The most common operation in SQL is the query, which is performed with the declarative SELECT statement. SELECT retrieves data from one or more tables, or expressions. Standard SELECT statements have no persistent effects on the database.

Queries allow the user to describe desired data, leaving the database management system (DBMS) responsible for planning, optimizing, and performing the physical operations necessary to produce that result as it chooses.

A query includes a list of columns to be included in the final result immediately following the SELECT keyword. An asterisk ("*") can also be used to specify that the query should return all columns of the queried tables. SELECT is the most complex statement in SQL, with optional keywords and clauses that include:

- The **FROM** clause which indicates the table(s) from which data is to be retrieved. The FROM clause can include optional JOIN sub-clauses to specify the rules for joining tables.

- The **WHERE** clause includes a comparison predicate, which restricts the rows returned by the query. The WHERE clause eliminates all rows from the result set for which the comparison predicate does not evaluate to True.

- The **GROUP BY** clause is used to project rows having common values into a smaller set of rows. GROUP BY is often used in conjunction with SQL aggregation functions or to eliminate duplicate rows from a result set. The WHERE clause is applied before the GROUP BY clause.

- The **HAVING** clause includes a predicate used to filter rows resulting from the GROUP BY clause. Because it acts on the results of the GROUP BY clause, aggregation functions can be used in the HAVING clause predicate.

The **ORDER BY** clause identifies which columns are used to sort the resulting data, and in which direction they should be sorted (options are ascending or descending). Without an ORDER BY clause, the order of rows returned by an SQL query is undefined.

The following is an example of a SELECT query that returns a list of expensive books. The query retrieves all rows from the Book table in which the price column contains a value greater than 100.00.

The result is sorted in ascending order by title. The asterisk (*) in the select list indicates that all columns of the Book table should be included in the result set.

SELECT *

FROM Book

WHERE price > 100.00

ORDER BY title;

The example below demonstrates a query of multiple tables, grouping, and aggregation, by returning a list of books and the number of authors associated with each book.

SELECT Book.title,count(*) AS Authors

FROM Book

JOIN Book_author ON Book.isbn = Book_author.isbn

GROUP BY Book.title

Example output might resemble the following:

Title	Authors
SQL Examples and Guide	4
The Joy of SQL	1
An Introduction to SQL	2
Pitfalls of SQL	1

5.4.2 Data manipulation

The **Data Manipulation Language (DML)** is the subset of SQL used to add, update and delete data. The acronym **CRUD** refers to all of the major functions that need to be implemented in a relational database application to consider it complete. Each letter in the acronym can be mapped to a standard SQL statement, the table 5-1 is shown the DML operations.

Operation	SQL
Create	INSERT
Read (Retrieve)	SELECT
Update	UPDATE
Delete (Destroy)	DELETE

Table 5-1

Example 1: INSERT

INSERT adds rows to an existing table:

INSERT INTO My_table field1, field2, field3)

VALUES ('test', 'N', NULL)

Example 2: UPDATE

UPDATE modifies a set of existing table rows:

UPDATE My_table

SET field1 = 'updated value'

WHERE field2 = 'N'

Example 3: DELETE

DELETE removes existing rows from a table:

DELETE FROM My_table

WHERE field2 = 'N'

5.4.3 Data definition

The **Data Definition Language (DDL)** manages table and index structure. The most basic items of DDL are the CREATE, ALTER, RENAME and DROP statements:

- ☐ **CREATE** creates an object (a table, for example) in the database.

- ☐ **DROP** deletes an object in the database, usually irretrievably.

- ☐ **ALTER** modifies the structure an existing object in various ways—for example, adding a column to an existing table.

Example 4: CREATE

Create a Database Table:

CREATE TABLE My_table

(

      my_field1   INT,

      my_field2   VARCHAR(50),

      my_field3   DATE   NOT NULL,

      PRIMARY KEY (my_field1)

)

5.4.4 Data types

Each column in an SQL table declares the type(s) that column may contain. ANSI SQL includes the following data types.

1. Character strings

- ☐ CHARACTER(n) or CHAR(n) — fixed-width n-character string, padded with spaces as needed

- ☐ CHARACTER VARYING(n) or VARCHAR(n) — variable-width string with a maximum size of (n) characters

- ☐ NATIONAL CHARACTER(n) or NCHAR(n) — fixed width string supporting an international character set

- ☐ NATIONAL CHARACTER VARYING(n) or NVARCHAR(n) — variable-width NCHAR string

2. Bit strings

- o   BIT(n) — an array of n bits

- o   BIT VARYING(n) — an array of up to n bits

3. Numbers

- INTEGER and SMALLINT

- FLOAT, REAL and DOUBLE PRECISION

- NUMERIC(precision, scale) or DECIMAL(precision, scale)

4. Date and Time

- DATE

- TIME

- TIMESTAMP

- INTERVAL

5.5 Database Modeling

5.5.1 ER Diagram & Microsoft Visio

In software engineering, an Entity-Relationship Model (ERM) is an abstract and conceptual representation of data. Entity-relationship modeling is a database modeling method, used to produce a type of conceptual schema or semantic data model of a system, often a relational database, and its requirements in a top-down fashion. Diagrams created using this process are called entity-relationship diagrams, or ER diagrams or ERDs for short.

There are many ER diagramming tools. Some of the proprietary ER diagramming tools are ER win, Enterprise Architect and Microsoft Visio. Microsoft SQL Server has also a built-in tool for creating Database Diagrams.

Microsoft Visio is a diagramming program for creating different kinds of diagrams. Visio have a template for creating Database Model Diagrams.

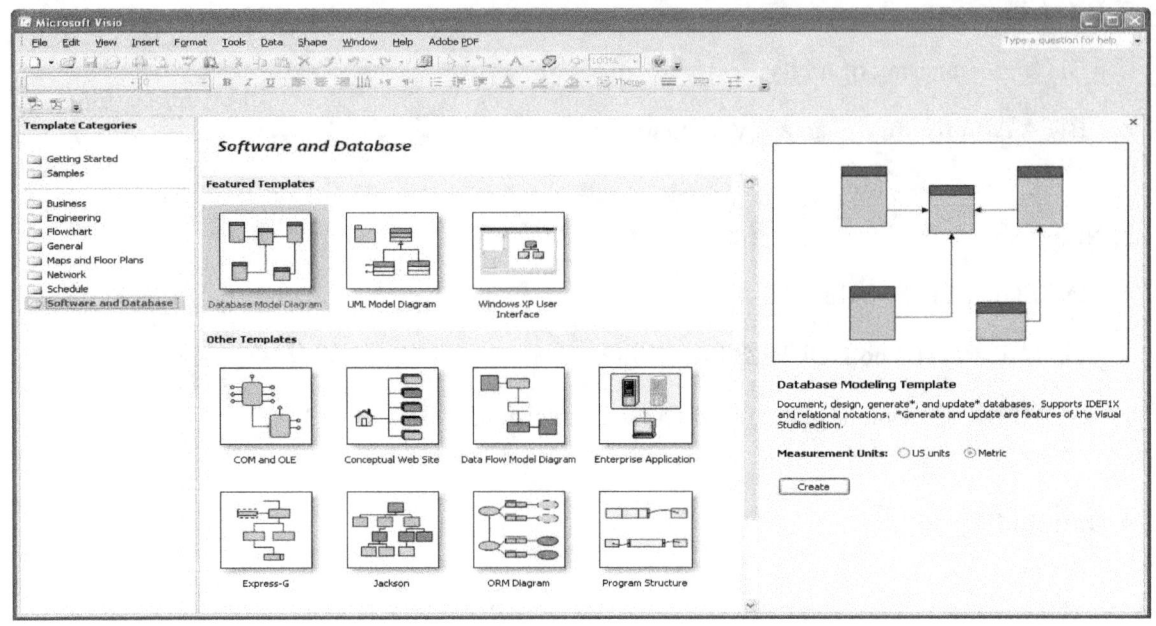

In the Database menu Visio offers lots of functionality regarding your database model.

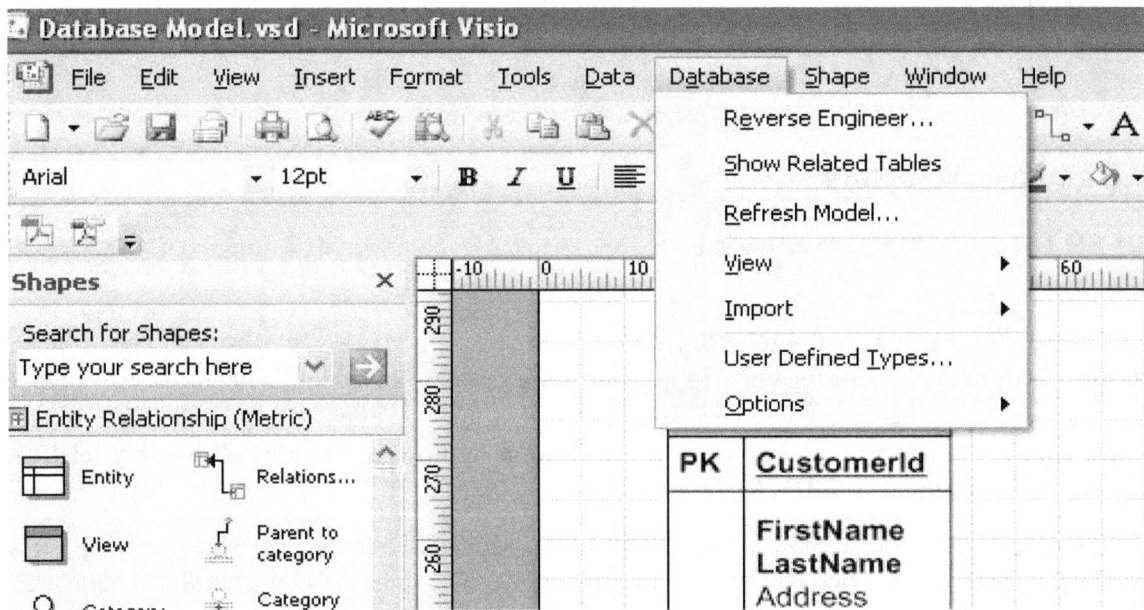

"Reverse Engineering" is the opposite procedure, i.e., extraction of a database schema from an existing database into a database model in Microsoft Visio.

5.5.2 EXERCISES

Exercise 1: Database Diagram

Create the following tables in an ER Diagram using MS Visio.

- CUSTOMER
    - **CustomerId (PK)**
    - FirstName
    - LastName
    - Address
    - Phone
    - PostCode
    - PostAddress
- PRODUCT
    - **ProductId (PK)**
    - ProductName
    - ProductDescription
    - Price
    - ProductCode
- ORDER
    - **OrderId (PK)**
    - OrderNumber
    - OrderDescription
    - **CustomerId (FK)**
- ORDER_DETAIL
    - OrderDetailId (PK)
    - **OrderId (FK)**
    - **ProductId (FK)**

Then the Database Diagram is shown in figure 5-2:

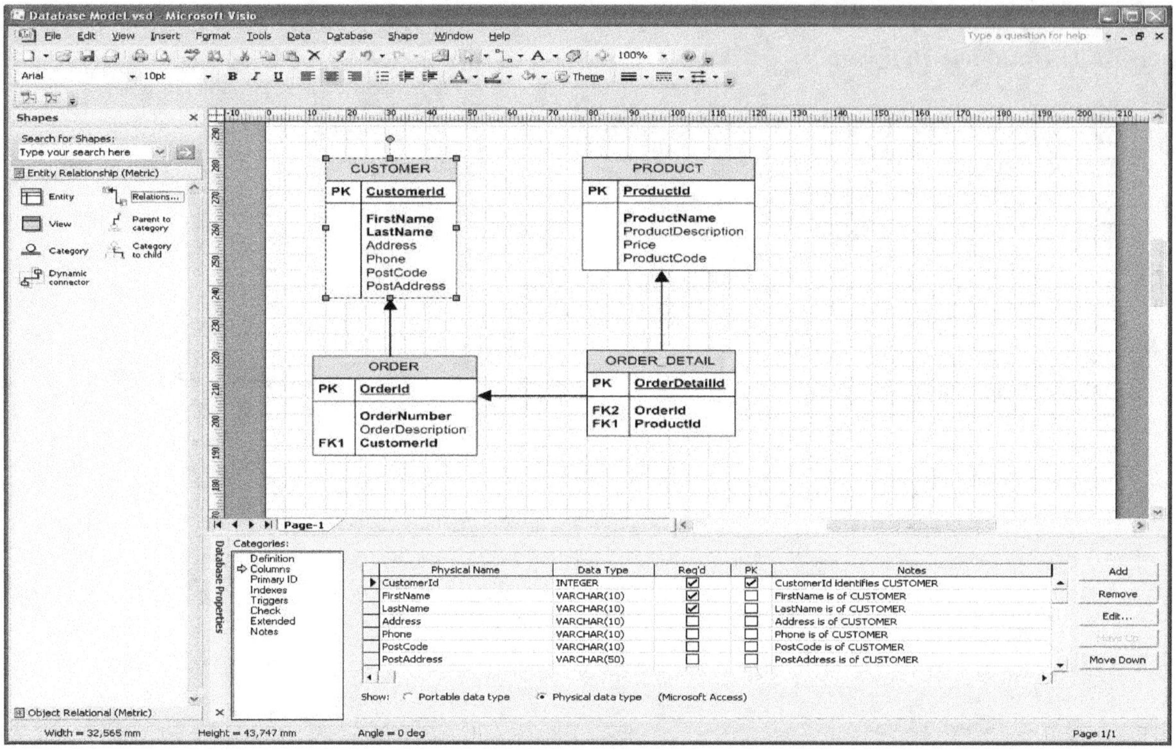

Figure 5-2

5.6 Microsoft SQL Server

5.6.1 Introduction

Microsoft SQL Server is a relational model database server produced by Microsoft. Its primary query languages are T-SQL and ANSI SQL. Microsoft SQL Server homepage: www.microsoft.com/sqlserver

The Microsoft SQL Server comes in different versions, such as:

☐ SQL Server Developer Edition

☐ SQL Server Enterprise Edition

☐ SQL Server Web Edition

☐ SQL Server Express Edition

The SQL Server Express Edition is a freely-downloadable and - distributable version.

5.6.2 SQL Server Express

The SQL Server Express Edition is a freely-downloadable and -distributable version. However, the Express edition has a number of technical restrictions which make it undesirable for large-scale deployments, including:

- Maximum database size of 4 GB per. The 4 GB limit applies per database (log files excluded); but in some scenarios users can access more data through the use of multiple interconnected databases.

- Single physical CPU, multiple cores

- 1 GB of RAM (runs on any size RAM system, but uses only 1 GB)

SQL Server Express offers a GUI tools for database management in a separate download and installation package, called **SQL Server Management Studio Express**.

5.6.3 Adventure Works

The **Adventure Works** is a sample Database with lots of examples. One should install this sample Database because some of the examples in this chapter will use the Adventure Works database.

5.6.4 SQL Server Management Studio

SQL Server Management Studio is a GUI tool included with SQL Server for configuring, managing, and administering all components within Microsoft SQL Server. The tool includes both script editors and graphical tools that work with objects and features of the server. As mentioned earlier, version of SQL Server Management Studio is also available for SQL Server Express Edition, for which it is known as SQL Server Management Studio Express.

A central feature of SQL Server Management Studio is the Object Explorer, which allows the user to browse, select, and act upon any of the objects within the server. It can be used to visually observe and analyze query plans and optimize the database performance, among others. SQL Server Management Studio can also be used to create a new database, alter any existing database schema by adding or modifying tables and indexes, or analyze performance. It includes the query windows which provide a GUI based interface to write and execute queries.

107

5.6.5 Create a new Database

It is quite simple to create a new database in Microsoft SQL Server. Just right-click on the "Databases" node and select "New Database...". There are lots of settings you may set regarding your database, but the only information you must fill in is the name of your database, see the figure 5-3:

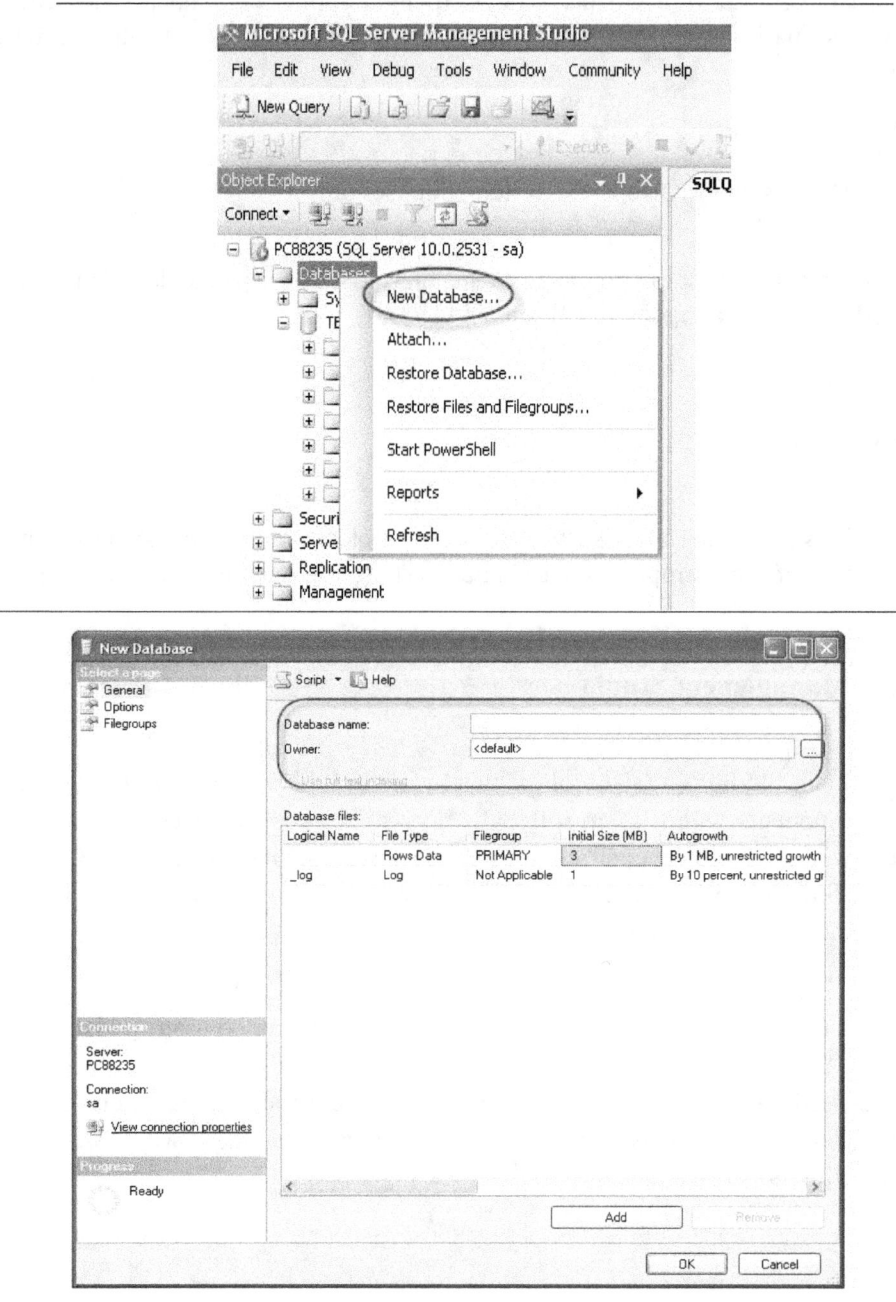

Figure 5-3

5.6.6 Backup/Restore

An important task in database systems is to take backup of the database with regular intervals, for example; during the night when the system is not in use. Database backup and Restore:

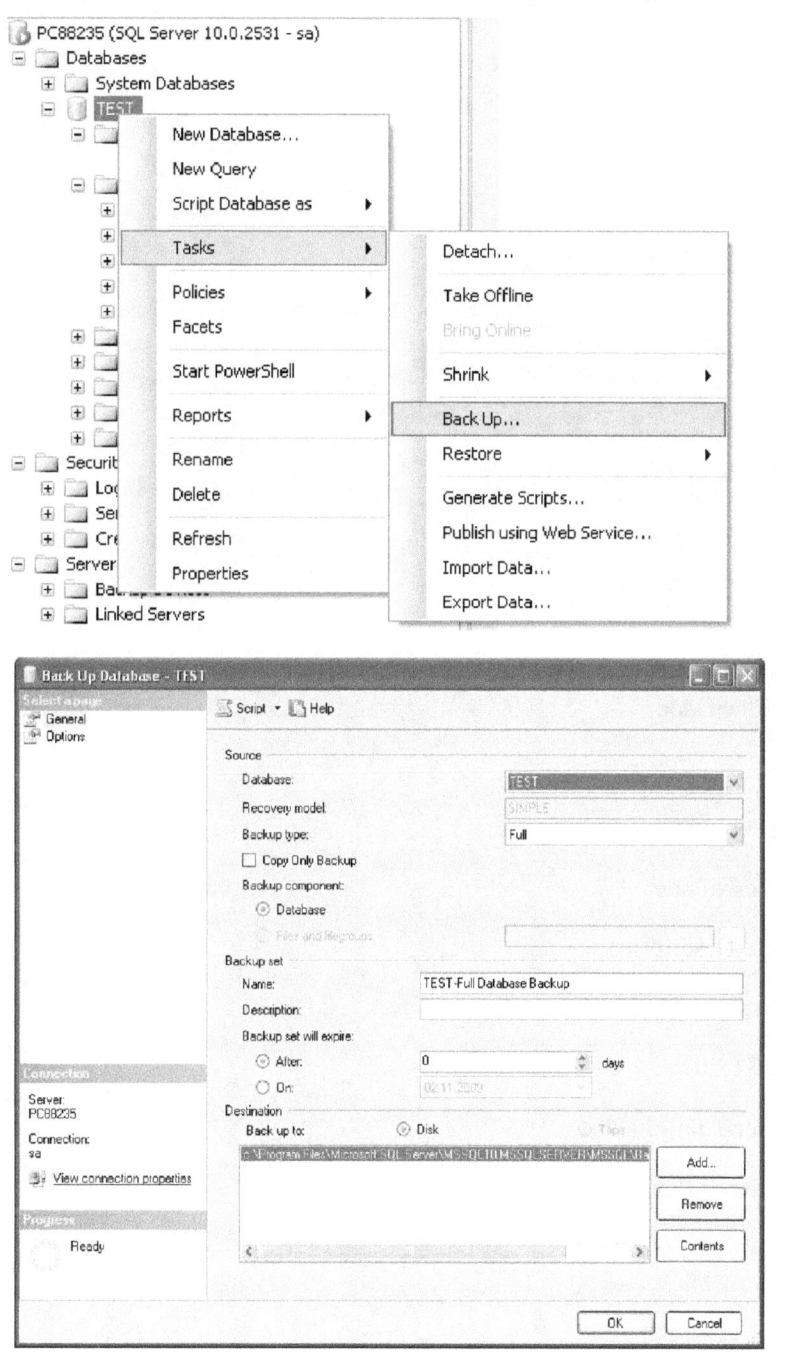

5.6.7 Databases Examples

Examples and exercises in this chapter are based on some basic tables. The Database's Example consists of the following Tables:

- CUSTOMER
    - **CustomerId (PK)**
    - FirstName
    - LastName
    - Address
    - Phone
    - PostCode
    - PostAddress

- PRODUCT
    - **ProductId (PK)**
    - ProductName
    - ProductDescription
    - Price
    - ProductCode

- ORDER
    - **OrderId (PK)**
    - OrderNumber
    - OrderDescription
    - **CustomerId (FK)**

- ORDER_DETAIL
    - OrderDetailId (PK)
    - **OrderId (FK)**
    - **ProductId (FK)**

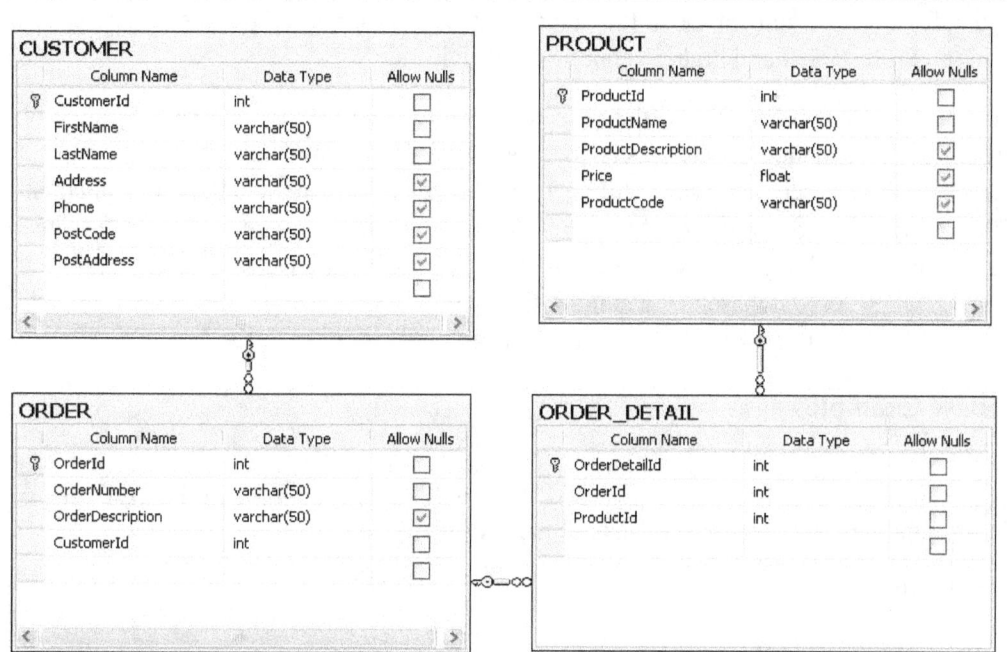

5.6.8 Exercises

Exercise - 1: New Database

Create a new Database in MS SQL Server called TEST_SQLSERVER.

Exercise - 2: Database Diagram

Create the tables in the Example Database using the Diagram Designer Tool in Microsoft SQL Server.

Exercise - 3: Database Script

Create the tables in the Example Database Tables using SQL Code. Save the Tables as a SQL Script file (.sql). Use The Query Tool in Microsoft SQL Server.

Exercise - 4: ODBC

Create an ODBC connection for the Database.

5.7 Microsoft Office Access

5.7.1 Introduction

Microsoft Office Access, previously known as Microsoft Access, is a relational database management system from Microsoft that combines the relational Microsoft Jet Database Engine with a graphical user interface and software development tools. It is a member of the Microsoft Office suite of applications and is included in the Professional and higher versions for Windows.

Access stores data in its own format based on the Access Jet Database Engine. Also, Microsoft Access is used by programmers and non-programmers to create their own simple database solutions. Microsoft Access is a file server-based database. Unlike client-server relational database management systems (RDBMS), e.g., Microsoft SQL Server, Microsoft Access does not implement database triggers, stored procedures, or transaction logging. All database tables, queries, forms, reports, macros, and modules are stored in the Access Jet database as a single file. This makes Microsoft Access useful in small applications, teaching, etc. because it is easy to move from one computer to another.

5.7.2 Database Example

The present example database in Microsoft Access 2007 will be used in some of the examples and exercises in this chapter. The database consists of the following table:

- CUSTOMER

    - **CustomerId (PK)**

    - FirstName

    - LastName

    - Address

    - Phone

    - PostCode

    - PostAddress

- PRODUCT

    - **ProductId (PK)**

    - ProductName
    - ProductDescription

    - Price

    - ProductCode

- ORDER

    - **OrderId (PK)**

    - OrderNumber

    - OrderDescription

    - **CustomerId (FK)**

112

☐ ORDER_DETAIL

  o OrderDetailId (PK)

  o **OrderId (FK)**
  o **ProductId (FK)**

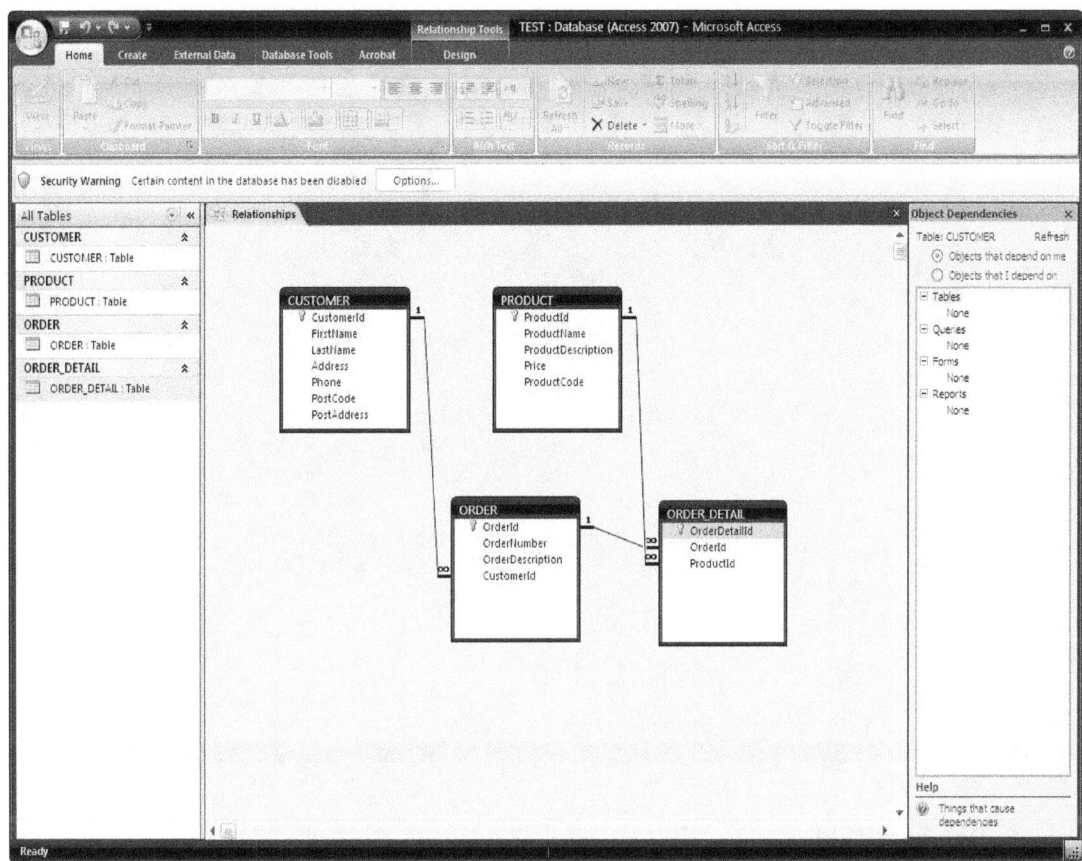

ODBC Connection:

Administrative Tools → Data Sources (ODBC)

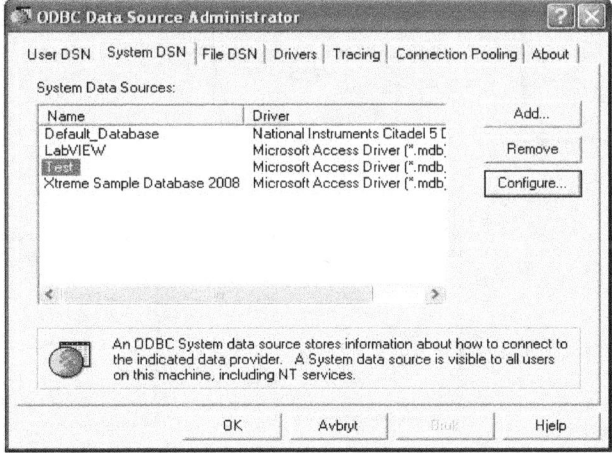

5.7.3 More Exercises

Exercise -5: Database

Create a new Database in MS Access called TEST.

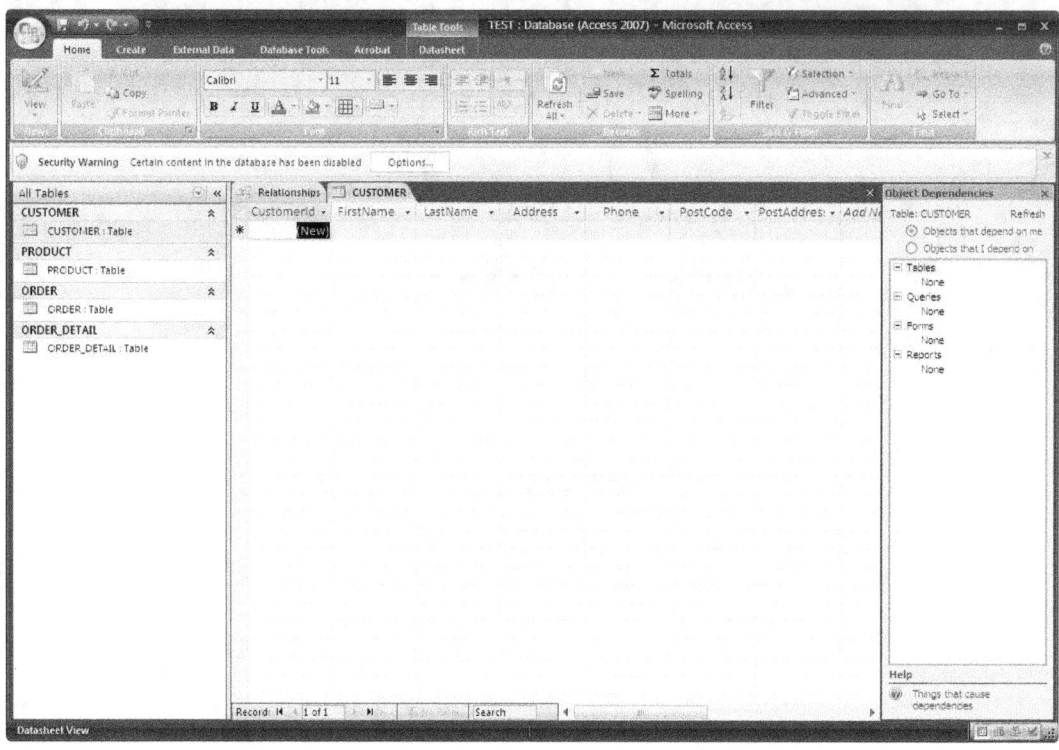

Exercise - 6: Database Tables

Create the tables in the Example Database Tables using the Diagram Designer Tool in Microsoft SQL Server.

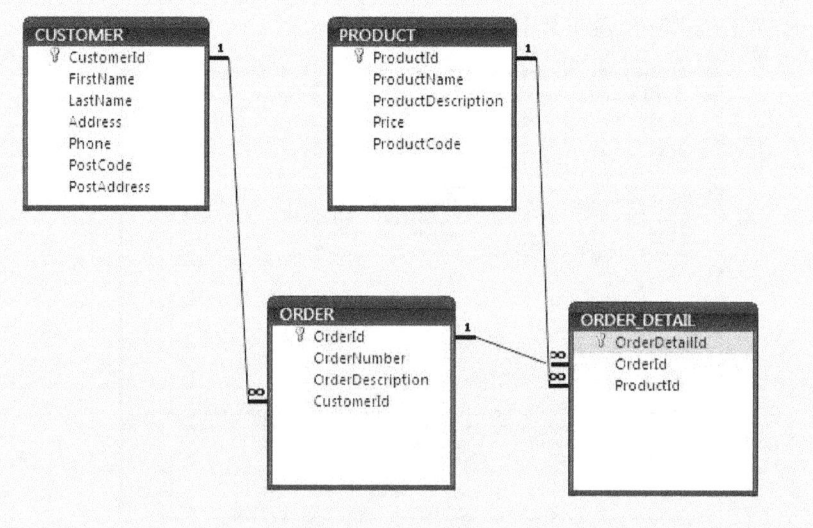

Exercise -7: ODBC

Create an ODBC connection for the Database. For answering this exercise, one should know more about the ODBC database connection.

5.7.4 What is ODBC

In computing, Open Database Connectivity (ODBC) provides a standard software API method for using database management systems (DBMS). The designers of ODBC aimed to make it independent of programming languages, database systems, and operating systems.

Create an ODBC Connection in "ODBC Data Source Administrator": Follow these steps:

Add a new Data Source and select the SQL Server driver:

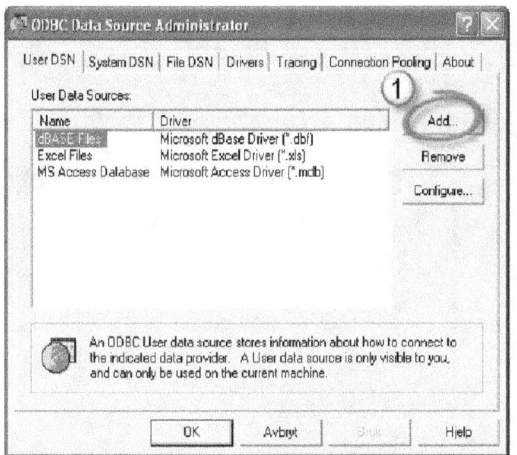

Type a Name for your Connection and your SQL Server Name. You find your Server name as shown below:

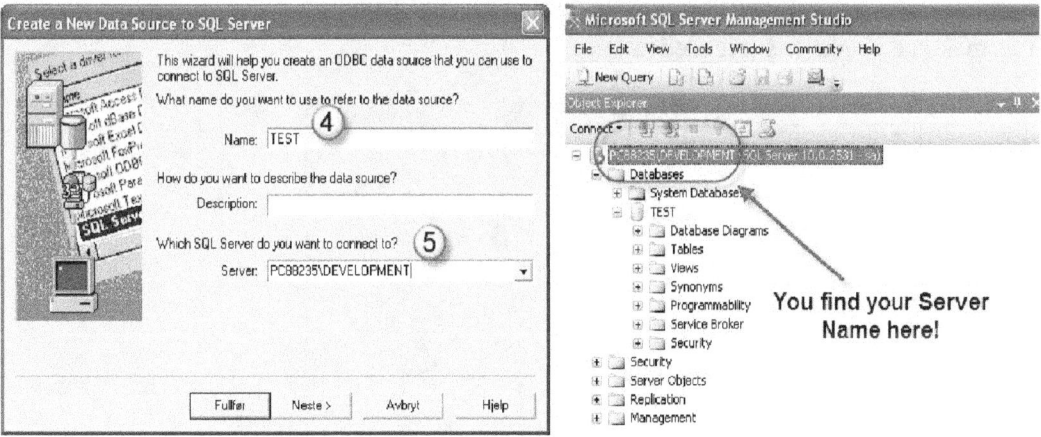

115

Select SQL Server authentication and type the password of the (System Administrator). One defined the password for the user during the setup procedure of SQL Server:

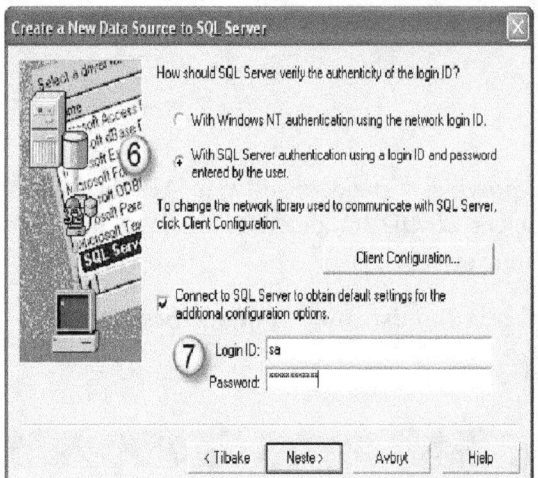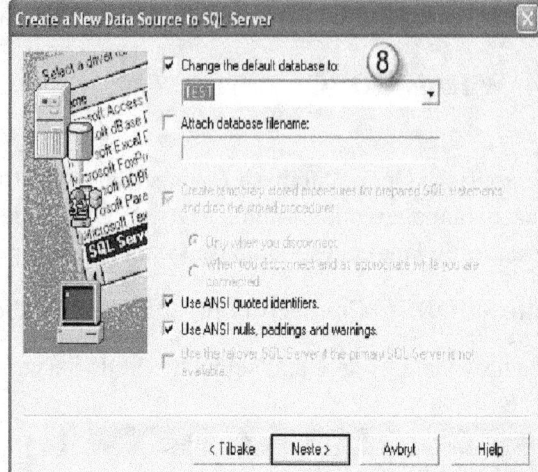

Complete your configuration and Test your data source to see if it's OK:

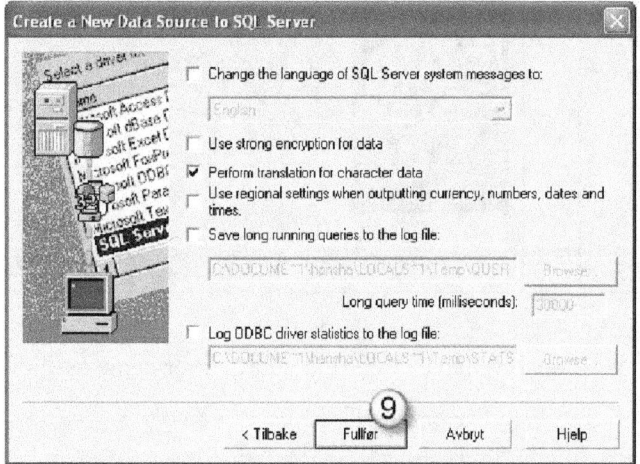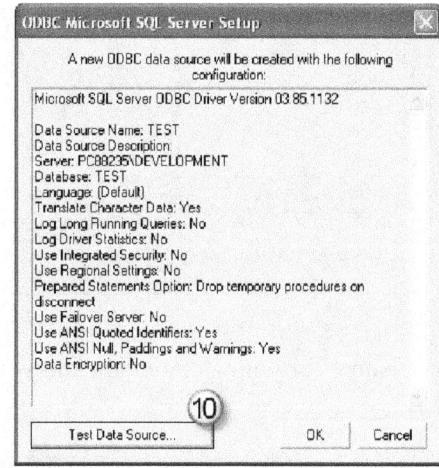

If you get this message you have succeeded:

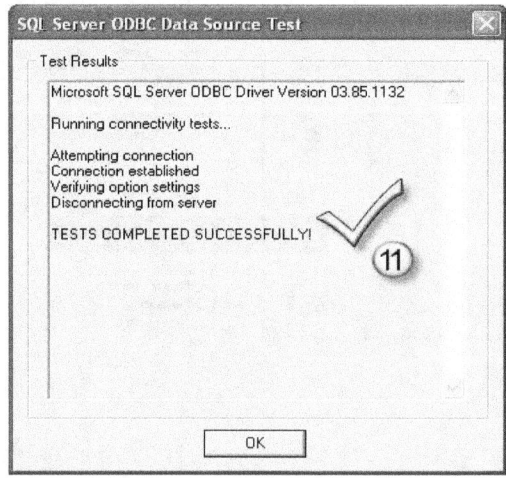

5.7.5 Get data into Excel using your ODBC Connection

The purpose is to use Excel as a client and get data into Excel from your SQL Server.

Step 1: Open Excel and go to the Data section:

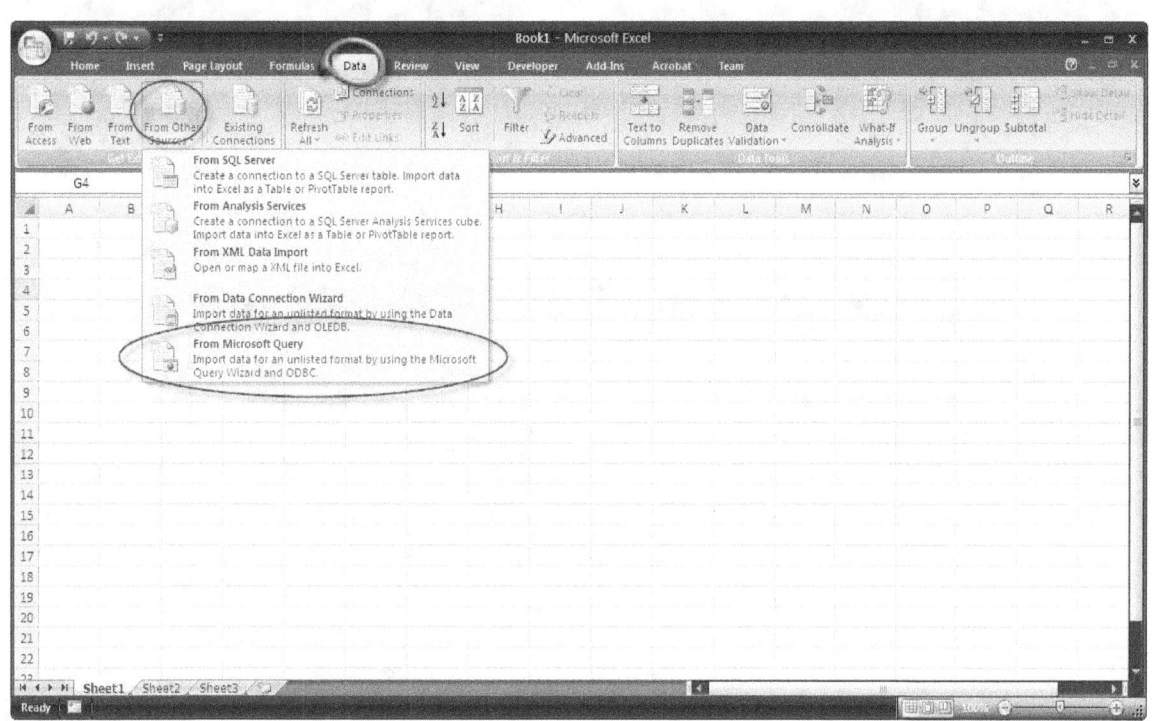

Step 2: Select your ODBC connection

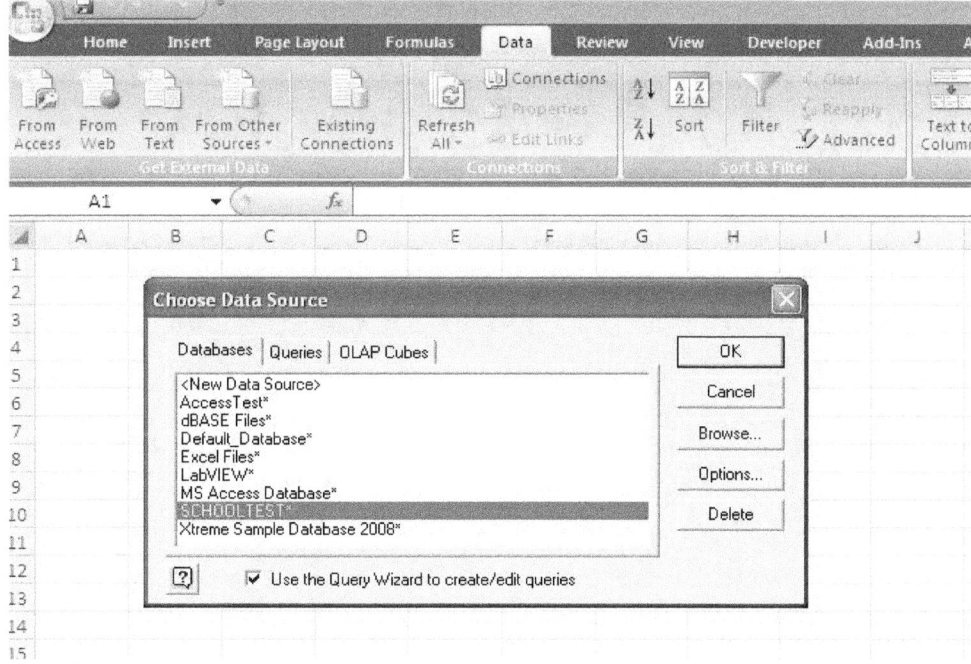

Step 3: Select your Table(s)

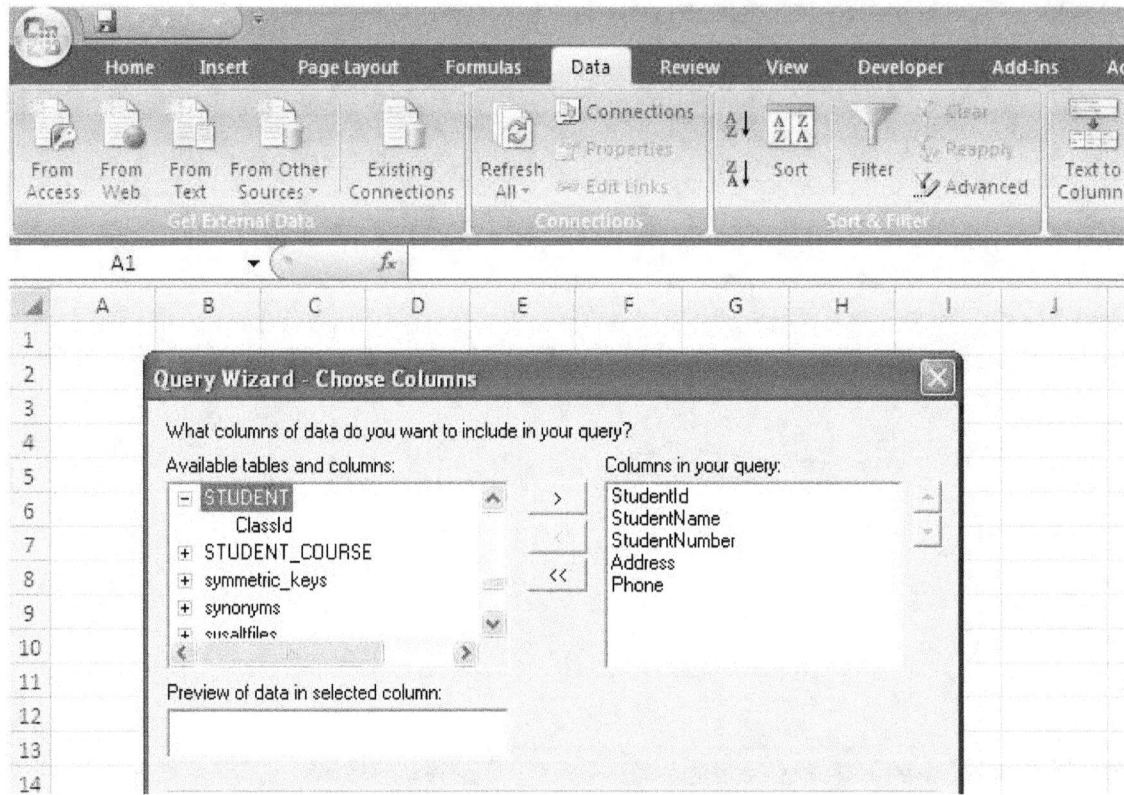

Step 4: Insert Data into Excel

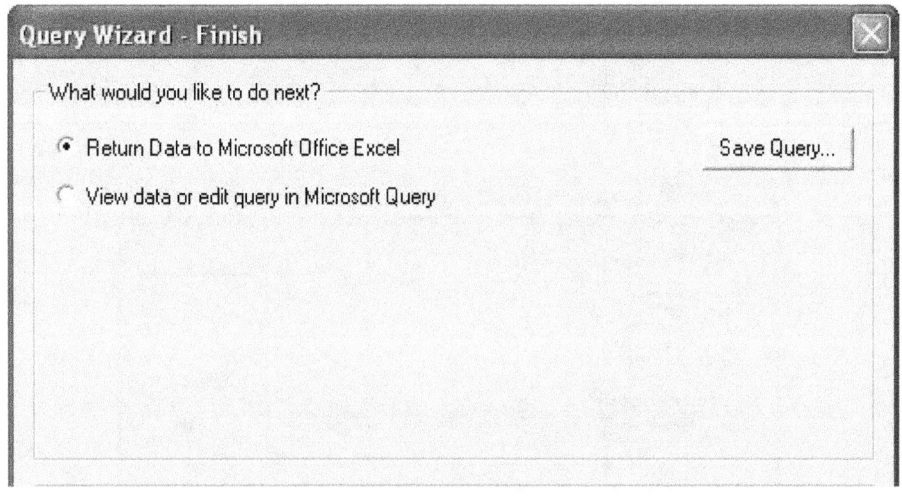

The results should look something like the figure 5-4:

118

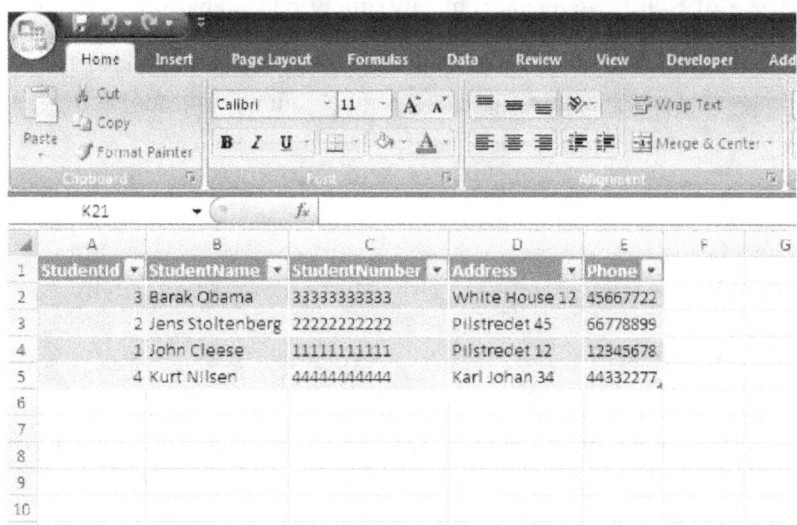

Figure 5-4

5.8 LabVIEW Database Connectivity Toolkit

LabVIEW offers an additional Toolkit called "LabVIEW Database Connectivity Toolkit". With this toolkit you can communicate with different databases, such as SQL Server, Oracle, etc.

Functions Palette: Connectivity → Database

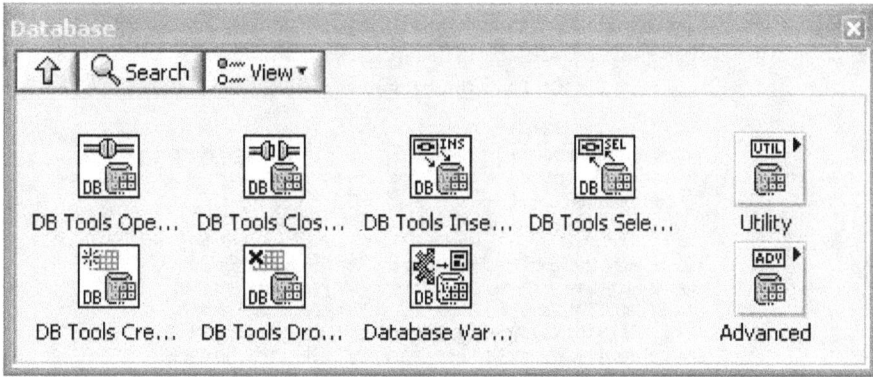

The following list describes the main features of the Database Connectivity Toolkit:

- Works with any provider that adheres to the Microsoft ActiveX Data Object (ADO) standard.

- Works with any database driver that complies with ODBC or OLE DB.

- Maintains a high level of portability. In many cases, you can port an application to another database by changing the connection information you pass to the DB Tools Open Connection VI.

- Converts database column values from native data types to standard Database Connectivity Toolkit data types, further enhancing portability.

119

- Permits the use of SQL statements with all supported database systems, even non-SQL systems.

- Includes VIs to retrieve the name and data type of a column returned by a SELECT statement.

- Creates tables and selects, inserts, updates, and deletes records without using SQL statements.

Some of the text in this chapter is based on the "LabVIEW Database Connectivity Toolkit User Manual".

5.8.1 Connect to the Database

Before you can access data in a table or execute SQL statements, you must establish a connection to a database. You may use different methods in order to connect to the database:

- ➤ ODBC Data Source Name (DSN)

- ➤ Universal Data Link (UDL)

- ➤ Connection String

These different methods are explained, and you will use the same VI as seen below:

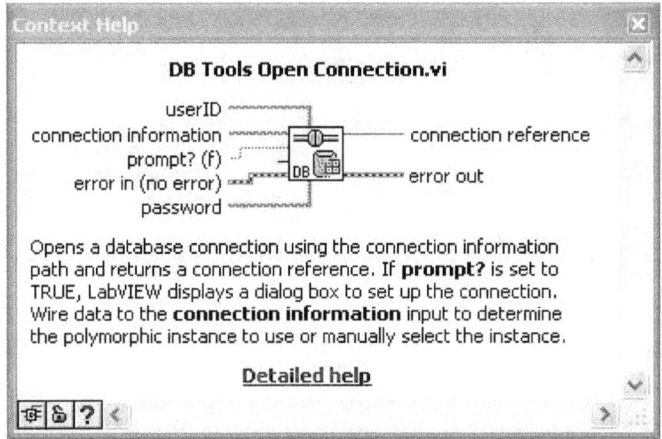

Connecting to a database is where most errors occur because each database management system (DBMS) uses different parameters for the connection and different levels of security. The different standards also use different methods of connecting to databases. For example, ODBC uses Data Source Names (DSN) for the connection, whereas the Microsoft ActiveX Data Object (ADO) standard uses Universal Data Links (UDL) for the connection. The "DB Tools Open Connection.vi" VI supports all these methods for connecting to a database. When one has finished with reading from the database and writing to the

120

database, it should always close the Connection. Use the "DB Tools Close Connection.vi". See the following figure:

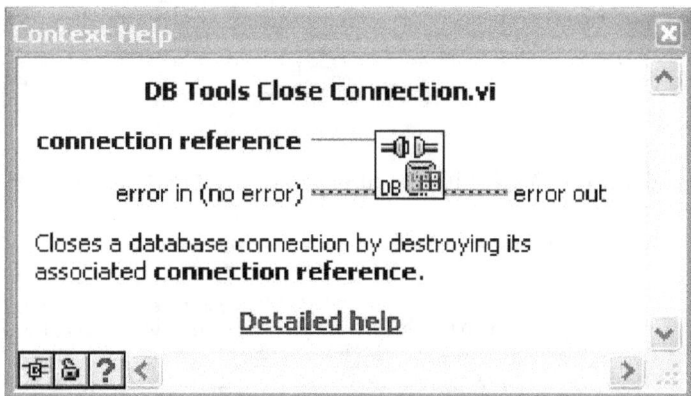

A) DSN

A **DSN (ODBC Data Source Name (DSN))** is the name of the data source, or database, to which you are connecting. The DSN also contains information about the ODBC driver and other connection attributes including paths, security information, and read-only status of the database.

Two main types of DSNs exist: machine DSNs and file DSNs. Machine DSNs are in the system registry and apply to all users of the computer system or to a single user. DSNs that apply to all users of a computer system are system DSNs. DSNs that apply to single users are user DSNs.

A file DSN is a text file with a .dsn extension and is accessible to anyone with proper permissions. File DSNs are not restricted to a single user or computer system. Use the ODBC Data Source Administrator to create and configure DSNs. In the Control Panel, Administrative Tools, one may find the ODBC Data Source Administrator tool.

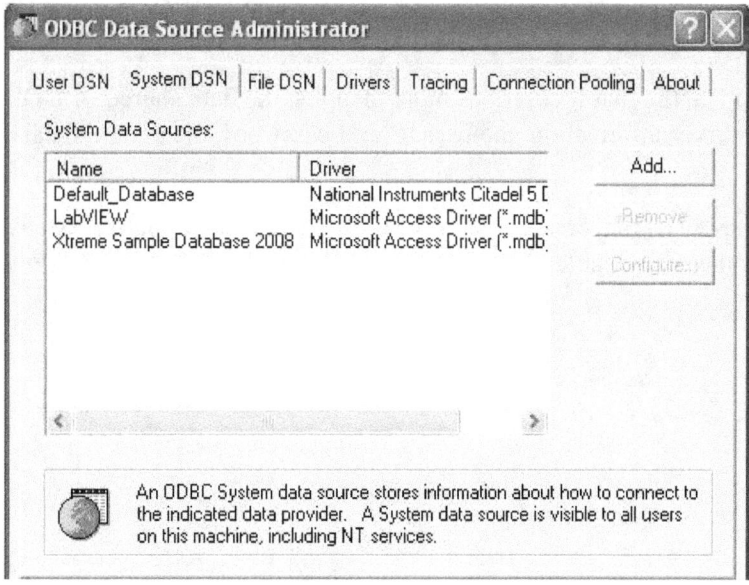

Example -1 : DSN

This Example specifies a DSN called MS Access to open a connection to that specific database.

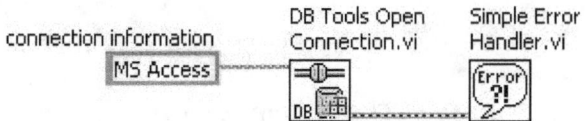

Example -2: DSN from File

You can use a path to specify a file DSN. This example specifies a path to a file DSN called "access.dsn" to open a connection to the database.

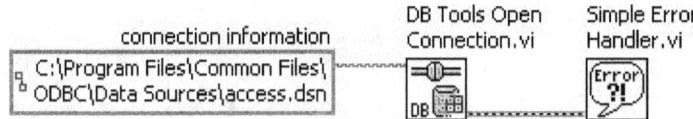

Example -3: DSN with User Id and Password

Most Database systems (DBMS – Database Management Systems) also require a User Id and a Password.

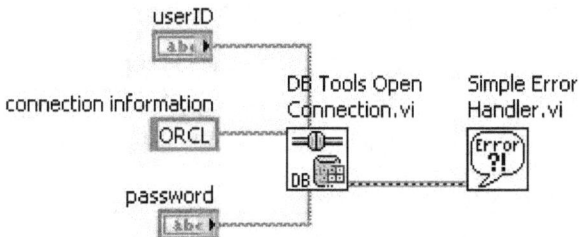

B) UDL

Whereas you must create a DSN to connect to a database using ODBC, you use **UDL (Universal Data Link)** to connect to databases that use ADO and OLE DB.

A UDL is similar to a DSN in that it describes more than just the data source. A UDL specifies what OLE DB provider is used, server information, the user ID and password, the default database, and other related information.

In order to create a new UDL file, create an empty text file and change the file extension of this document from .txt to .udl. You then can double-click the UDL file to display the Data Link Properties dialog box.

Example - 4: UDL

Connect to a Database using UDL:

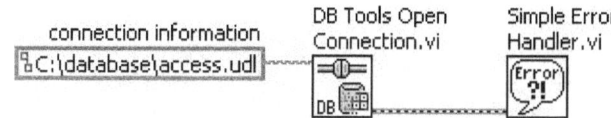

C) Connection String

Rather than including an existing UDL in an application, you also can use an ODBC connection string with the Microsoft ActiveX Data Object (ADO) standard.

A connection string is written like this:

PROVIDER=SQLOLEDB;DATA

SOURCE=server_name;UID=user_name;PWD=password;DATABASE=database_name;

One could use more parameters, but the parameters used above are the most common ones.

5.8.2 Reading Data from the Database

Reading data from a database table is similar to writing data to the database. You open a connection to the database, select the data from a table, and then close the connection. The "DB Tools Select Data.vi" is used to read data from the Database:

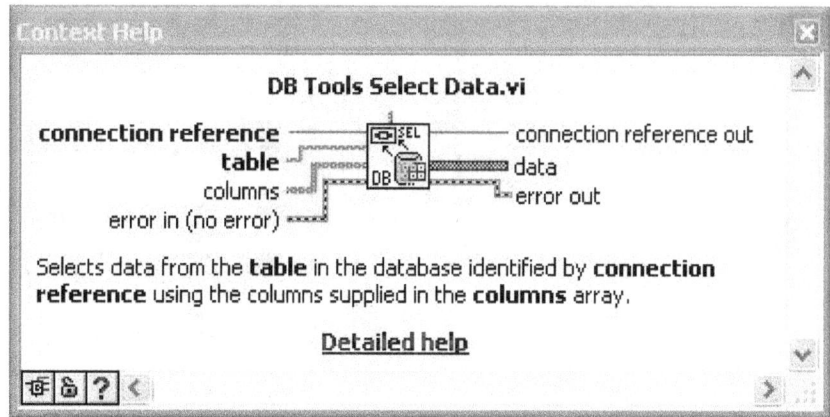

Example -1: Select Data from MS Access

The following example gets data from the CUSTOMER table in MS Access.

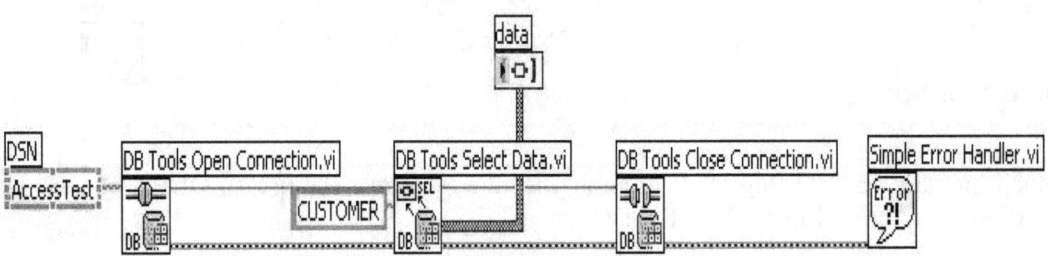

As the name implies, the Microsoft ActiveX Data Object (ADO) standard is based on ActiveX, which defines variants as its data types. Variants work well in languages such as Visual Basic that are not strongly typed. Because LabVIEW is strongly typed, you must use the Database Variant to Data function to convert the variant data to a LabVIEW data type before you can display the data in standard indicators such as graphs, charts, and LEDs.

Example -2: Select Data from MS Access

The following example gets data from the CUSTOMER table in MS Access and converts the data to text.

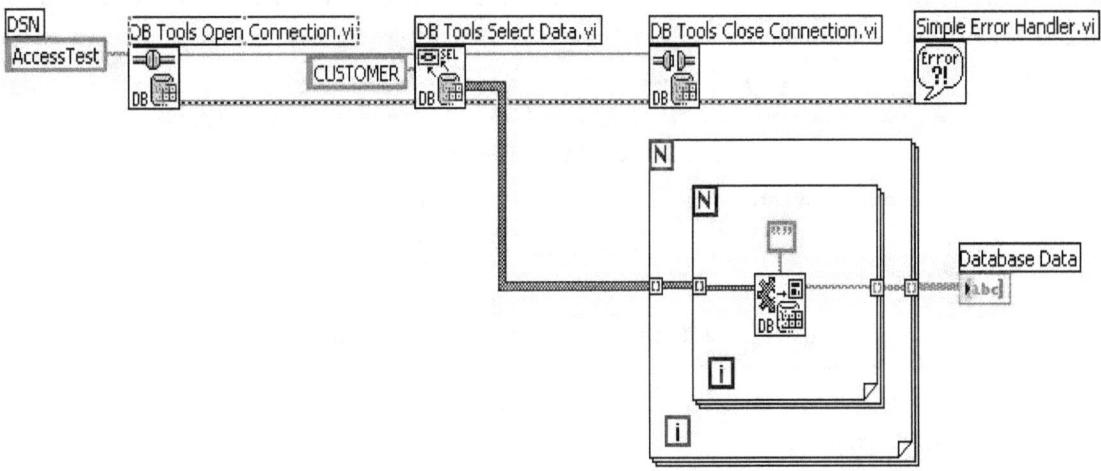

One may read from more than one table if you use a comma-delimited string to specify multiple table names:

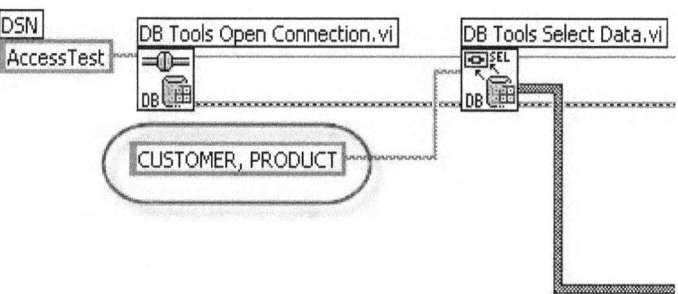

One may select which columns you want to read by using the "Columns" input:

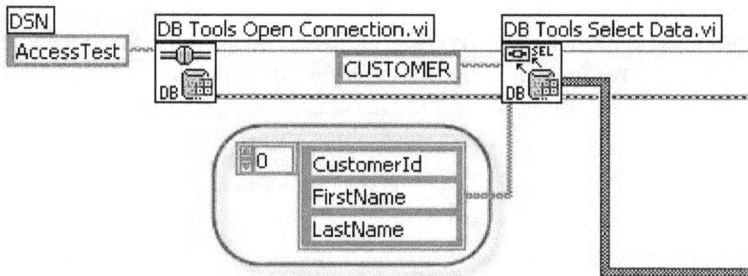

One may also restrict which data to receive using the "optional Clause" input:

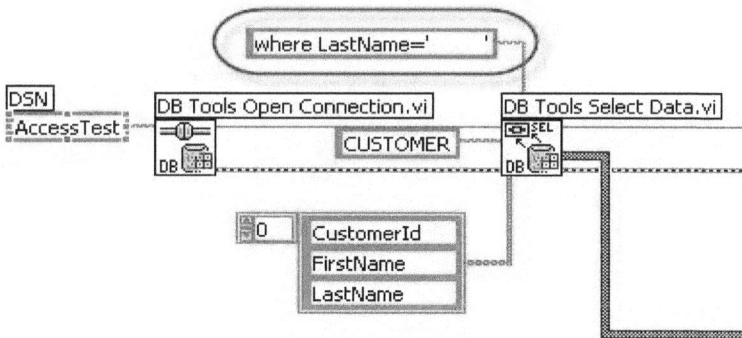

Example -3: Read Data

Using some VIs from the "Advanced" palette, create the following example:

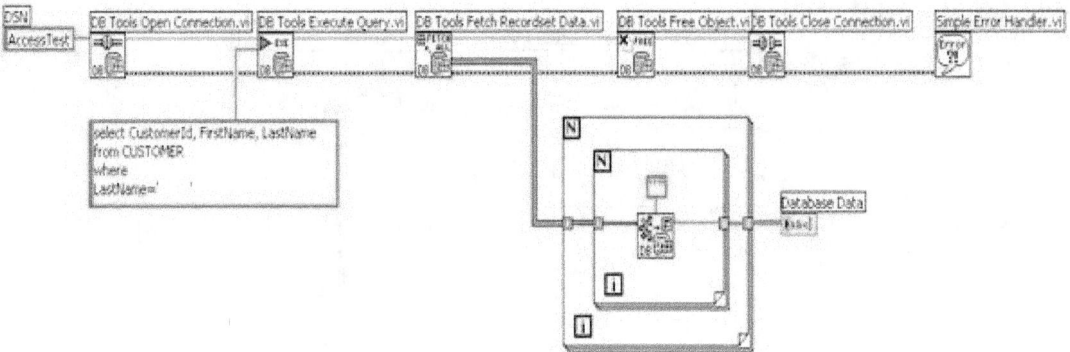

5.8.3 Writing Data to the Database

Writing data to a database with the LabVIEW Database Connectivity Toolkit is similar to reading data to a file. You open a connection, insert the data, and close the connection when you are finished. The "DB Tools Insert Data.vi" is used to write data to the Database:

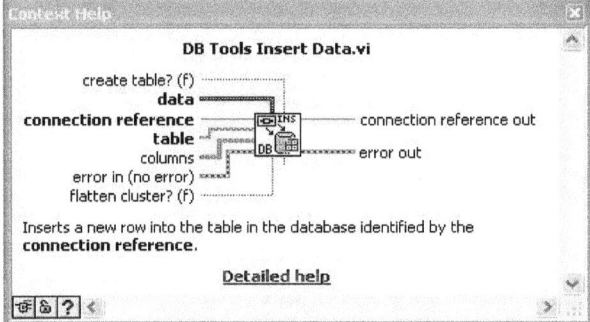

Example -4: Write Data

Create the following block diagram:

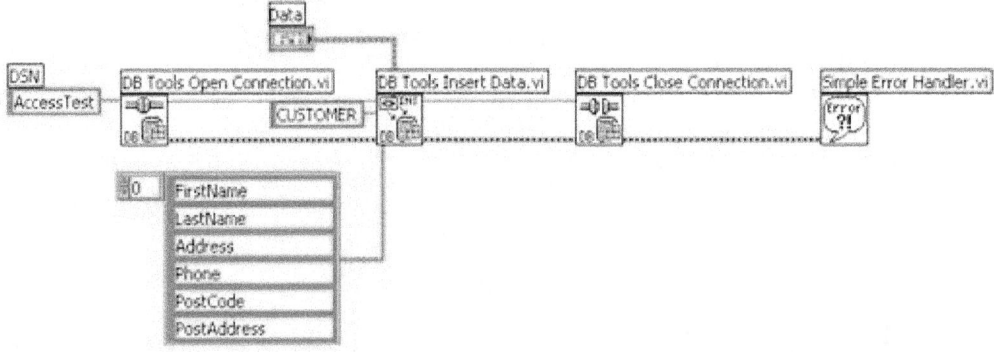

Example -5: Write Data

Create the following block diagram using some VIs from the "Advanced" palette.

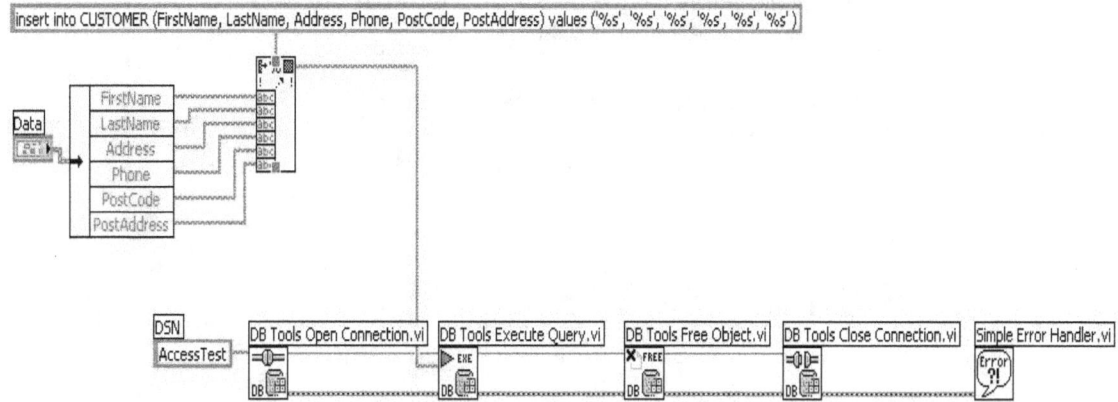

5.8.4 Creating and Dropping Tables

It could be use the standard SQL syntax in order to create a table:

CREATE TABLE <TableName> (…)

Or you may use the "DB Tools Create Table.vi" in order to create a table.

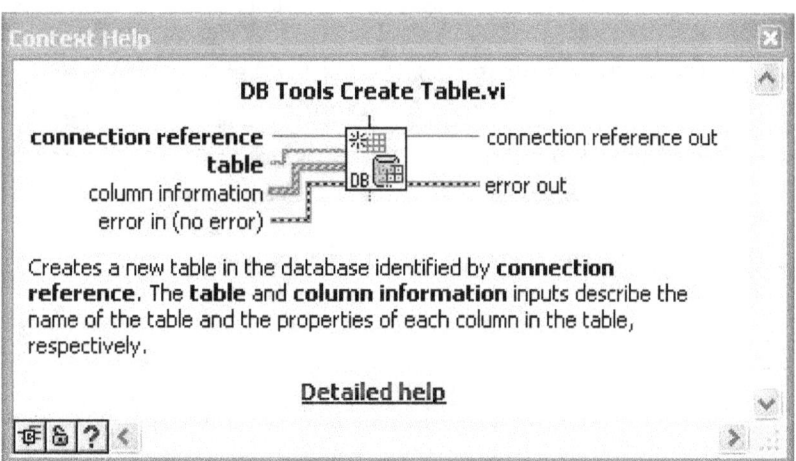

You may use standard SQL syntax in order to drop tables (delete tables):

DROP TABLE <TableName>

Or you may use the "DB Tools Drop Table.vi" in order to drop/delete a table.

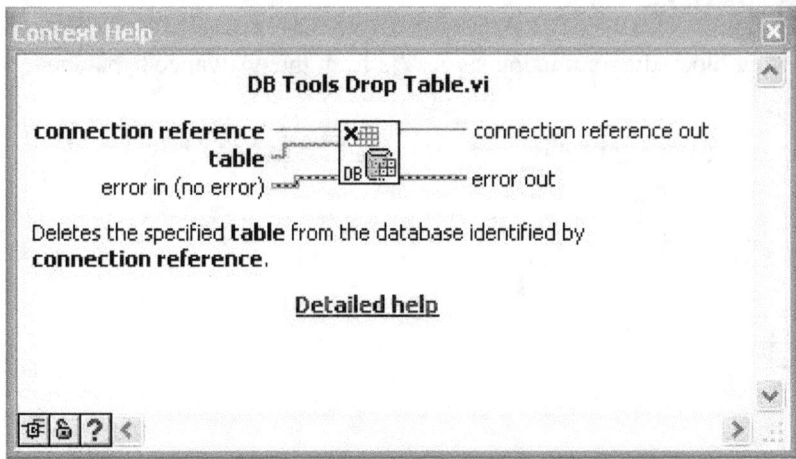

5.8.5 Using the Database Connectivity Toolkit Utility VIs

In the "**Utility**" palette there are several useful VIs for getting more information about tables, saving to text files.

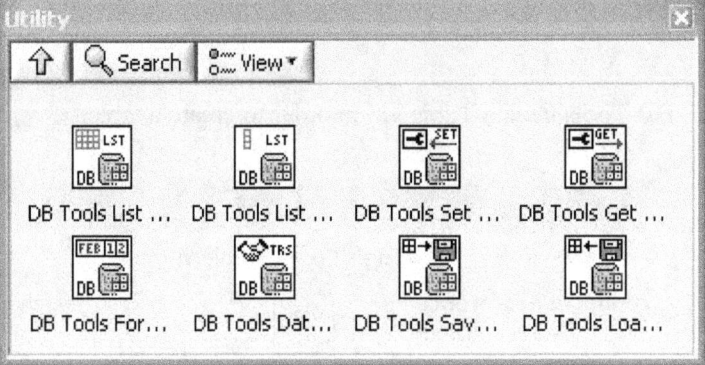

Here is a short description of the VIs located in the "Utility" palette:

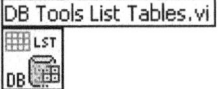
This VI lists the tables in the database identified by connection reference.

This VI lists the columns present in table. The column information includes the name, the data type, and the defined size of the column.

DB Tools Set Properties.vi

This VI sets properties on the object as determined by the inputs.

DB Tools Get Properties.vi

This VI gets properties of the object as determined by the inputs.

DB Tools Format Datetime Str.vi

This VI Returns a string containing the formatted date and time, and identifies the string as a date/time string so other VIs can interpret it.

DB Tools Database Transaction.vi

This VI begins, commits, rolls back a transaction for any type of reference.

DB Tools Save Recordset To File.vi

This VI saves the record set identified by the record set reference to either an XML or ADTG file. The ADTG file format is a proprietary format that only the LabVIEW Database Connectivity Toolkit can interpret. The ADTG format results in a smaller file than the XML format.

DB Tools Load Recordset From File.vi

This VI loads a record set from a file and returns a record set reference that identifies this record set. You can retrieve data from this record set like any other record set, but some properties might not be available on this record set.

5.8.6 Performing Advanced Database Operations

When creating real programs you will soon need some of the VIs in the "**Advanced**" palette.

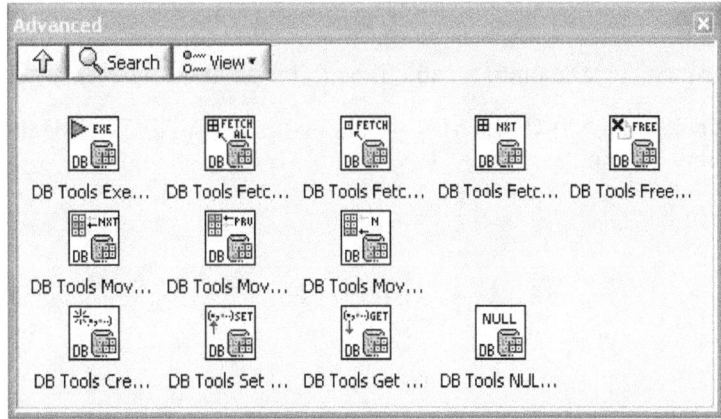

Here is a short description of some of the VIs located in the "Advanced" palette:

DB Tools Execute Query.vi

This VI Executes an SQL query and returns a record set reference that you must eventually free with the DB Tools Free Object VI.

DB Tools Fetch Recordset Data.vi

This VI retrieves the data in the record set identified by the record set reference input. You can convert each element in the array to its native LabVIEW data type using the "Database Variant To Data function".

DB Tools Free Object.vi

This VI frees an object by destroying its associated reference and returns a different reference object.

5.9 Creating and Using Tables

The SQL syntax for creating a Table is as follows:

CREATE TABLE <TableName>

(

<ColumnName> <datatype>

…

)

The SQL syntax for inserting Data into a Table is as follows:

INSERT INTO <TableName> (<Column1>, <Column2>, …) VALUES(<Data for Column1>, <Data for Column2>, …)

Example -6: Insert Data into Tables

We will insert some data into our tables:

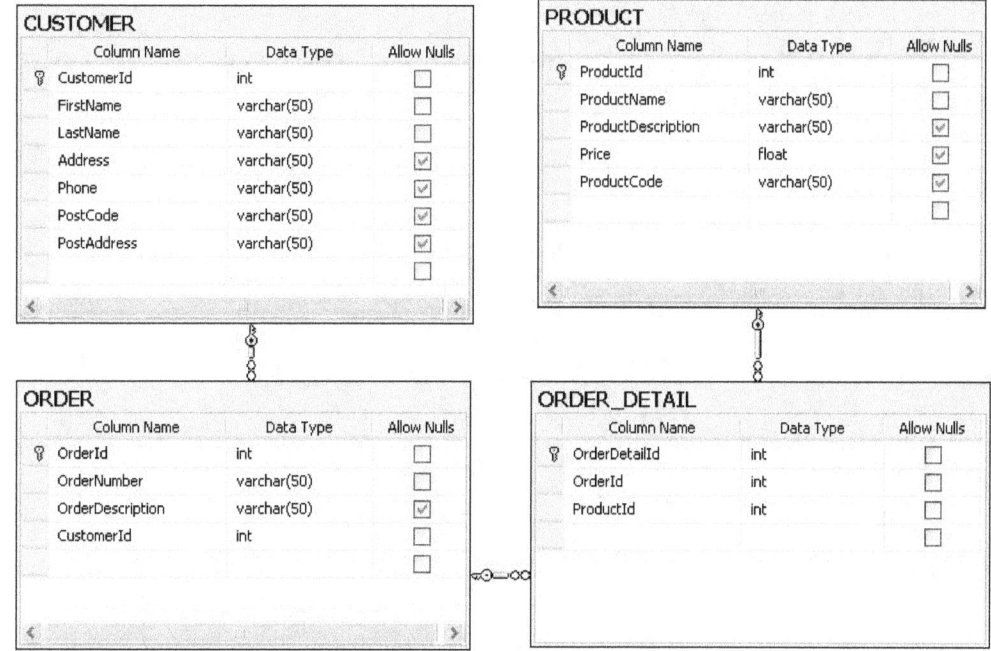

The following SQL Query inserts some example data into these tables:

```
--CUSTOMER


INSERT INTO [CUSTOMER]
([FirstName],[LastName],[Address],[Phone],[PostCode],[PostAddress]) VALUES
('NN', 'MM', 'LL', '12345678', '222', 'koko')


GO


INSERT INTO [CUSTOMER]
([FirstName],[LastName],[Address],[Phone],[PostCode],[PostAddress]) VALUES
('VV', 'GG', 'FF', '323439', '003', 'pp')


GO


INSERT INTO [CUSTOMER]
([FirstName],[LastName],[Address],[Phone],[PostCode],[PostAddress]) VALUES
('AA', 'BB', 'CC', '12345778', '656', 'jpo')

GO



--PRODUCT
```

131

```sql
INSERT INTO [PRODUCT]
([ProductName],[ProductDescription],[Price],[ProductCode]) VALUES ('Product
A', 'This is product A', 1000, 'A-1234')


GO


INSERT INTO [PRODUCT]
([ProductName],[ProductDescription],[Price],[ProductCode]) VALUES ('Product
B', 'This is product B', 1000, 'B-1234')


GO


INSERT INTO [PRODUCT]
([ProductName],[ProductDescription],[Price],[ProductCode]) VALUES ('Product
C', 'This is product C', 1000, 'C-1234')


GO



--ORDER


INSERT INTO [ORDER] ([OrderNumber],[OrderDescription],[CustomerId]) VALUES
('10001', 'This is Order 10001', 1)


GO


INSERT INTO [ORDER] ([OrderNumber],[OrderDescription],[CustomerId]) VALUES
('10002', 'This is Order 10002', 2)

GO


INSERT INTO [ORDER] ([OrderNumber],[OrderDescription],[CustomerId]) VALUES
('10003', 'This is Order 10003', 3)

GO

--ORDER_DETAIL

INSERT INTO [ORDER_DETAIL] ([OrderId],[ProductId]) VALUES (1, 1)

GO


INSERT INTO [ORDER_DETAIL] ([OrderId],[ProductId]) VALUES (1, 2)

GO

INSERT INTO [ORDER_DETAIL] ([OrderId],[ProductId]) VALUES (1, 3)


GO

INSERT INTO [ORDER_DETAIL] ([OrderId],[ProductId]) VALUES (2, 1)


GO
```

```
INSERT INTO [ORDER_DETAIL] ([OrderId],[ProductId]) VALUES (2, 2)
GO

INSERT INTO [ORDER_DETAIL] ([OrderId],[ProductId]) VALUES (3, 3)
GO

INSERT INTO [ORDER_DETAIL] ([OrderId],[ProductId]) VALUES (3, 1)
GO

INSERT INTO [ORDER_DETAIL] ([OrderId],[ProductId]) VALUES (3, 2)

GO

INSERT INTO [ORDER_DETAIL] ([OrderId],[ProductId]) VALUES (3, 3)

GO
```

Executing the following Queries then gives:

select * from CUSTOMER

select * from PRODUCT

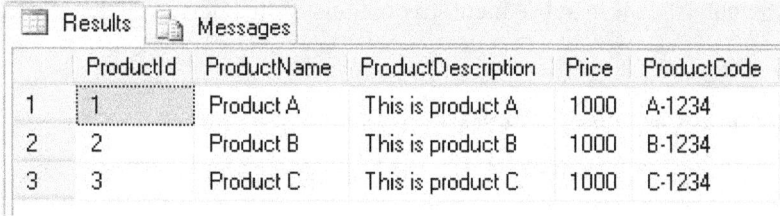

	ProductId	ProductName	ProductDescription	Price	ProductCode
1	1	Product A	This is product A	1000	A-1234
2	2	Product B	This is product B	1000	B-1234
3	3	Product C	This is product C	1000	C-1234

select * from [ORDER]

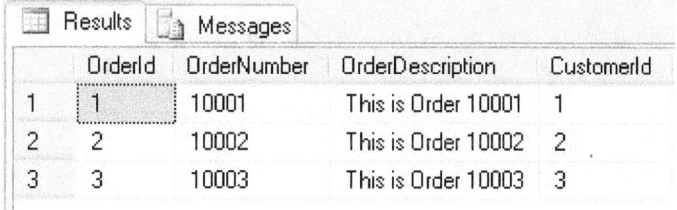

	OrderId	OrderNumber	OrderDescription	CustomerId
1	1	10001	This is Order 10001	1
2	2	10002	This is Order 10002	2
3	3	10003	This is Order 10003	3

133

select * from ORDER_DETAIL

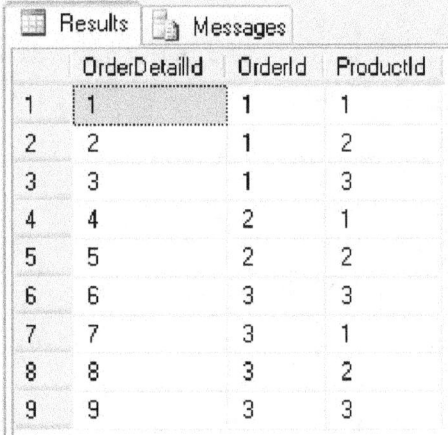

5.10 Creating and Using Views

In database theory, a view consists of a stored query accessible as a virtual table composed of the result set of a query. Unlike ordinary tables in a relational database, a view does not form part of the physical schema: it is a dynamic, virtual table computed or collated from data in the database. Changing the data in a table alters the data shown in subsequent invocations of the view.

Views can provide advantages over tables:

- Views can represent a subset of the data contained in a table

- Views can join and simplify multiple tables into a single virtual table

- Views can act as aggregated tables, where the database engine aggregates data (sum, average etc) and presents the calculated results as part of the data

- Views can hide the complexity of data; for example a view could appear as Sales2000 or Sales2001, transparently partitioning the actual underlying table

- Views take very little space to store; the database contains only the definition of a view, not a copy of all the data it presents

- Depending on the SQL engine used, views can provide extra security

- Views can limit the degree of exposure of a table or tables to the outer world

Just as functions (in programming) can provide abstraction, so database users can create abstraction by using views. In another parallel with functions, database users can manipulate nested views, thus one view can aggregate data from other views.

Syntax:

CREATE VIEW <ViewName>
AS

…

Create a VIEW:

Step 1: Create a new View

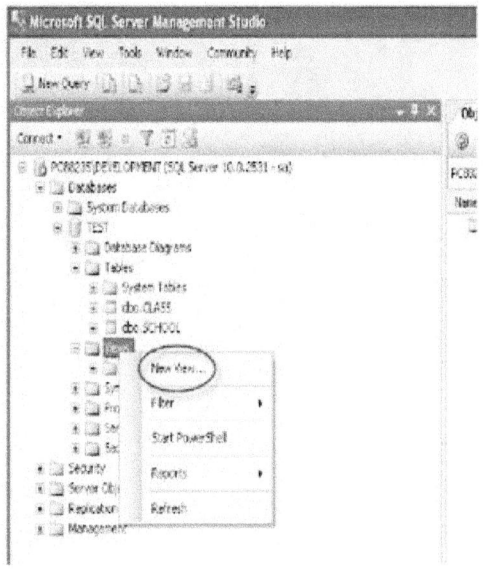

Step 2: Add your tables

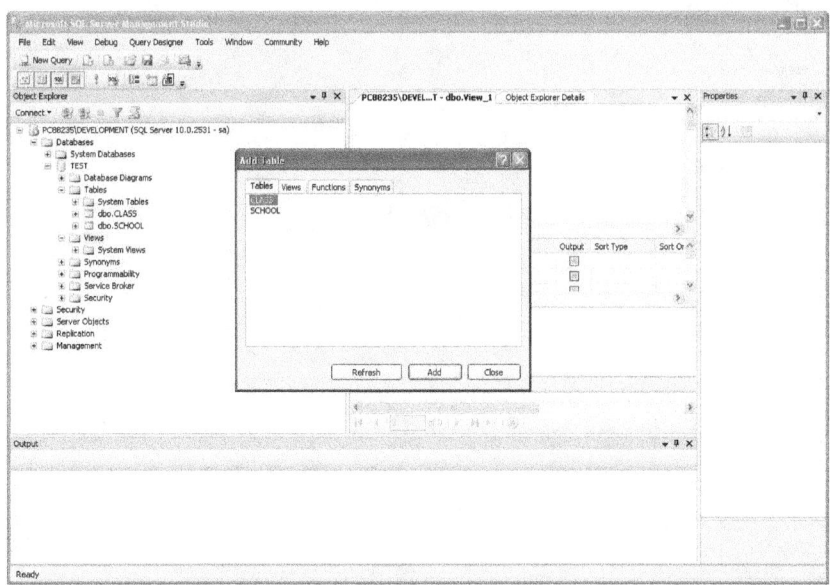

Step 3: Add your columns

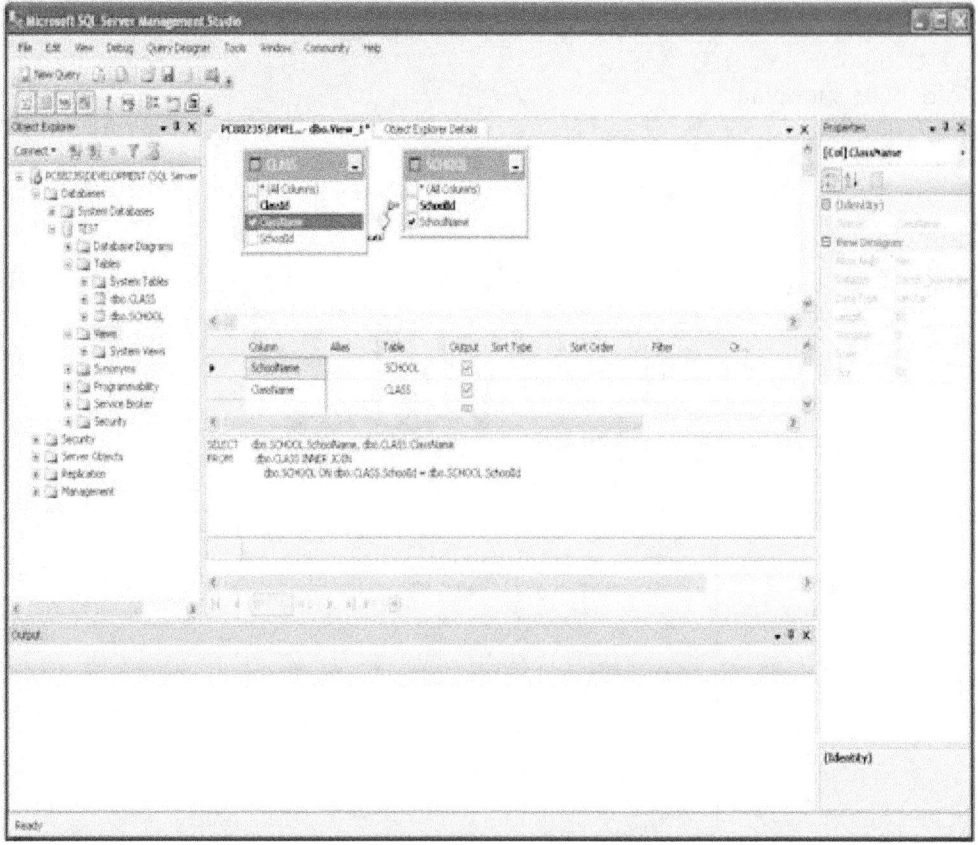

Step 4: Save it

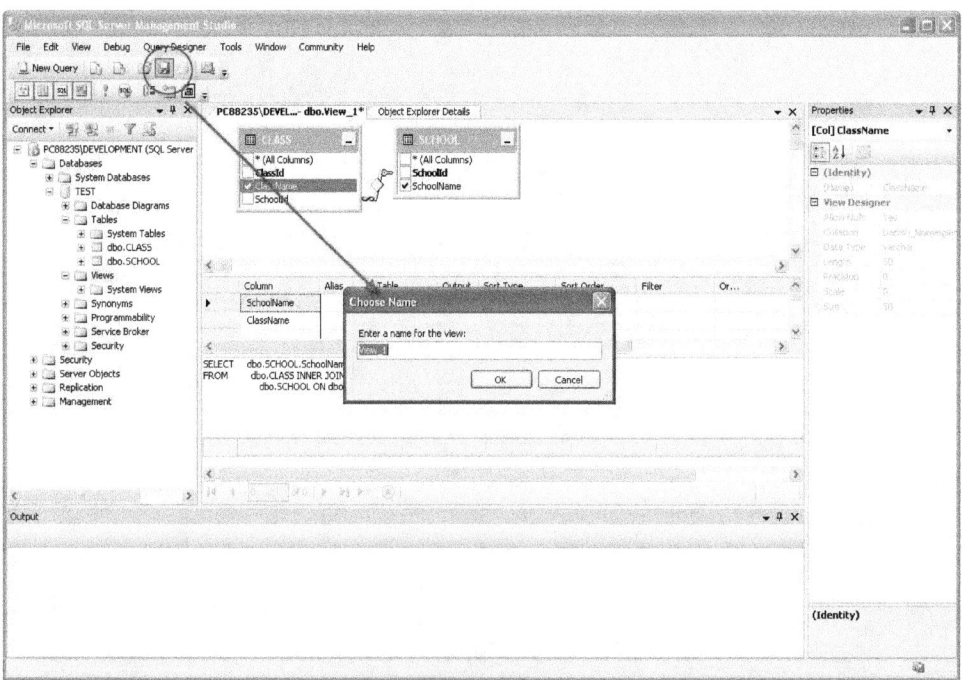

5.11 Creating and using Stored Procedures

A stored procedure is a subroutine available to applications accessing a relational database system. Typical uses for stored procedures include data validation (integrated into the database) or access control mechanisms. Furthermore, stored procedures are used to consolidate and centralize logic that was originally implemented in applications. Large or complex processing that might require the execution of several SQL statements is moved into stored procedures, and all applications call the procedures only.

A stored procedure is a precompiled collection of SQL statements and optional control-of-flow statements, similar to a macro. Each database and data provider supports stored procedures differently. Stored procedures offer the following benefits to your database applications:

Performance—Stored Procedures are usually more efficient and faster than regular SQL queries because SQL statements are parsed for syntactical accuracy and precompiled by the DBMS when the stored procedure is created. Also, combining a large number of SQL statements with conditional logic and parameters into a stored procedure allows the procedures to perform queries, make decisions, and return results without extra trips to the database server.

Maintainability—Stored Procedures isolate the lower-level database structure from the application. As long as the table names, column names, parameter names, and types do not change from what is stated in the stored procedure, you do not need to modify the procedure when changes are made to the database schema. Stored procedures are also a way to support modular SQL programming because after you create a procedure, you and other users can reuse that procedure without knowing the details of the tables involved.

Security—When creating tables in a database, the Database Administrator can set EXECUTE permissions on stored procedures without granting SELECT, INSERT, UPDATE, and DELETE permissions to users. Therefore, the data in these tables is protected from users who are not using the stored procedures.

Stored procedures are similar to user-defined functions. The major difference is that functions can be used like any other expression within SQL statements, whereas stored procedures must be invoked using the CALL statement.

The syntax for creating a Stored Procedure is as follows:

```
CREATE  PROCEDURE  <ProcedureName>
@<Parameter1> <datatype>
```

5.12 Creating and Using Triggers

A database trigger is procedural code that is automatically executed in response to certain events on a particular table or view in a database. The trigger is mostly used for keeping the integrity of the information on the database. For example, when a new record (representing a new worker) added to the employees table, new records should be created also in the tables of the taxes, vacations, and salaries.

Triggers are commonly used to:

- prevent changes (e.g. prevent an invoice from being changed after it's been mailed out)

- log changes (e.g. keep a copy of the old data)

- audit changes (e.g. keep a log of the users and roles involved in changes)

- enhance changes (e.g. ensure that every change to a record is time-stamped by the server's clock, not the client's)

- enforce business rules (e.g. require that every invoice have at least one line item)

- execute business rules (e.g. notify a manager every time an employee's bank account number changes)

- replicate data (e.g. store a record of every change, to be shipped to another database later)

- enhance performance (e.g. update the account balance after every detail transaction, for faster queries)

The major features of database triggers, and their effects, are:

- do not accept parameters or arguments (but may store affected-data in temporary tables)

- cannot perform commit or rollback operations because they are part of the triggering SQL statement

- can cancel a requested operation

- can cause mutating table errors, if they are poorly written.

Microsoft SQL Server supports triggers either after or instead of an insert, update, or delete operation.

The syntax is as follows:

```
CREATE TRIGGER <TriggerName> on <TableName>
FOR INSERT, UPDATE, DELETE

AS

… Create your Code here

GO
```

- Replace <TriggerName> with the Name of your Trigger

- Replace <TableName> with the Name of your Table

Define when the Trigger should be execute

- If the Trigger should be executed only when you insert data into the table: `FOR INSERT`

- If the Trigger should be executed only when you update data into the table: `FOR UPDATE`

- If the Trigger should be executed only when you delete data into the table: `FOR DELETE`

- If the Trigger should be executed when you insert and update data into the table: `FOR INSERT, UPDATE`

Example 1: Trigger

The Example above change the "below" in the Table "SCHOOL" from 'OSU' to 'Oklahoma State University'

```
CREATE TRIGGER CheckSchoolData on SCHOOL

FOR INSERT, UPDATE

AS

DECLARE

@SchoolName varchar(50)

select @SchoolName=SchoolName from INSERTED

If @SchoolName='OSU'

    update SCHOOL set SchoolName='Oklahoma State University' where

SchoolName=@SchoolName

GO
```

Note: the use of a temporary table called **"INSERTED"**. This temporary table contains the last inserted record into the SCHOOL table , and SQL you define a variable like this

```
DECLARE

@myVariable <datatype>
```

5.13 Creating and using Functions

In SQL databases, a user-defined function provides a mechanism for extending the functionality of the database server by adding a function that can be evaluated in SQL statements. The SQL standard distinguishes between scalar and table functions. A scalar function returns only a single value (or NULL), whereas a table function returns a (relational) table comprising zero or more rows, each row with one or more columns.

Stored Procedures vs. Functions:

- Only functions can return a value (using the RETURN keyword).

- Stored procedures can use RETURN keyword but without any value being passed.

- Functions could be used in SELECT statements, provided they don't do any data manipulation and also should not have any OUT or IN OUT parameters.

- Functions must return a value, but for stored procedures this is not compulsory.

- A function can have only IN parameters, while stored procedures may have OUT or IN OUT parameters.

- A function is a subprogram written to perform certain computations and return a single value.

- A stored procedure is a subprogram written to perform a set of actions, and can return multiple values using the OUT parameter or return no value at all.

User-defined functions in SQL are declared using the CREATE FUNCTION statement.

5.14 SQL Toolkit

The SQL Toolkit palette is available in LabVIEW:

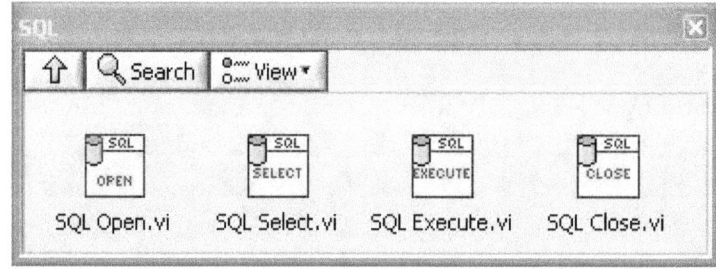

The SQL Toolkit contains the following VIs:

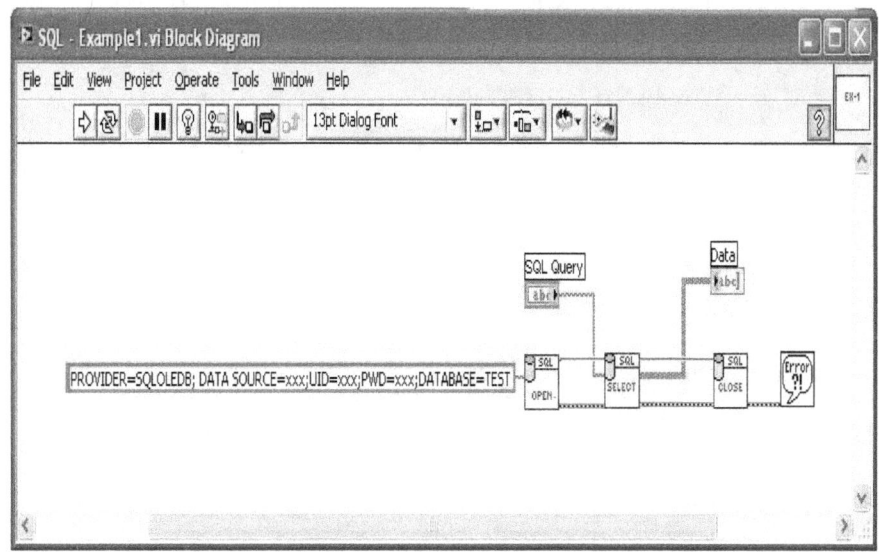 **"SQL Open.vi"** - This VI open a connection to the database specified in the Connection string. The connection string may be as follows:

"PROVIDER=SQLOLEDB; DATA SOURCE=xxx;UID=xxx;PWD=xxx;DATABASE=xxx"

You need to replace the "xxx" with the parameters from your database.

"SQL Select.vi" -This VI get data from the database specified in the SQL Query. The output is a 2D string array with data.

"SQL Execute.vi" - This VI executes a Query with no return Data, e.g., an INSERT statement

"SQL Close.vi" - This VI Close the connection to the database opened by "SQL Open.vi"

Two examples are also included:

SQL – Example 1.vi – This example selects data from a table. The example uses "SQL Select.vi" in order to get data from the database. The following are:

1) Block Diagram.

2) Front panel.

SQL – Example 2.vi – This example inserts data into a table. The example uses "SQL Execute.vi" in order to insert data into the database. No data is returned.

The Block Diagram is:

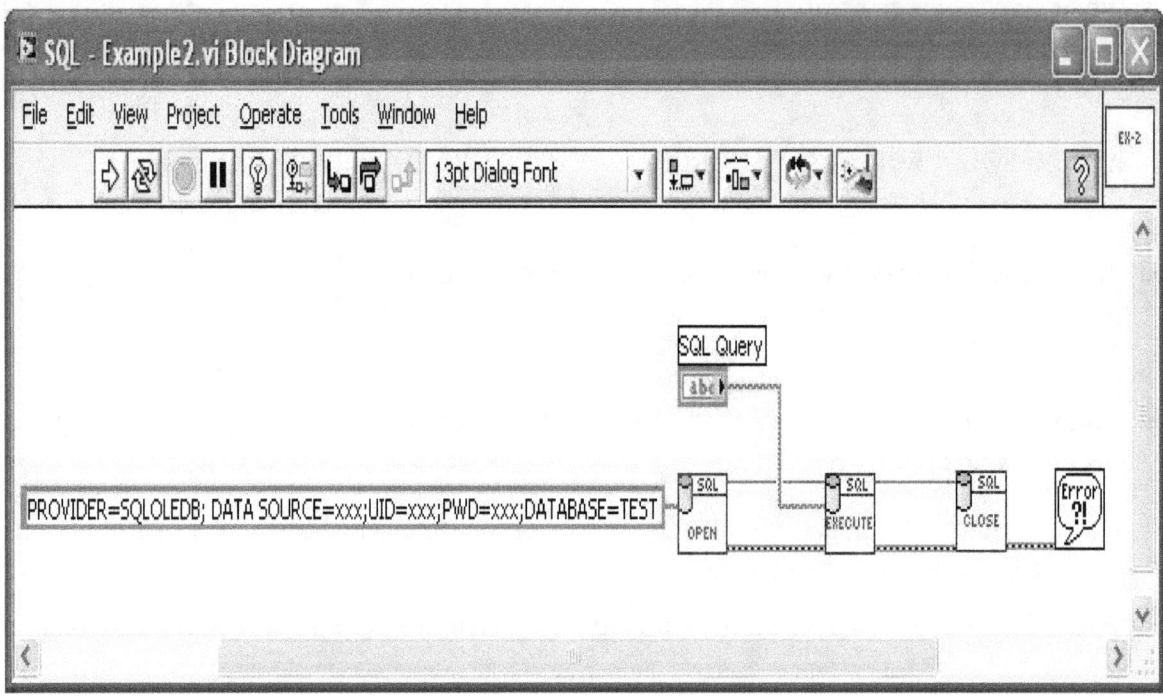

Get Data into LabVIEW using SQL Toolkit

Download the SQL toolkit from the **Homepage of the Database Lab (ni.com)** and follow the instructions in the ReadMe file.

On the Functions palette on your Block Diagram the following palette should appear:

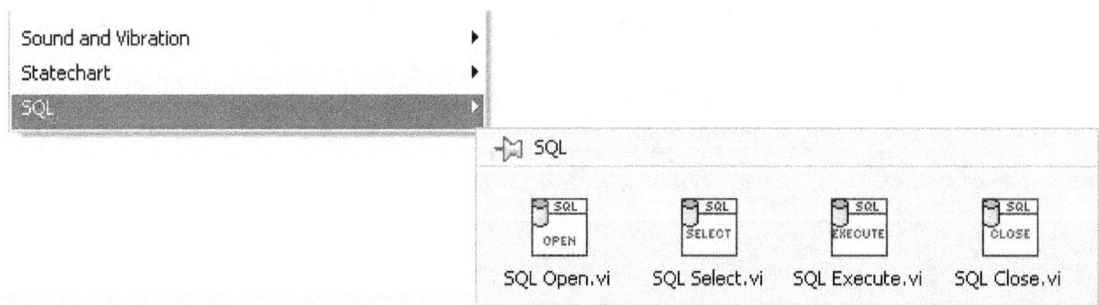

Here is a simple example of how you get data from the database into LabVIEW, the procedure is as follows:

Step 1:

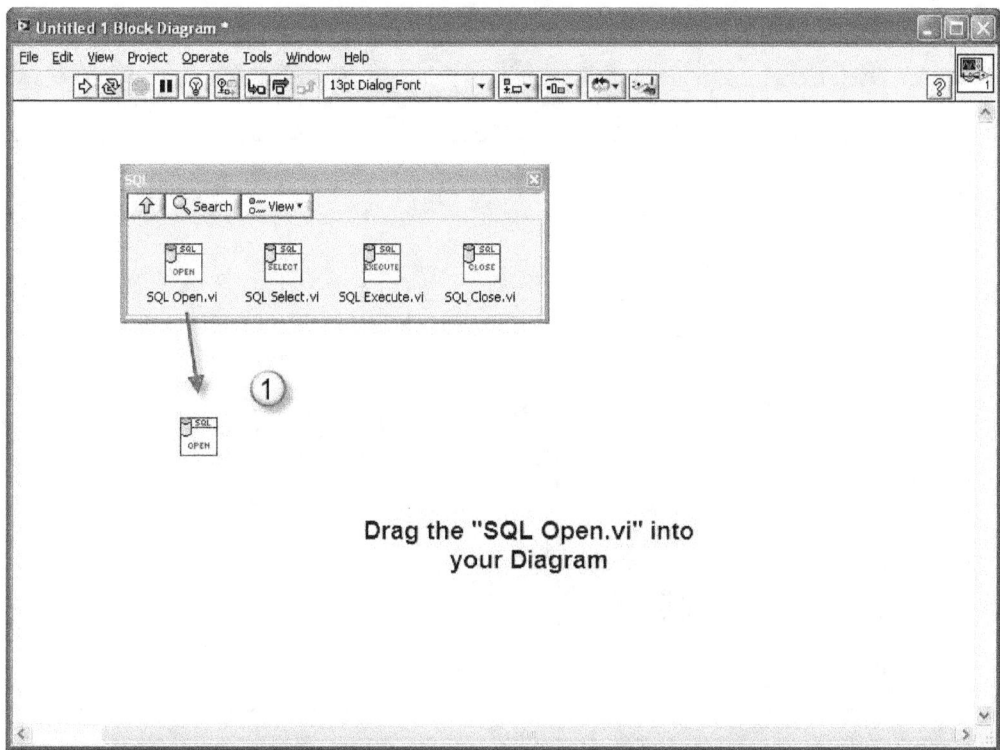

Drag the "SQL Open.vi" into
your Diagram

Step 2:

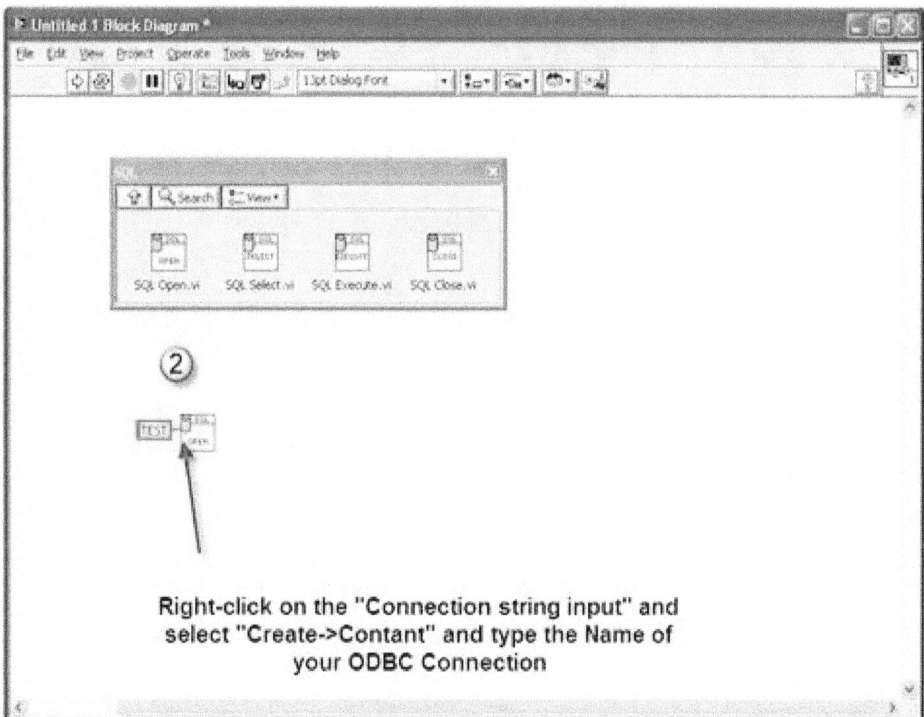

Right-click on the "Connection string input" and
select "Create->Contant" and type the Name of
your ODBC Connection

Step 3 and 4:

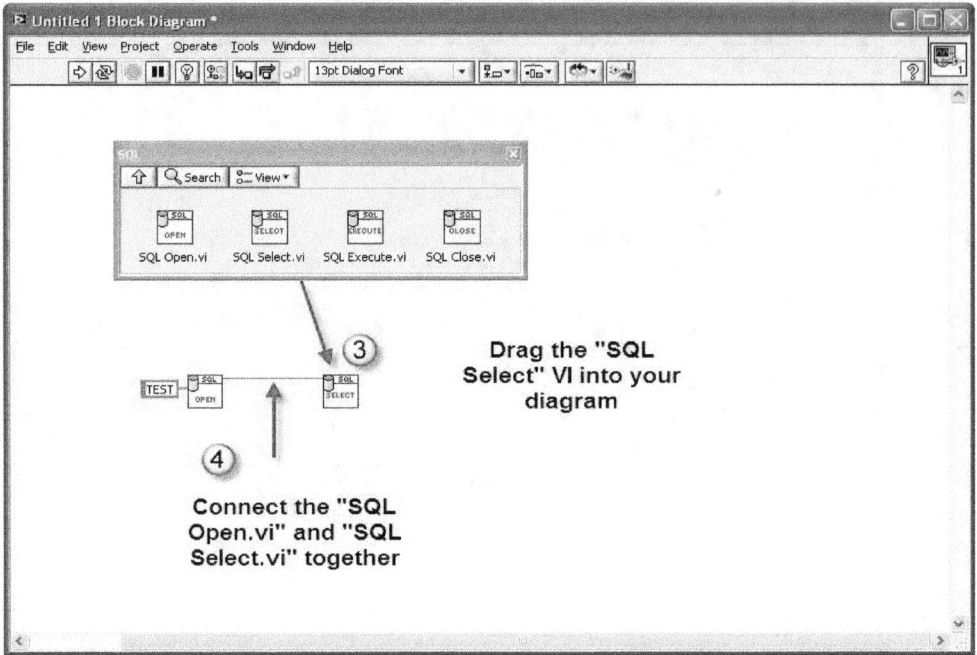

Step 5:

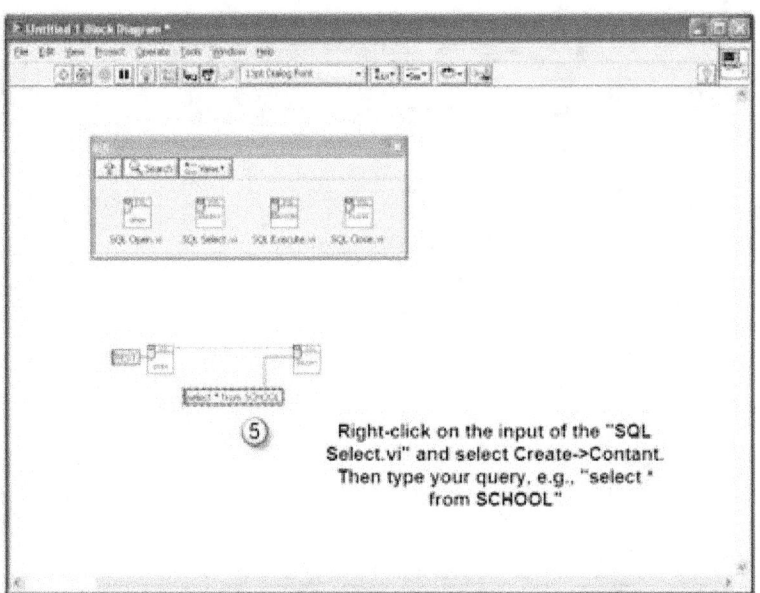

Step 6:

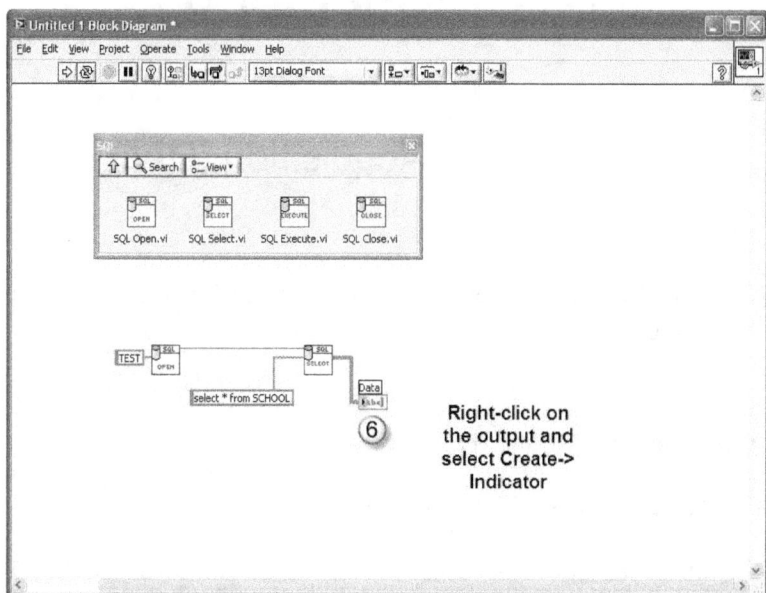

Step 7:

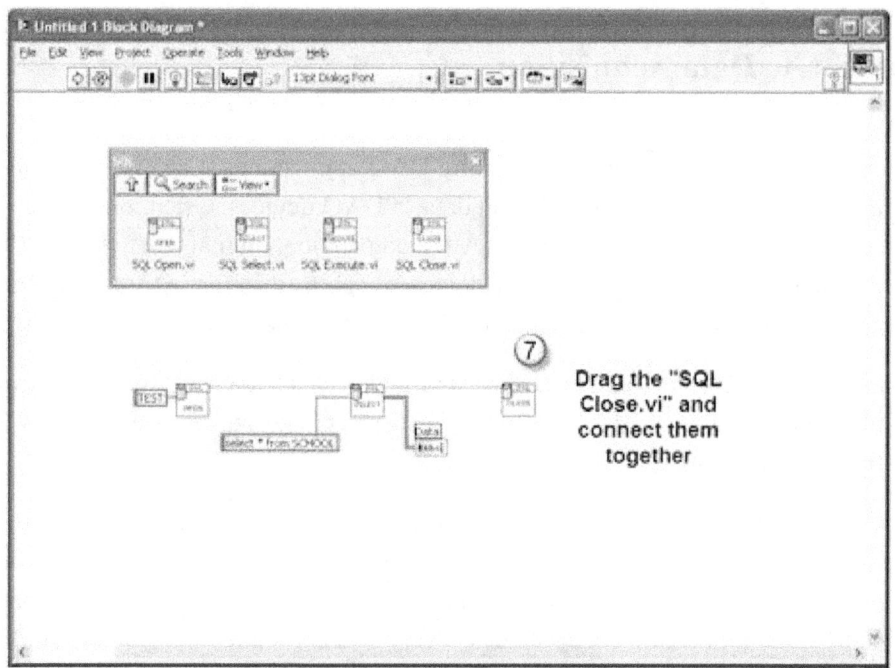

Chapter 6

Data Acquisition (DAQ)

This chapter explains the basic concepts of using DAQ in LabVIEW. And it covers the most powerful topics, as the following ;

- Introduction to DAQ - Data Acquisition

- MAX – Measurement and Automation Explorer

- NI-DAQmx

6.1 Introduction to Data Acquisition

LabVIEW is very powerful when it comes to creating DAQ applications. LabVIEW includes a set of VIs that let you configure, acquire data from, and send data to DAQ devices. Often, one device can perform a variety of functions, such as analog-to-digital (A/D) conversion, digital-to-analog (D/A) conversion, digital I/O, and counter/timer operations. Each device supports different DAQ and signal generation speeds. Also, each DAQ device is designed for specific hardware, platforms and operating systems.

National Instruments, the inventor of LabVIEW, also make DAQ devices, so the integration with the DAQ devices from NI and the LabVIEW software is seamless and makes it easy to do I/O operations from the LabVIEW environment.

The purpose of data acquisition is to measure an electrical or physical phenomenon such as voltage, current, temperature, pressure, or sound. PC-based data acquisition uses a combination of modular hardware, application software, and a computer to take measurements. While each data acquisition system is defined by its application requirements, every system shares a common goal of acquiring, analyzing, and presenting information. Data acquisition systems incorporate signals, sensors, actuators, signal conditioning, data acquisition devices, and application software.

Summing up, the Data Acquisition (DAQ) is the process of:

- o Acquiring signals from real-world phenomena

- o Digitizing the signals

- o Analyzing, presenting and saving the data

The DAQ system has the following parts involved, as shown in Figure 6-1:

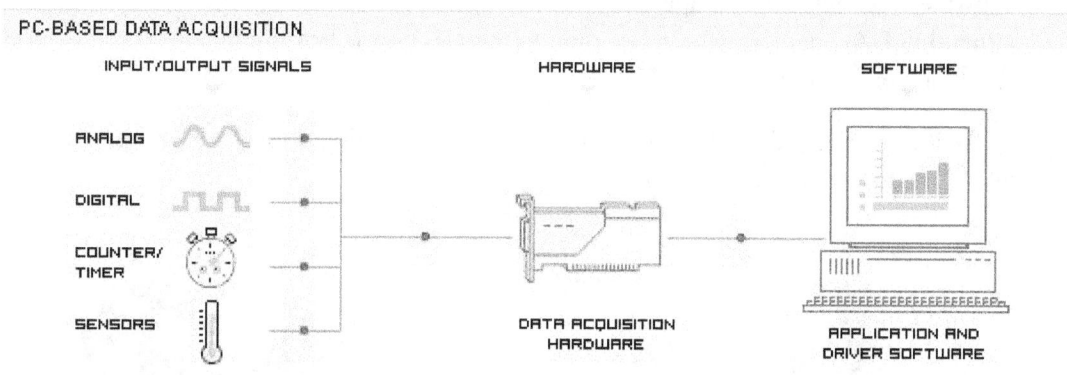

Figure 6-1

Those parts are:

- Physical input/output signals

- DAQ device/hardware

- Driver software

- Your software application (Application software)

6.1.1 Physical input/output signals

A physical input/output signal is typically a voltage or current signal. A voltage signal can typically be a 0-5V signal, while a current signal can typically be a 4-20mA signal.

6.1.2 DAQ device/hardware

DAQ hardware acts as the interface between the computer and the outside world. It primarily functions as a device that digitizes incoming analog signals so that the computer can interpret them. A DAQ device (Data Acquisition Hardware) usually has these functions:

- Analog input

- Analog output

- Digital I/O

- Counter/timers

We have different DAQ devices, such as:

- "**Desktop** DAQ devices" where you need to plug a PCI DAQ board into your computer. The software is running on a computer.
- "**Portable** DAQ devices" for connection to the USB port, Wi-Fi connections, etc. The software is running on a computer

- "**Distributed** DAQ devices" where the software is developed on your computer and then later downloaded to the distributed DAQ device.

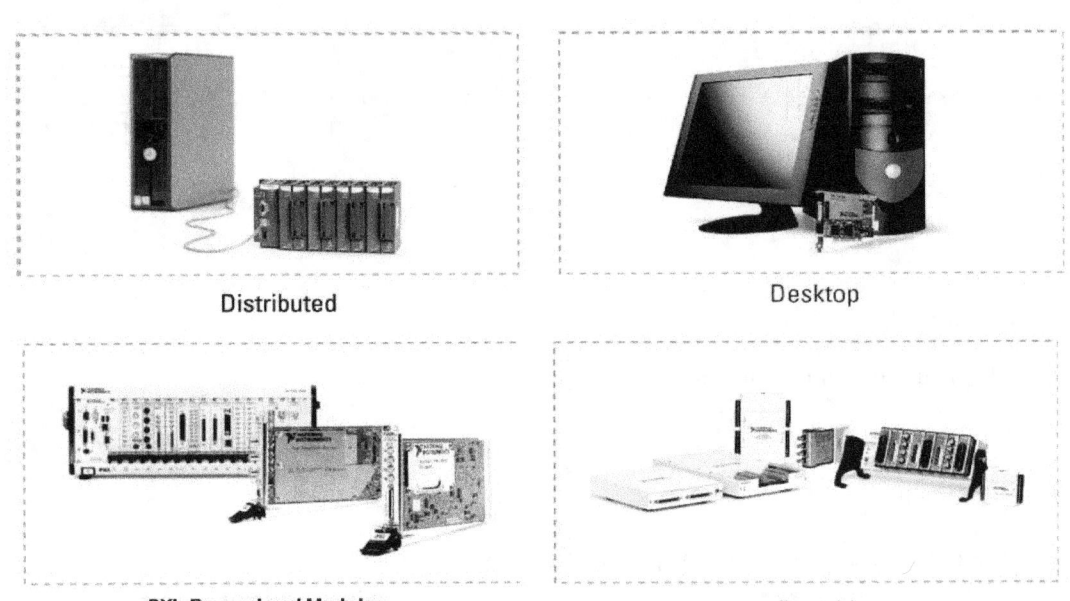

Distributed   Desktop

PXI: Rugged and Modular   Portable

Figure 6-2

6.1.3 Driver Software

Driver software is the layer of software for easily communicating with the hardware. It forms the middle layer between the application software and the hardware. Driver software also prevents a programmer from having to do register-level programming or complicated commands in order to access the hardware functions.

Driver software from National Instruments:

- ➢ NI-DAQmx

- ➢ NI-DAQmx Base

The **DAQ Assistant**, included with NI-DAQmx, is a graphical, interactive guide for configuring, testing, and acquiring measurement data. With a single click, you can even generate code based on your configuration, making it easier and faster to develop complex operations.

Because DAQ Assistant is completely menu-driven, you will make fewer programming errors and drastically decrease the time from setting up your DAQ system to taking your first measurement. NI-DAQmx Base offers a subset of NI-DAQmx functionality on Windows and Linux, Mac OS X, Windows Mobile and Windows CE.

6.1.4 Application Software

Application software adds analysis and presentation capabilities to the driver software. Your software application normally does such tasks as:

- Real-time monitoring

- Data analysis

- Data logging

- Control algorithms

- Human machine interface (HMI)

In order to create your DAQ application you need a programming development tool, such as LabVIEW.

6.2 MAX - Measurement and Automation Explorer

Measurement & Automation Explorer (MAX) provides access to your National Instruments devices and systems, see figure 6-3.

With MAX, you can:

- ❖ Configure your National Instruments hardware and software

- ❖ Create and edit channels, tasks, interfaces, scales, and virtual instruments

- ❖ Execute system diagnostics

- ❖ View devices and instruments connected to your system

- ❖ Update your National Instruments software

In addition to the standard tools, MAX can expose item-specific tools you can use to configure, diagnose, or test your system, depending on which NI products you install. As you navigate through MAX, the contents of the application menu and toolbar change to reflect these new tools. The following Figure 6-3 are shown both front page and explorer page of MAX.

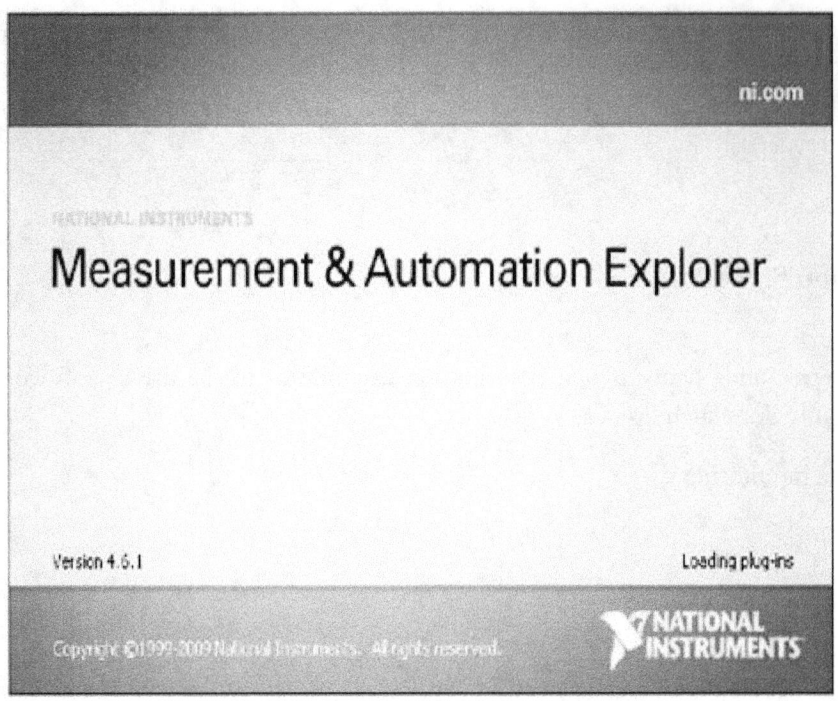

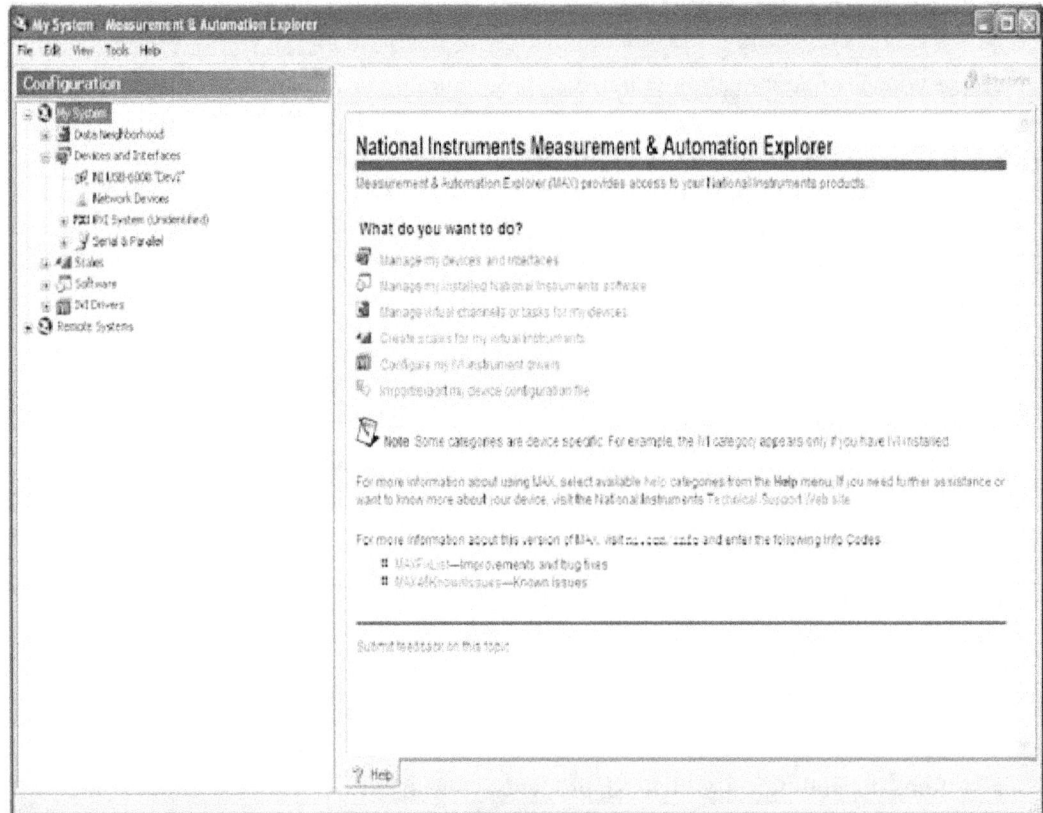

Figure 6-3

6.3 NI-DAQmx

The NI-DAQmx Driver software is the layer of software for easily communicating with the hardware. It forms the middle layer between the application software and the hardware. Driver software also prevents a programmer from having to do register-level programming or complicated commands in order to access the hardware functions. The "DAQ Assistant" is an easy way to start using the DAQ features in LabVIEW. We'll learn more about the "DAQ Assistant" in a later chapter. The DAQmx palette in LabVIEW is shown like figure 6-4:

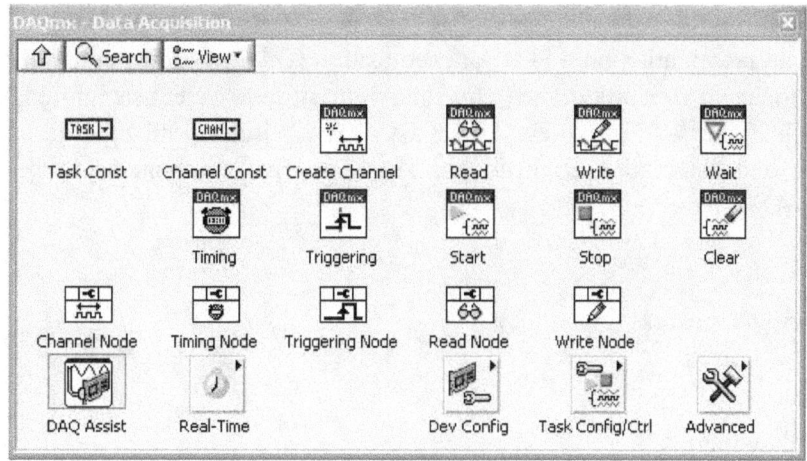

Figure 6-4

6.3.1 DAQ Assistant

The **DAQ Assistant**, included with NI-DAQmx, is a graphical, interactive guide for configuring, testing, and acquiring measurement data. With a single click, you can even generate code based on your configuration, making it easier and faster to develop complex operations. Because DAQ Assistant is completely menu-driven, you will make fewer programming errors and drastically decrease the time from setting up your DAQ system to taking your first measurement.

6.4 NI USB-6008

NI USB-6008 is a simple and low-cost multifunction I/O device from National Instruments. The NI USB-6008 is well suited for education purposes due to its small size and easy USB connection. The device has the following specifications:

- 8 analog inputs (12-bit, 10 kS/s)

- 2 analog outputs (12-bit, 150 S/s)

- 12 digital I/O

- USB connection, No extra power-supply needed

- Compatible with LabVIEW, Lab Windows/CVI, and Measurement Studio for Visual Studio .NET

- NI-DAQmx driver software

6.5 Physical input/output signals

Data acquisition involves gathering signals from measurement sources and digitizing the signal for storage, analysis, and presentation on a PC. Data acquisition (DAQ) systems come in many different PC technology forms for great flexibility when choosing your system. Scientists and engineers can choose from PCI, PXI, PCI Express, PXI Express, PCMCIA, USB, Wireless and Ethernet data acquisition for test, measurement, and automation applications. There are five components to be considered when building a basic DAQ system:

- Transducers and sensors

- Signals

- Signal conditioning

- DAQ hardware

- Driver and application software

In this chapter we focus on Transducers, sensors and Signals.

6.5.1 Transducers

The transducer is a device that converts a physical phenomenon into a measurable electrical signal, such as voltage or current. The ability of a DAQ system to measure different phenomena depends on the transducers to convert the physical phenomena into signals measurable by the DAQ hardware. Transducers are synonymous with sensors in DAQ systems. There are specific transducers for many different applications, such as measuring temperature, pressure, or fluid flow. Below we've seen some common phenomena and the transducers used to measure them.

Whereby Data acquisition begins with the physical phenomenon to be measured. This physical phenomenon could be the temperature of a room, the intensity of a light source, the pressure inside a chamber, the force applied to an object, or many other things. An effective DAQ system can measure all of these different phenomena. The following table 6-1 shows some of these phenomena matched with specific transducer.

Phenomenon	Transducer
Temperature	Thermocouple, RTD, Thermostat
Light	Photo Sensor
Sound	Microphone
Force and Pressure	Strain Gage, Piezoelectric Transducer
Position and Displacement	Potentiometer, LVDT, Optical Encoder
Acceleration	Accelerometer
pH	pH Electrode

Table 6-1

Different transducers have different requirements for converting phenomena into a measurable signal. Some transducers may require excitation in the form of voltage or current. Other transducers may require additional components and even resistive networks to produce a signal.

6.5.2 Signals

The appropriate transducers convert physical phenomena into measurable signals. However, different signals need to be measured in different ways. For this reason, it is important to understand the different types of signals and their corresponding attributes. Signals can be categorized into two groups:

A) Analog Signals

B) Digital Signals

A) Analog Signals

Analog input is the process of measuring an analog signal and transferring the measurement to a computer for analysis, display, or storage. An analog signal is a signal that varies continuously. Analog input is most commonly used to measure voltage or current. You can use many types of devices to perform analog input, such as multifunction DAQ (MIO) devices, high-speed digitizers, digital multi meters, and Dynamic Signal Acquisition (DSA) devices.

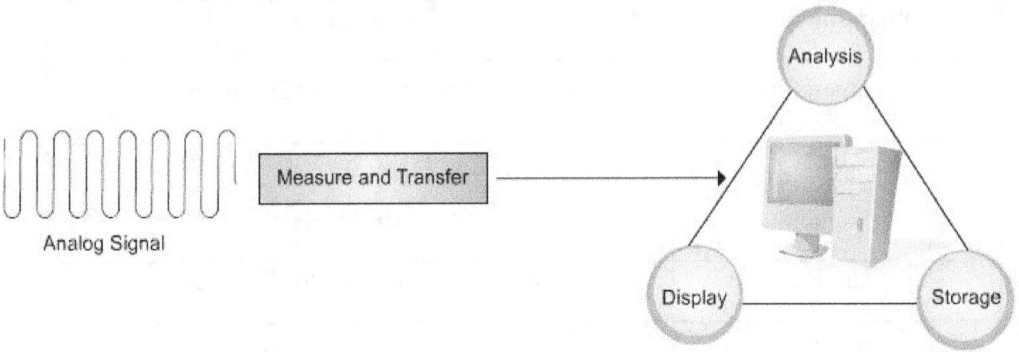

An analog signal can be at any value with respect to time. A few examples of analog signals include voltage, temperature, pressure, sound, and load. The three primary characteristics of an analog signal is:

- Level

- Shape

- Frequency

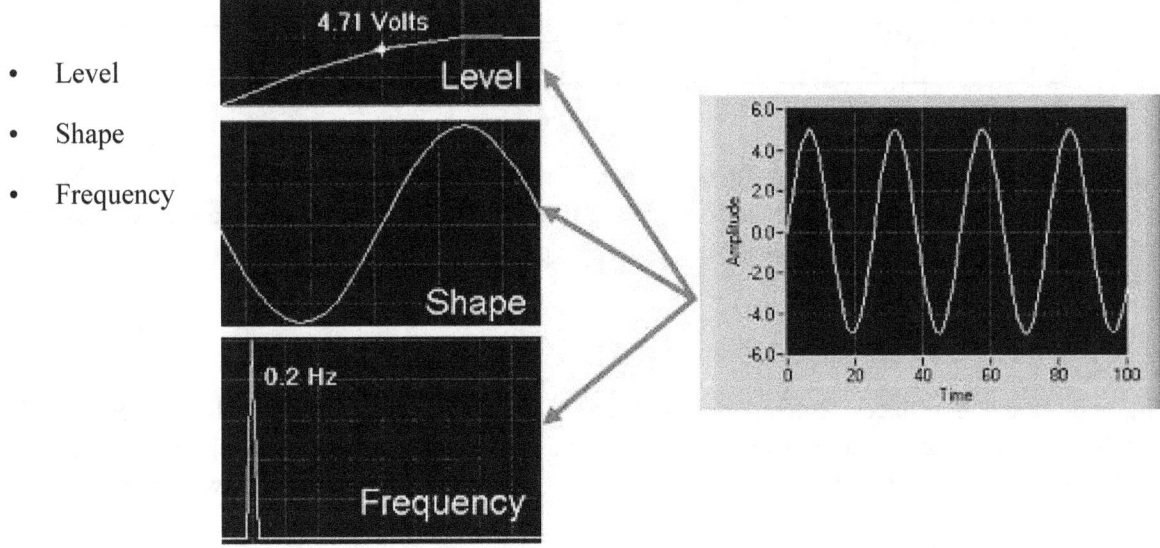

> **Level**

Because analog signals can take on any value, the level gives vital information about the measured analog signal. The intensity of a light source, the temperature in a room, and the pressure inside a chamber are all examples that demonstrate the importance of the level of a signal. When measuring the level of a signal, the signal generally does not change quickly with respect to time. The accuracy of the measurement, however, is very important. A DAQ system that yields maximum accuracy should be chosen to aid in analog level measurements.

> **Shape**

Some signals are named after their specific shape - sine, square, saw tooth, and triangle. The shape of an analog signal can be as important as the level, because by measuring the shape of an analog signal, you can further analyze the signal, including peak values, DC values, and slope. Signals where shape is of interest generally change rapidly with respect to time, but system accuracy is still important. The analysis of heartbeats, video signals, sounds, vibrations, and circuit responses are some applications involving shape measurements.

➢ Frequency

All analog signals can be categorized by their frequency. Unlike the level or shape of the signal, frequency cannot be directly measured. The signal must be analyzed using software to determine the frequency information. This analysis is usually done using an algorithm known as the Fourier transform.

When frequency is the most important piece of information, it is important to consider including both accuracy and acquisition speed. Although the acquisition speed for acquiring the frequency of a signal is less than the speed required for obtaining the shape of a signal, the signal must still be acquired fast enough that the pertinent information is not lost while the analog signal is being acquired. The condition that stipulates this speed is known as the Nyquist Sampling Theorem. Speech analysis, telecommunication, and earthquake analysis are some examples of common applications where the frequency of the signal must be known.

B) Digital Signals

A digital signal cannot take on any value with respect to time. Instead, a digital signal has two possible levels: high and low. Digital signals generally conform to certain specifications that define characteristics of the signal. Digital signals are commonly referred to as transistor-to-transistor logic (TTL). TTL specifications indicate a digital signal to be low when the level falls within 0 to 0.8 V, and the signal is high between 2 to 5 V. The useful information that can be measured from a digital signal includes the state and the rate.

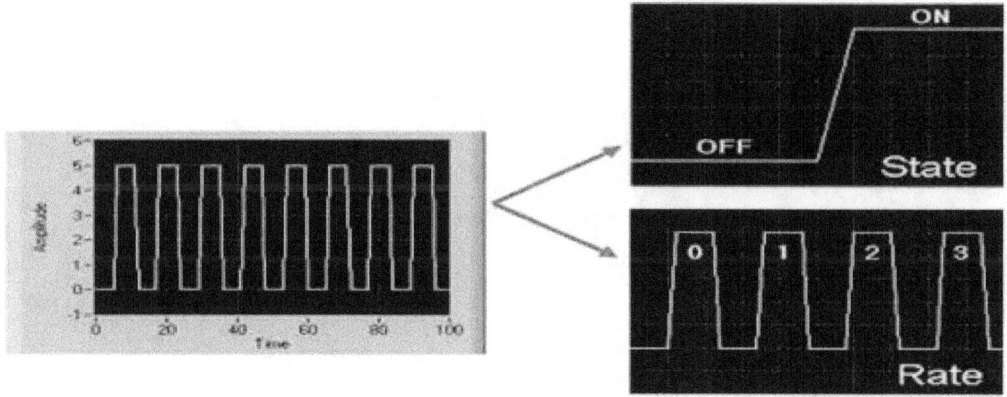

➢ State

Digital signals cannot take on any value with respect to time. The state of a digital signal is essentially the level of the signal - on or off, high or low. Monitoring the state of a switch - open or closed - is a common application showing the importance of knowing the state of a digital signal.

➢ Rate

The rate of a digital signal defines how the digital signal changes state with respect to time. An example of measuring the rate of a digital signal includes determining how fast a motor shaft spins. Unlike frequency, the rate of a digital signal measures how often a portion of a signal occurs. A software algorithm is not required to determine the rate of a signal.

6.6 Measurement & Automation Explorer (MAX) continue

As we have seen before the Measurement & Automation Explorer (MAX) provides access to your National Instruments devices and systems. In addition to the standard tools, MAX can expose item-specific tools you can use to configure, diagnose, or test your system, depending on which NI products you install. As you navigate through MAX, the contents of the application menu and toolbar change to reflect these new tools.

LabVIEW installs MAX to establish all devices and channel configuration parameters. MAX reads the information the Device Manager records in the Windows Registry and assigns a logical device number to each DAQ device. You use the device number to refer to the device in LabVIEW. It can be access MAX by selecting Tools» Measurement & Automation Explorer in LabVIEW. This displays the primary MAX window.

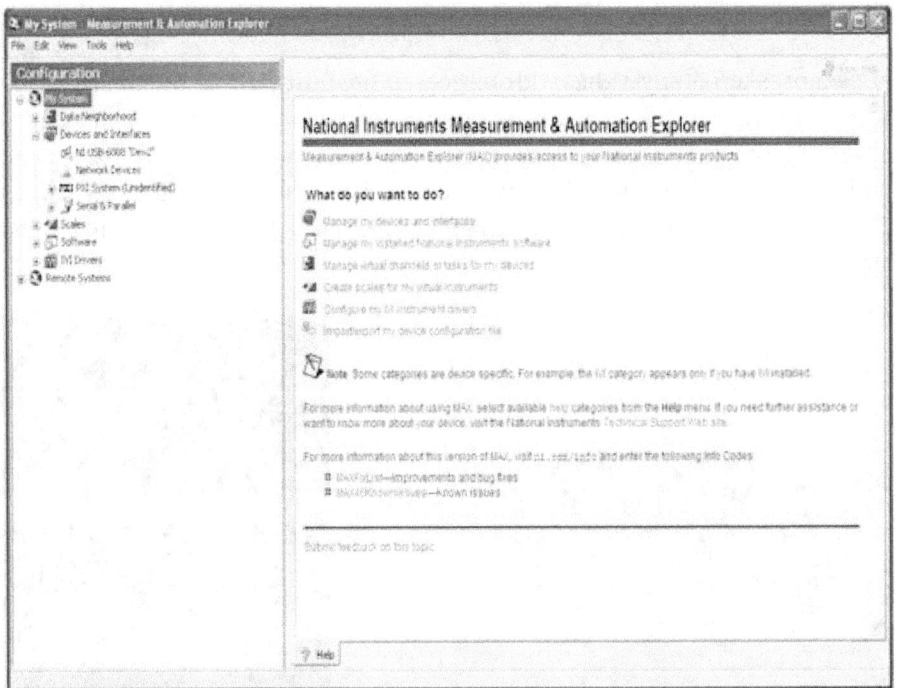

Before using a data acquisition board, you must confirm that the software can communicate with the board by configuring the devices. For Windows, the Windows Configuration Manager keeps track of all the hardware installed in the computer, including National Instruments DAQ devices. The Windows Configuration Manager automatically detects and configures Plug & Play (PnP) devices.

> ➤ **Windows Configuration Manager**

If one have a PnP device, such as an E Series MIO device, the Windows Configuration Manager automatically detects and configures the device. If you have a non-PnP device, or legacy device, you must configure the device manually using the Add New Hardware option in the Control Panel. You can verify the Windows Configuration by accessing the Device Manager.

6.7 NI-DAQmx

Driver software is the layer of software for easily communicating with the hardware. It forms the middle layer between the application software and the hardware. Driver software also prevents a programmer from having to do register-level programming or complicated commands in order to access the hardware functions.

Driver software from National Instruments:

- NI-DAQmx

- NI-DAQmx Base

The **DAQ Assistant**, included with NI-DAQmx, is a graphical, interactive guide for configuring, testing, and acquiring measurement data. With a single click, you can even generate code based on your configuration, making it easier and faster to develop complex operations. Because DAQ Assistant is completely menu-driven, you will make fewer programming errors and drastically decrease the time from setting up your DAQ system to taking your first measurement. NI-DAQmx Base offers a subset of NI-DAQmx functionality on Windows and Linux, Mac OS X, Windows Mobile and Windows CE.

National Instruments DAQ boards have a driver engine that communicates between the board and the application software. There are two driver engines, NI-DAQmx and Traditional NI-DAQ. You can also use the DAQ Assistant, an Express VI that communicates with NI-DAQmx, in LabVIEW to communicate with the DAQ board. In addition, National Instruments provides Measurement & Automation Explorer (MAX) for configuring DAQ boards.

The NI-DAQmx Driver software is the layer of software for easily communicating with the hardware. It forms the middle layer between the application software and the hardware. Driver software also prevents a programmer from having to do register-level programming or complicated commands in order to access the hardware functions. The DAQmx palette as seen in the LabVIEW is:

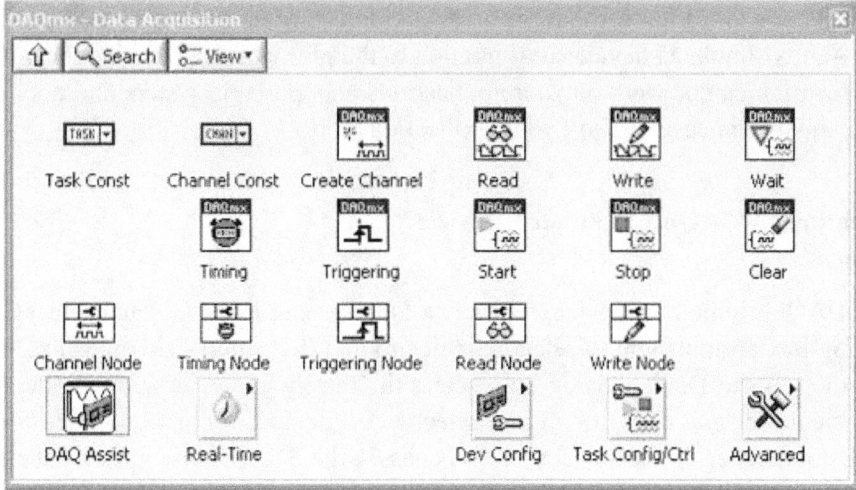

6.7.1 DAQ Assistant

The **DAQ Assistant**, included with NI-DAQmx, is a graphical, interactive guide for configuring, testing, and acquiring measurement data. With a single click, you can even generate code based on your configuration, making it easier and faster to develop complex operations. Because DAQ Assistant is completely menu-driven, you will make fewer programming errors and drastically decrease the time from setting up your DAQ system to taking your first measurement.

> ➢ **Scales**

One can configure custom scales for your measurements using MAX. This is very useful when working with sensors. It allows you to bring a scaled value into your application without having to work directly with the raw values. For example, you can use a temperature sensor that represents temperature with a voltage.

The conversion equation for the temperature is, Voltage x 100 = Celsius. After a scale is set, you can use it in your application program, providing the temperature value, rather than the voltage. When performing analog input, the task can be timed to:

- Acquire 1 Sample
- Acquire n Samples
- Acquire Continuously

6.7.2 Simulating a DAQ Device

It can create NI-DAQmx simulated devices in NI-DAQmx. Using NI-DAQmx simulated devices, and try NI products in your application without the hardware. Later, when one acquire the hardware, it can import the NI-DAQmx simulated device configuration to the physical device using the MAX Portable Configuration Wizard. It can be work on your applications on a portable system and upon returning to the original system, you can easily import your application work.

> ➢ **Creating NI-DAQmx Simulated Devices**

To create an NI-DAQmx simulated device, right-click Devices and Interfaces and select Create New. The Create New dialog box prompts you to select a device to add. Select NI-DAQmx Simulated Device and click Finish. In the Choose Device dialog box, select the family of devices for the device you want to simulate. Select the device and click OK. If you select a PXI device, you are prompted to select a chassis number and PXI slot number. If you select an SCXI chassis, the SCXI configuration panels open.

6.8 DAQ Devices

DAQ hardware acts as the interface between the computer and the outside world. It primarily functions as a device that digitizes incoming analog signals so that the computer can interpret them. A DAQ device (Data Acquisition Hardware) usually has these functions:

- Analog input
- Analog output
- Digital I/O
- Counter/timers

There are different DAQ devices, such as:

- **"Desktop** DAQ devices" where you need to plug a PCI DAQ board into your computer. The software is running on a computer.

- **"Portable** DAQ devices" for connection to the USB port, Wi-Fi connections, etc. The software is running on a computer

- **"Distributed** DAQ devices" where the software is developed on your computer and then later downloaded to the distributed DAQ device.

Most DAQ devices have four standard elements: analog input, analog output, digital I/O, and counters. The DAQ device transfers the measured signals to a computer through different bus structures. For example, you can plug a DAQ device into the PCI bus or the USB port of a computer or the Personal Computer Memory Card International Association (PCMCIA) socket of a laptop. You also can use PXI/Compact PCI to create a portable, versatile, and rugged measurement system.

6.8.1 Performing Analog to Digital Conversion

Analog-to-digital conversion is a process of acquiring and translating signals into digital data so that a computer can process it. Analog-to-digital converters (ADCs) are circuit components that convert a voltage level into a series of ones and zeroes. ADCs sample the analog signal on each rising or falling edge of a sample clock. In each cycle, the ADC takes a snapshot of the analog signal, measures and converts it into a digital value. The ADC obtains and approximates the signal with fixed precision and converts it into a series of digital values, see figure 6-5.

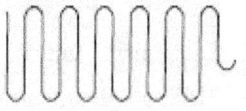

Analog Signal

Analog-to-Digital Converters

Digital Signal

Figure 6-5

6.8.2 Performing Digital to Analog Conversion

Digital-to-analog conversion is the opposite of analog-to-digital conversion. In digital-to-analog conversion, the computer generates the data.

A) Using Counters

A counter is a digital timing device. You typically use counters for event counting, frequency measurement, period measurement, position measurement, and pulse generation.

B) Using Digital I/O

- Digital signals are electrical signals that transfer digital data over a wire. These signals typically have only two states: on and off, also known as high and low, or 1 and 0. When sending a digital signal across a wire, the sender applies a voltage to the wire and the receiver uses the voltage level to determine the value being sent. The voltage ranges for each digital value depend on the voltage level standard being used.
- Digital signals have many users.
- Digital signals control or measure digital devices such as switches or LEDs.

6.9 NI-USB 6008

NI USB-6008 is a simple and low-cost multifunction I/O device from National Instruments. The device has the following specifications:

- 8 analog inputs (12-bit, 10 kS/s)
- 2 analog outputs (12-bit, 150 S/s)
- 12 digital I/O
- USB connection, No extra power-supply needed
- Compatible with LabVIEW, Lab Windows/CVI, and Measurement Studio for Visual Studio .NET
- NI-DAQmx driver software
- The NI USB-6008 is well suited for education purposes due to its small size and easy USB connection.

160

6.9.1 Connect USB-6008 to the PC

Configuring and testing: USB-6008 can be configured and tested using MAX (Measurement and Automation Explorer), which is installed with the NI-DAQmx Driver Software. The first time you connect the USB-6008 to the PC, the Windows Hardware Installer Wizard will open.

The wizard searches the PC for the necessary driver software for the USB-6008. This driver software was installed along with the installation of the NI-DAQ software. When the wizard has finished the installation of the driver software, the USB-6008 is ready for use.

6.9.2 Testing USB-6008 in MAX

Before you start to use the USB-6008 in an application, you should test the device in the Measurement and Automation Explorer (**MAX**).

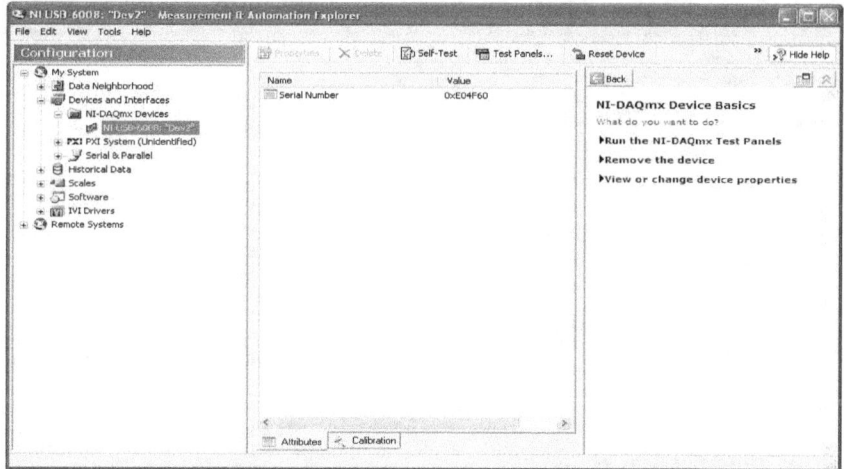

In the MAX window, expand the "Devices and Interfaces" node and then "NI DAQmx Devices". Right-click on the NI USB-6008 device and select "Self-Test".

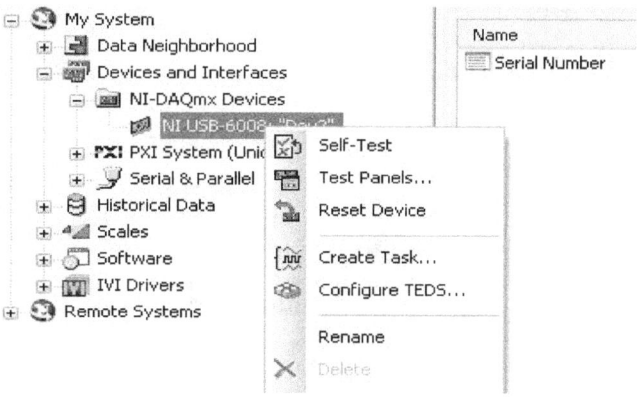

Hopefully the self-test passes without errors. Then, one should test the individual channels of the USB-6008 to check that the input signals are detected correctly by the USB-6008, and that the output signals generated by the USB-6009 have correct values. This I/O can be tested in several ways, depending on which channels you actually want to test. We will perform a simple **loopback test**:

> Here, let us test analog output channel 0 (AO0) and the analog input channel 0 (AI0) to see if they work correctly. We will perform a very simple test, which is sufficient if we are to check that both AO0 and AI0 work correctly. The test procedure, which is denoted loopback, is to connect the AI0 channel to the AO0 channel. Then we generate some legal voltage at AO0. If AI0 detects the same voltage, we know that both AO0 and AI0 work. (We may then repeat this procedure for other channels.) If for some reason AI0 detects some other voltage than the value we set for AO0, then there is an error in either the AI0 channel or in the AO0 channel, and further investigations are necessary.

To prepare for the loopback test, we wire together AI0 and AO0. To see the terminals of the USB-6000, select "Device Pin outs" from the right-click menu.

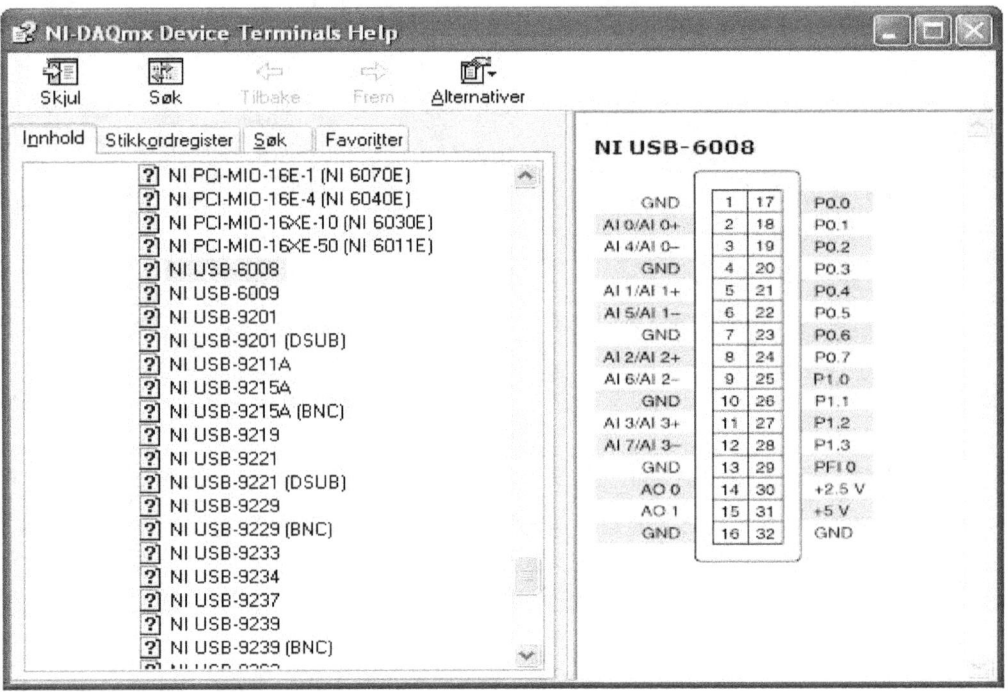

The figure below shows the AI0 and AO0 channels wired together.

162

As a matter of fact , to perform the loopback test, right-click on the NI USB-6008 device in MAX, and then select "Test Panels.." in to open the Test Panels. In the Test Panels window, select the Analog Output tab. In the Analog Output tab, select any voltage between 0V and 5V. Next, click the Analog Input tab in the Test Panels window. The following figures are shown these steps in some details and figure out the whole process.

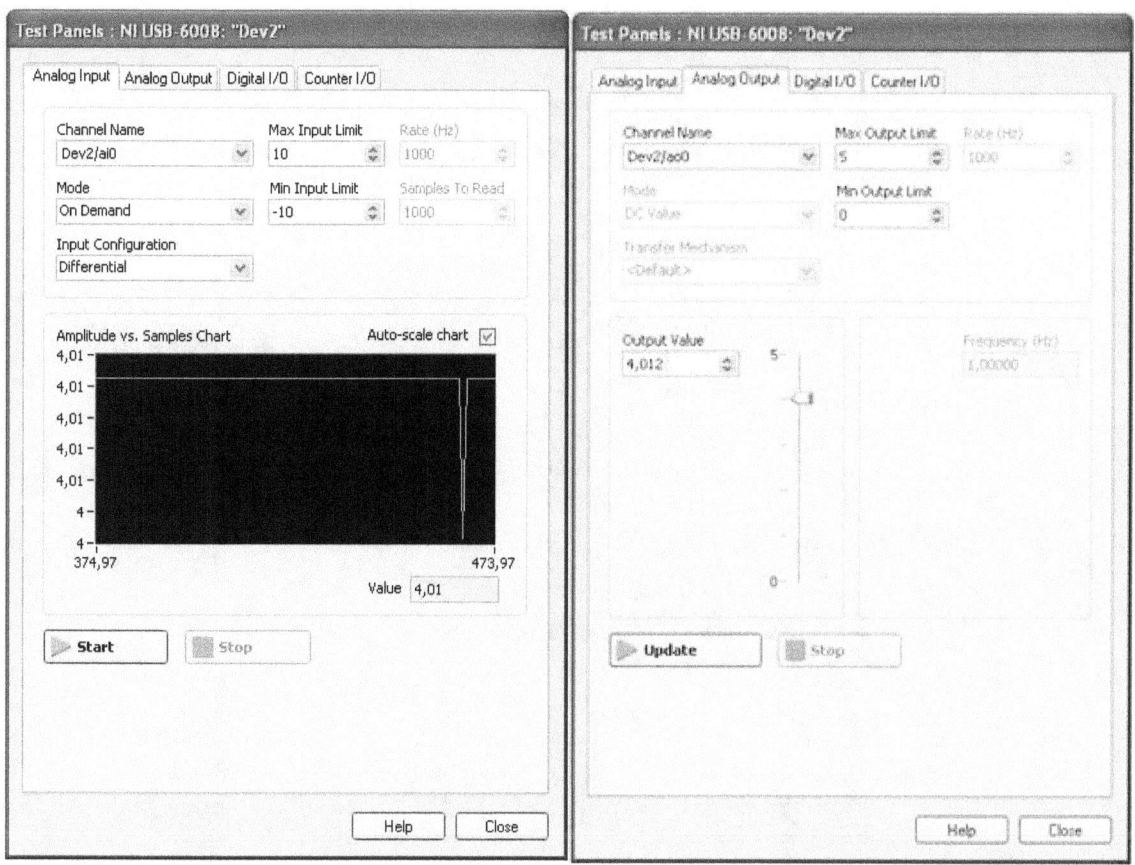

The Analog Input tab should indicate the same (or almost the same) voltage as is set out on AO0. There may be a small difference between the values due to the limited resolution in the DA-converter (digital-to-analog) and in the AD-converter (analog-to-digital). Also always use a **multi-meter** to check if the voltage levels on the output and input channels are correct according to your settings.

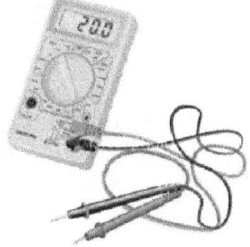

6.9.3 Using USB-6008 in LabVIEW

In order to use the NI USB-6008 in LabVIEW you need to use the DAQmx functions, see the Figure below. Then use DAQmx – Data Acquisition palette as seen it before.

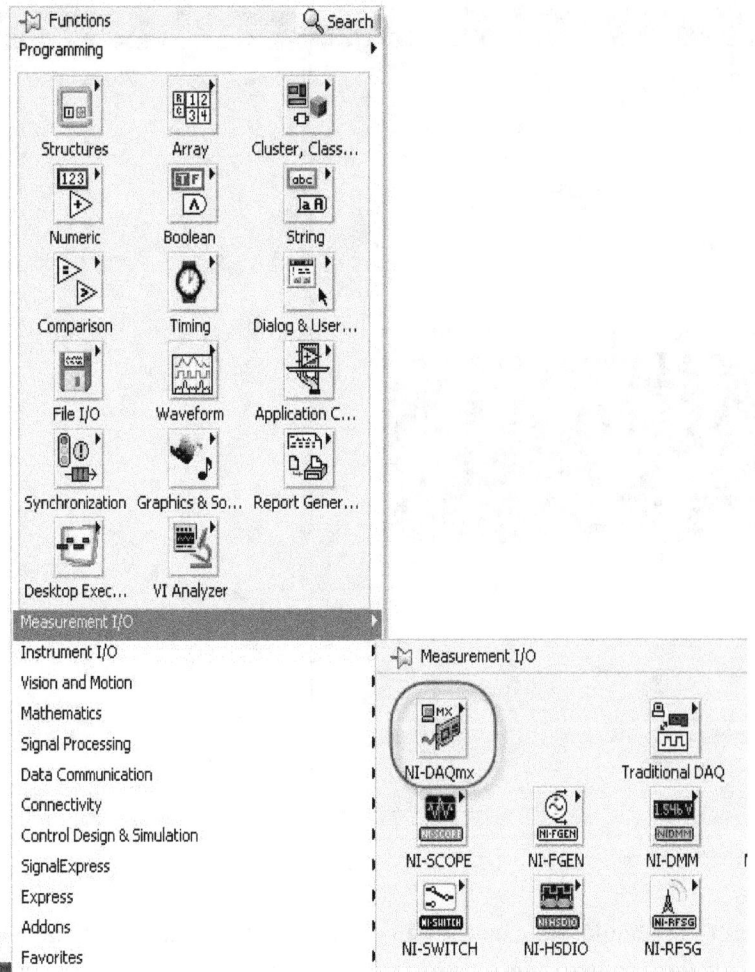

6.10 DAQ Assistant

The easiest ways is to use the DAQ Assistant by following the entire steps:

A) Analog Input: When you drag the DAQ Assistant icon on your Block Diagram, the following window appears:

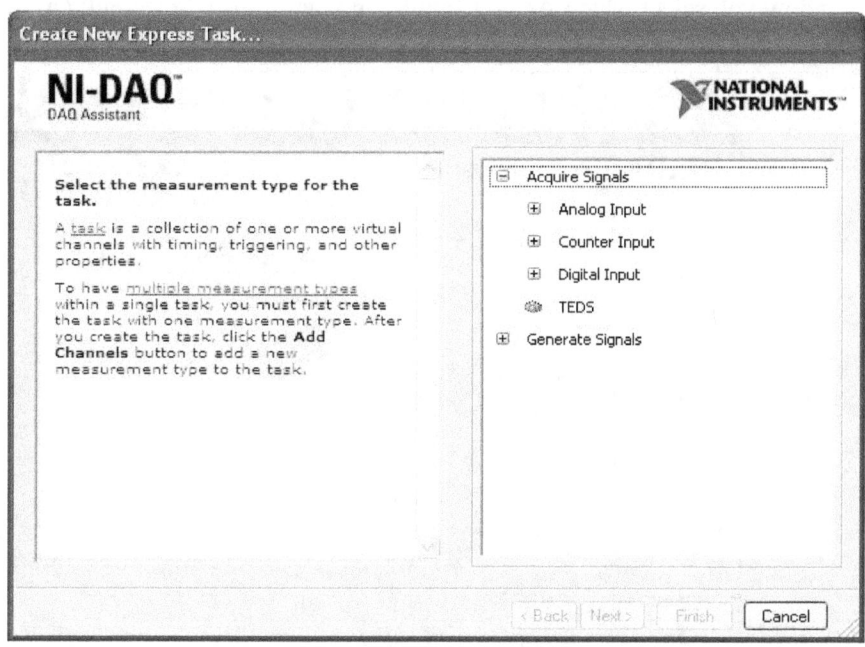

In this window you need to select either "Acquire Signals" (i.e., Input Signals) or "Generate Signals" (i.e., Output Signals). Select Acquire Signals → Analog Input → Voltage.

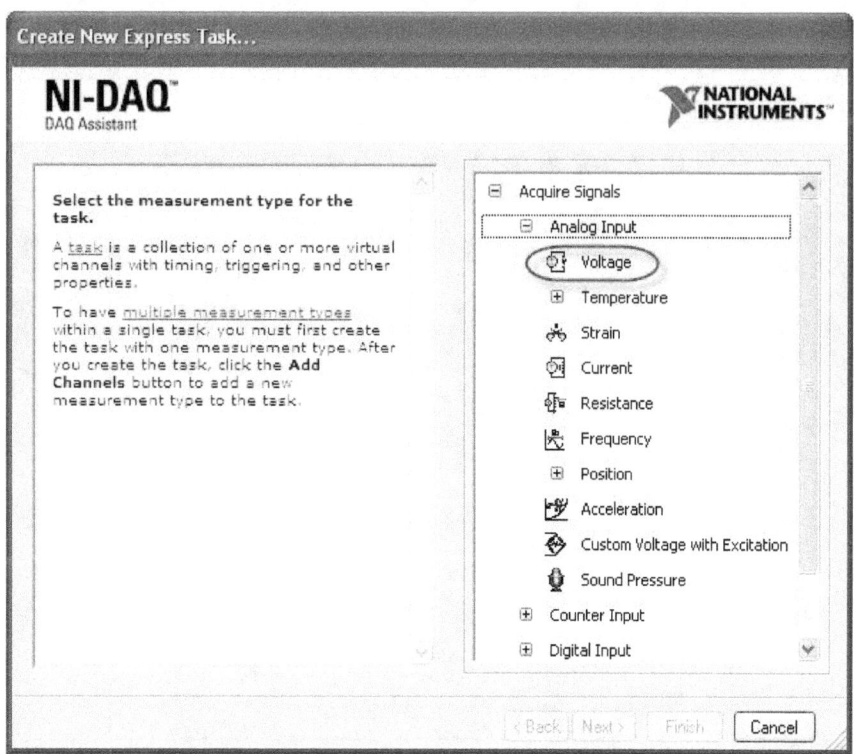

In the next window you select which Analog Input you want to use. Select ai0 (Analog Input channel 0) and click Finish.

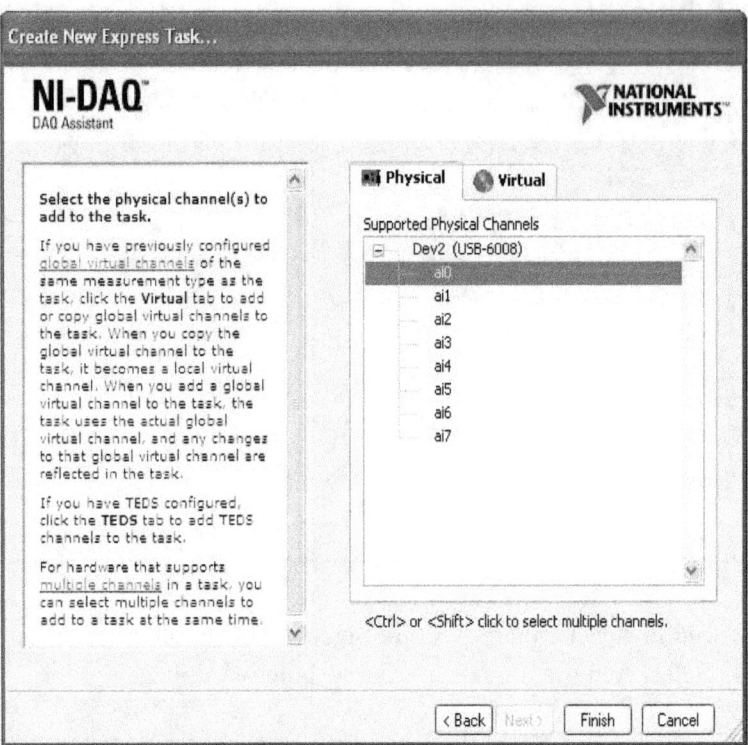

The following window appears:

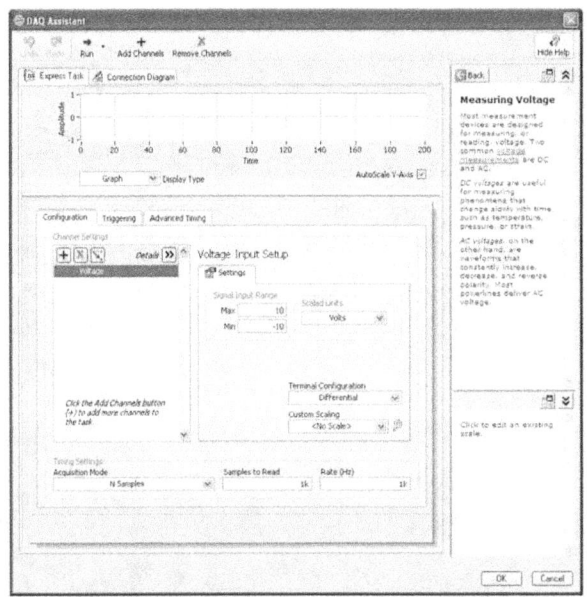

In the **Timing Settings** Select "**1 Sample (On Demand)**".

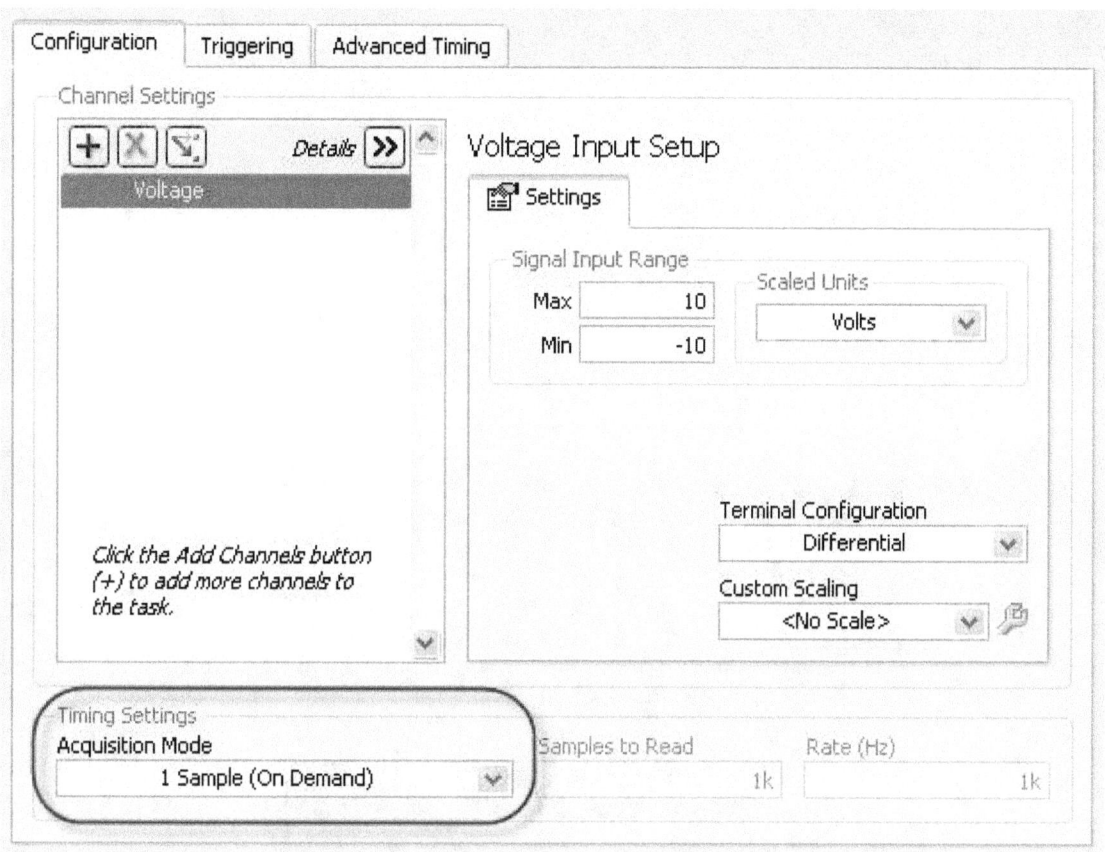

The next step is to select the Signal Input Range. A common signal is 0-5V.

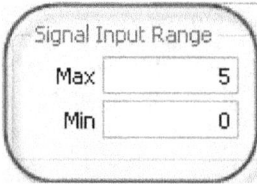

One may also rename the name of the channel (right-click on the name) as shown in the following figures.

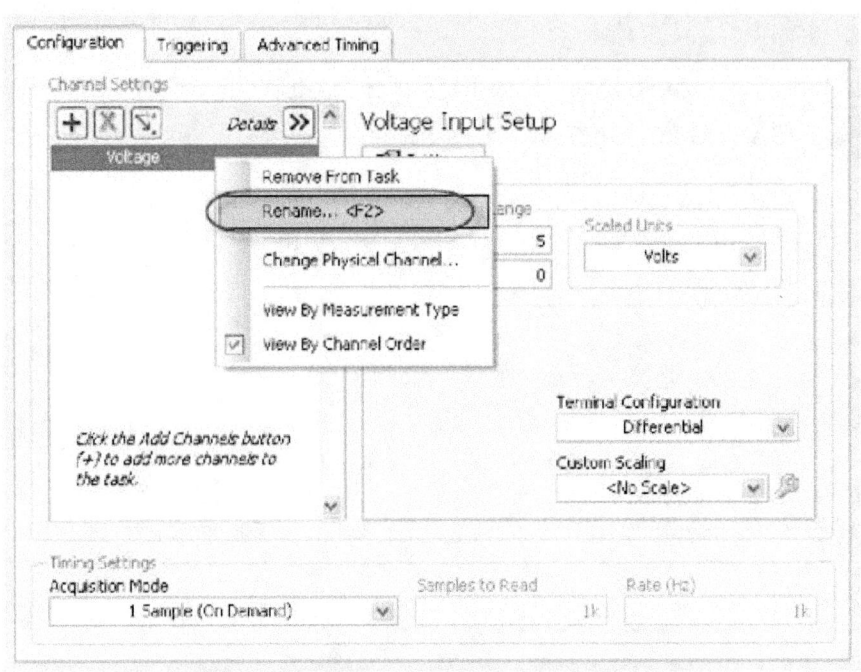

Now everything has finished with the configuration. Click OK in the DAQ Assistant window The DAQ Assistant icon appears on the Block Diagram:

Example1:

Wire the data output to a numeric indicator like shown in the figure below (and hit the Run button), you have seen the numeric indicator will show, the following value:

Example2:

If you want a "continuous" acquisition, put a While loop around the DAQ Assistant like the following figure:

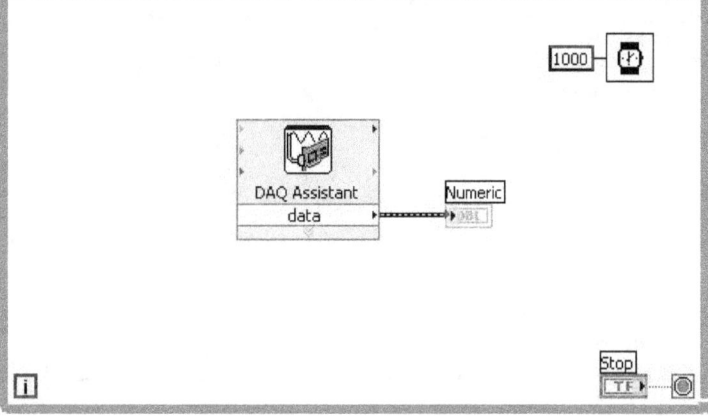

We can also communicate with the DAQ device without using the DAQ Assistant:

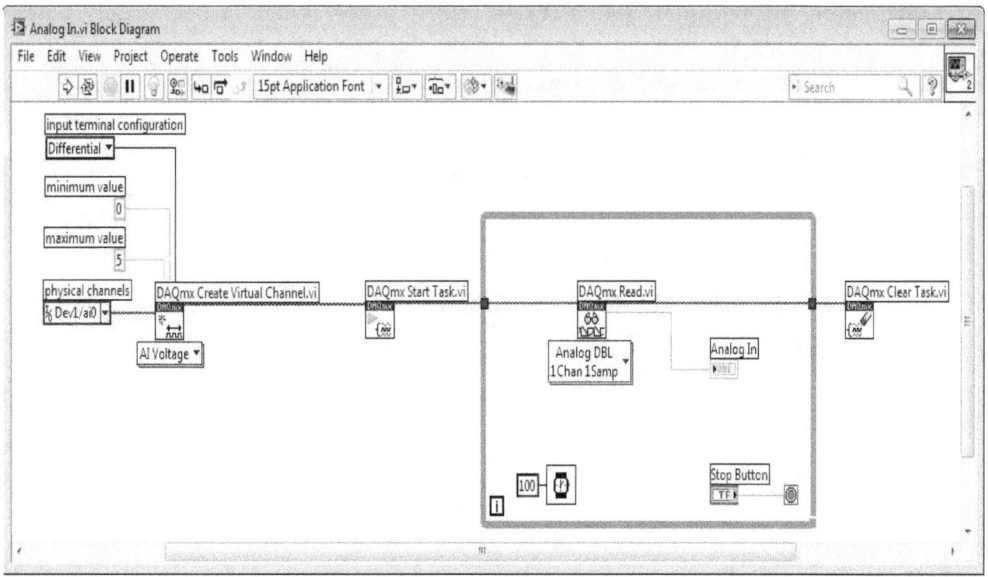

For more advanced applications this approach is recommended.

B) Analog output:

Analog Output is similar like the following Figures:

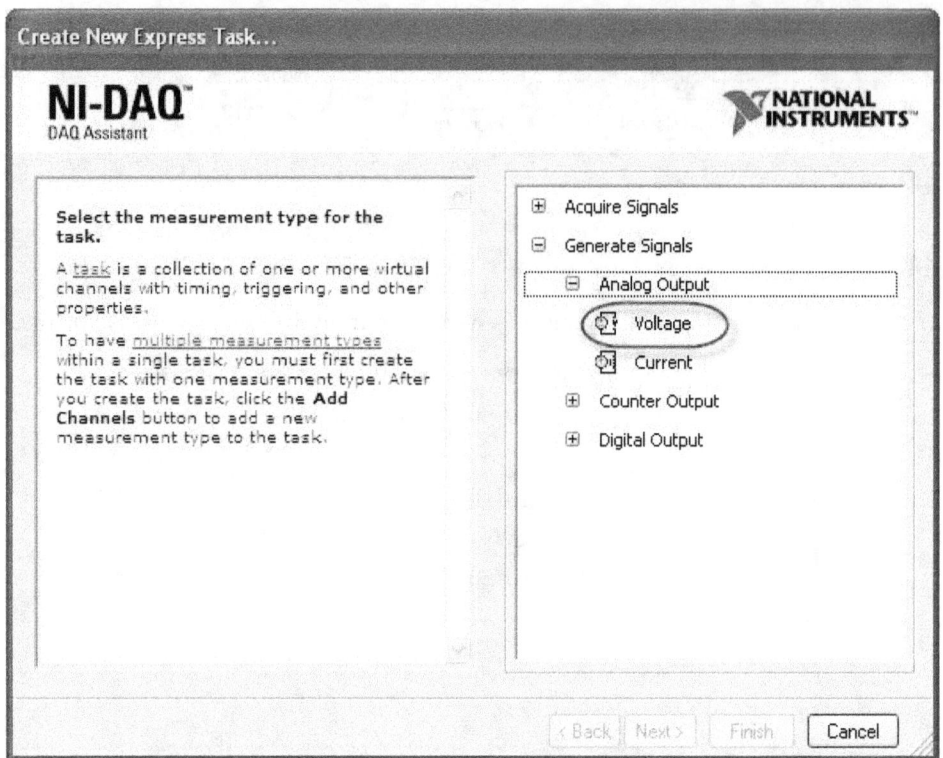

Example 3:

Inside a loop for "continuous" writing to the DAQ device:

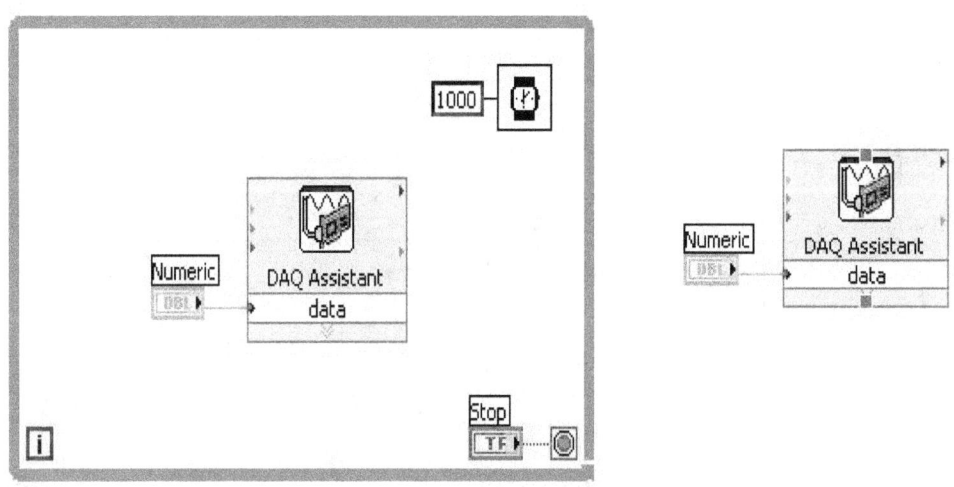

However you should not use the DAQ Assistant inside a loop because of the lack of performance. The following is therefore better:

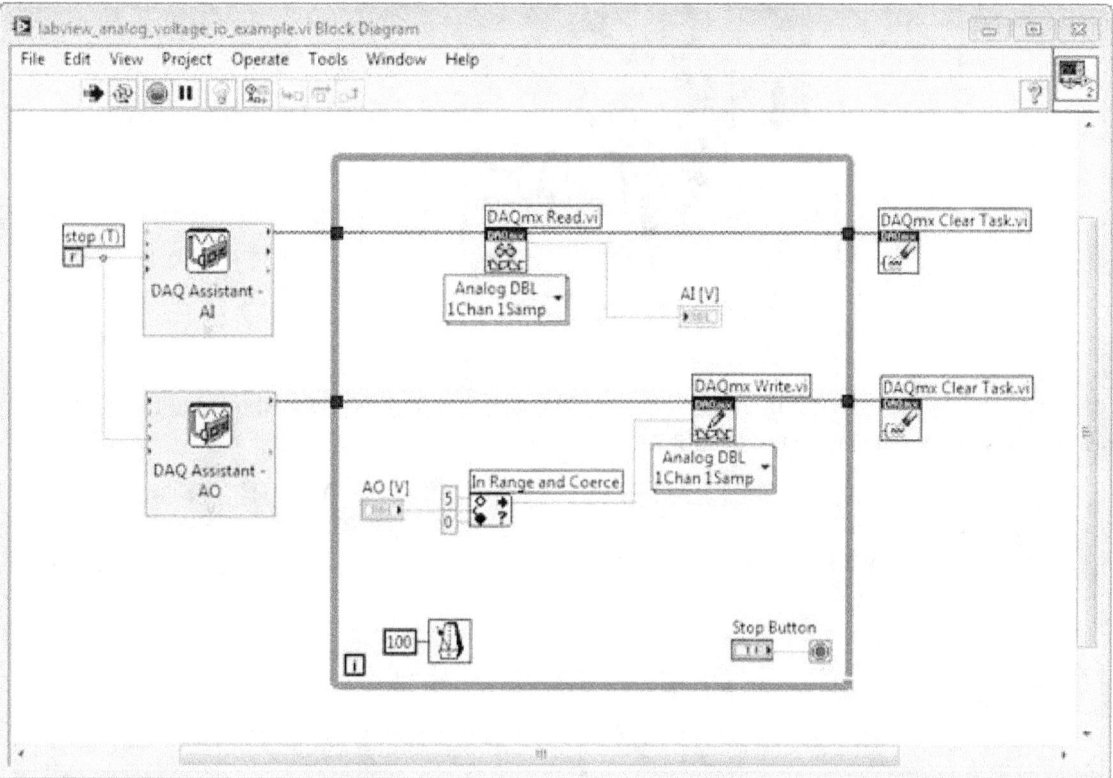

In this example we have used some of the other Vis in the DAQmx palette as well. The Front Panel may looks like the Figure below:

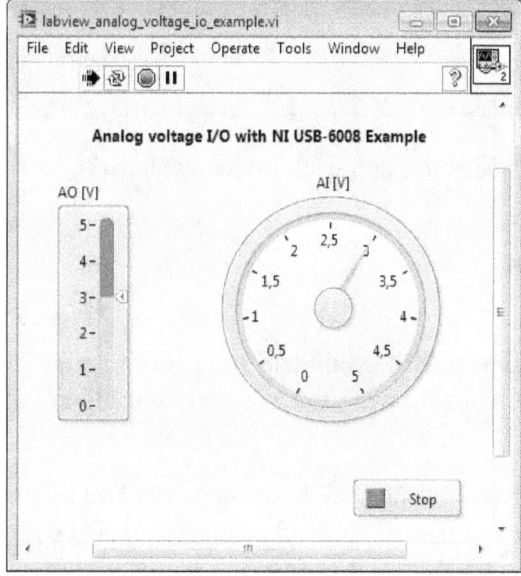

In this example we assume that we connect wires for Analog Out and Analog In together like this (a so called loopback test):

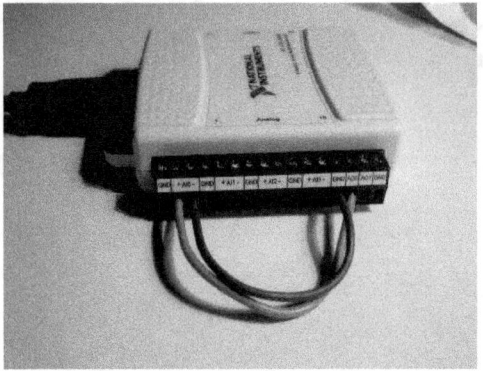

We can also communicate with the DAQ device without using the DAQ Assistant, see the next Figure that has shown the analog output block diagram:

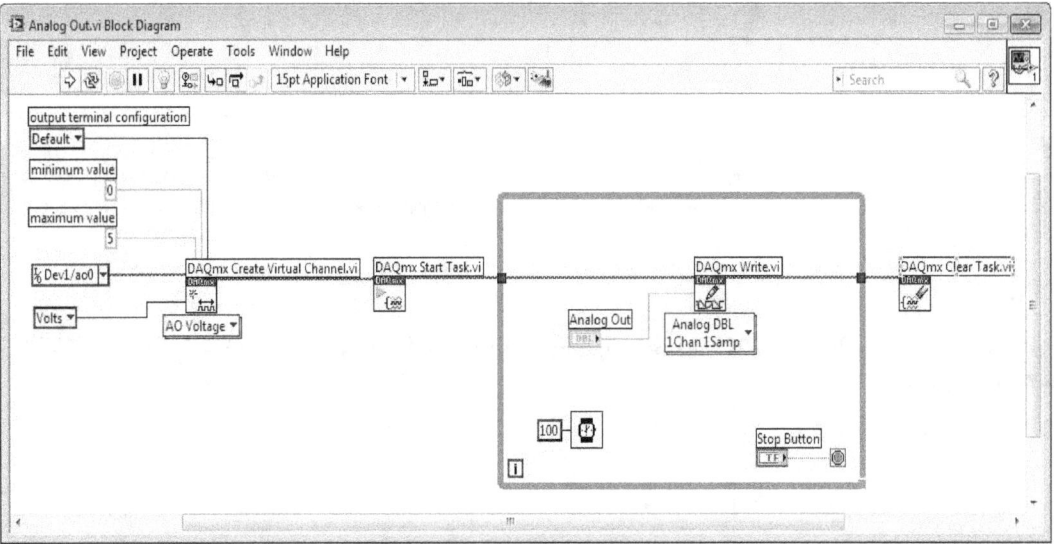

For more advanced applications this approach is recommended.

6.11 Logging to File

In many cases someone wants to write your data you get from the DAQ device to a text file for later use. In this chapter we will learn how to write to a measurement file in LabVIEW. We will also learn how to read the same file.

It can use the "Write to Measurement File" function on the File I/O palette in LabVIEW for writing data to text files You can save your data in a tab separated text (LVM data file format) or as a binary (TDMS file format) file. If we use the LVM, it is easy to open and view the data in Notepad. The following palette has shown the file I/O.

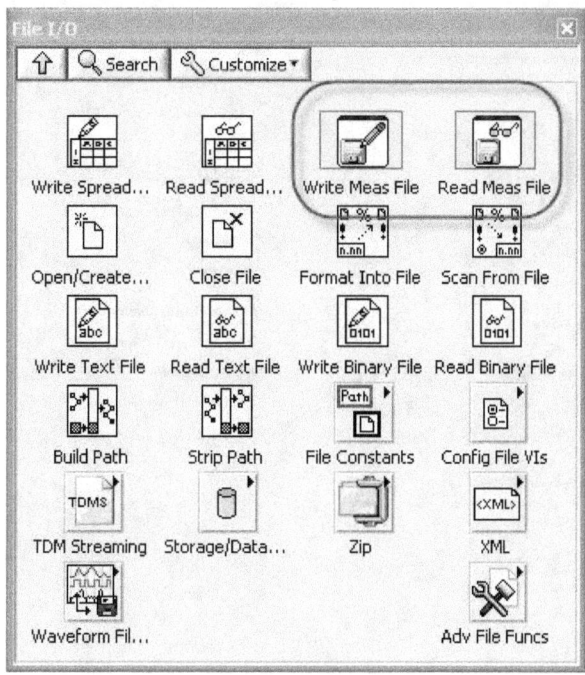

File I/O

6.11.1 Writing to Measurement File

We will use the "Write to Measurement File" function in the File I/O palette in LabVIEW for writing data to text files. We will also focus on the LVM data file format, not the TDMS file format which give binary files.

When you drag in the "Write to Measurement File", a configuration dialog window will automatically pop up. Recommended settings for the "Write To Measurement File" is as follows:

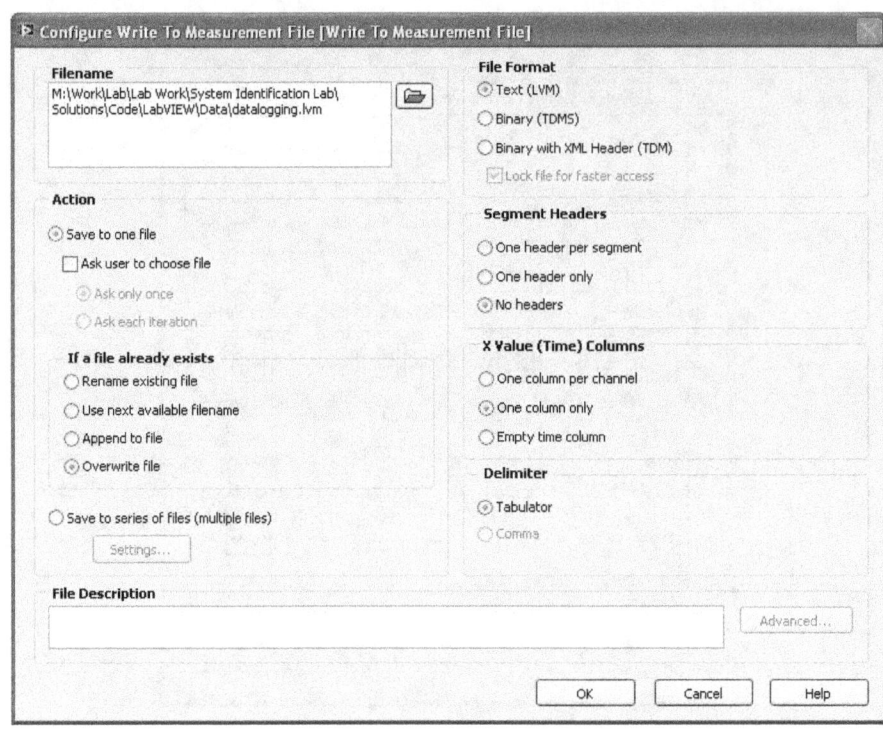

Configuration the measurement file

Example 4:

Example of LabVIEW Program that write data to a Measurement File.

See the Front Panel:

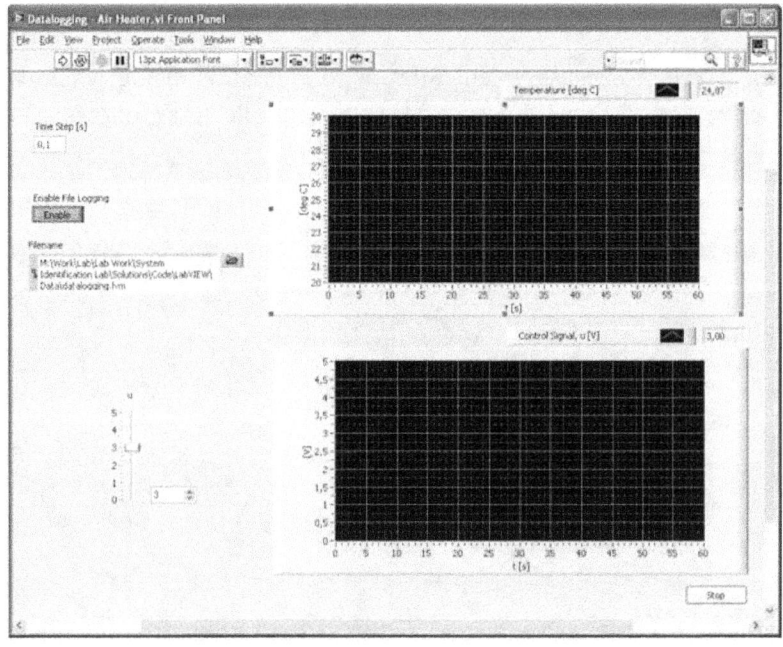

In this application we log data from a process based on a manual control signal. Both the input signal and the output signal (temperature) is saved to a Measurement File.

Block Diagram:

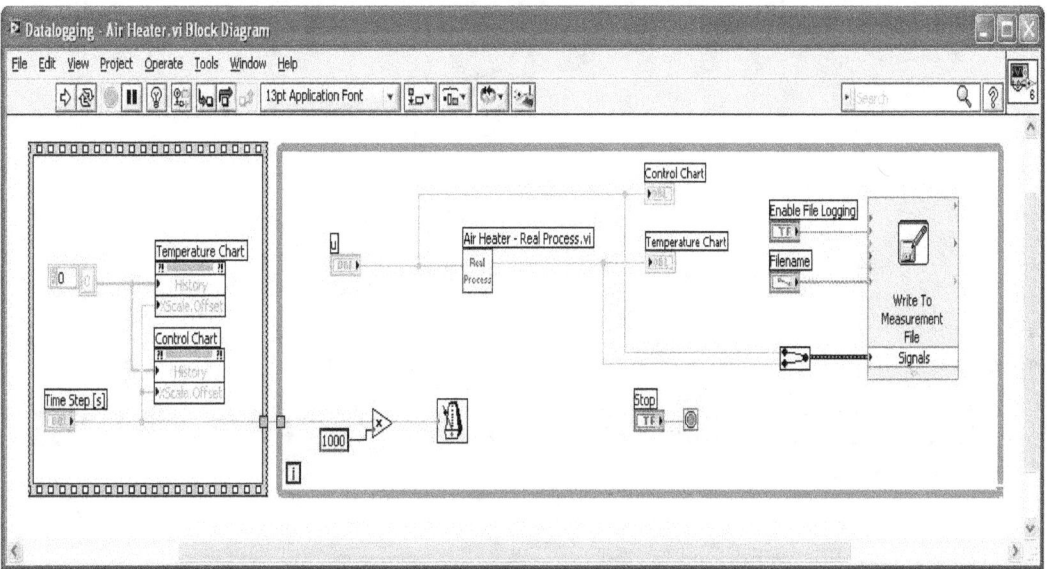

The LVM file may look something like the figure below:

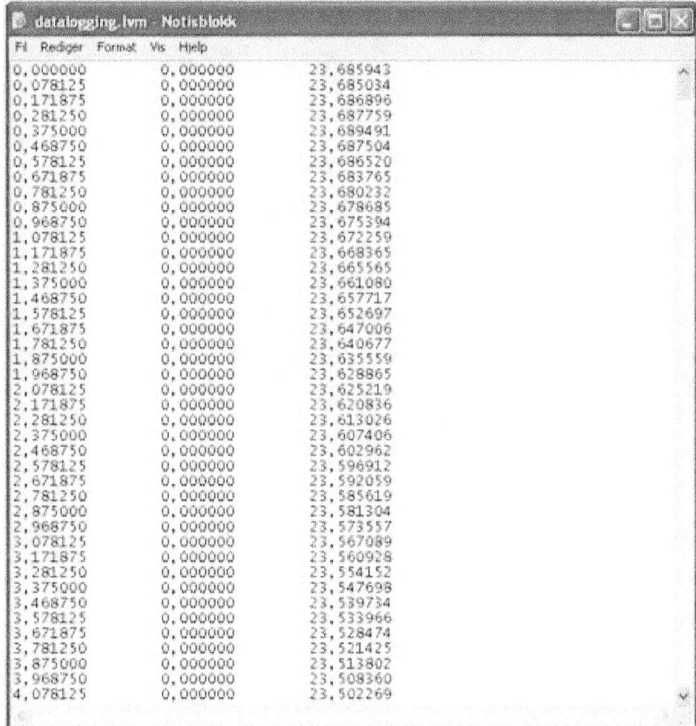

The first column is the time, the second column is the input signal, and the third column is the output signal.

6.11.2 Read from Measurement File

When you drag in the "Read from Measurement File", a configuration dialog window will automatically pop up.

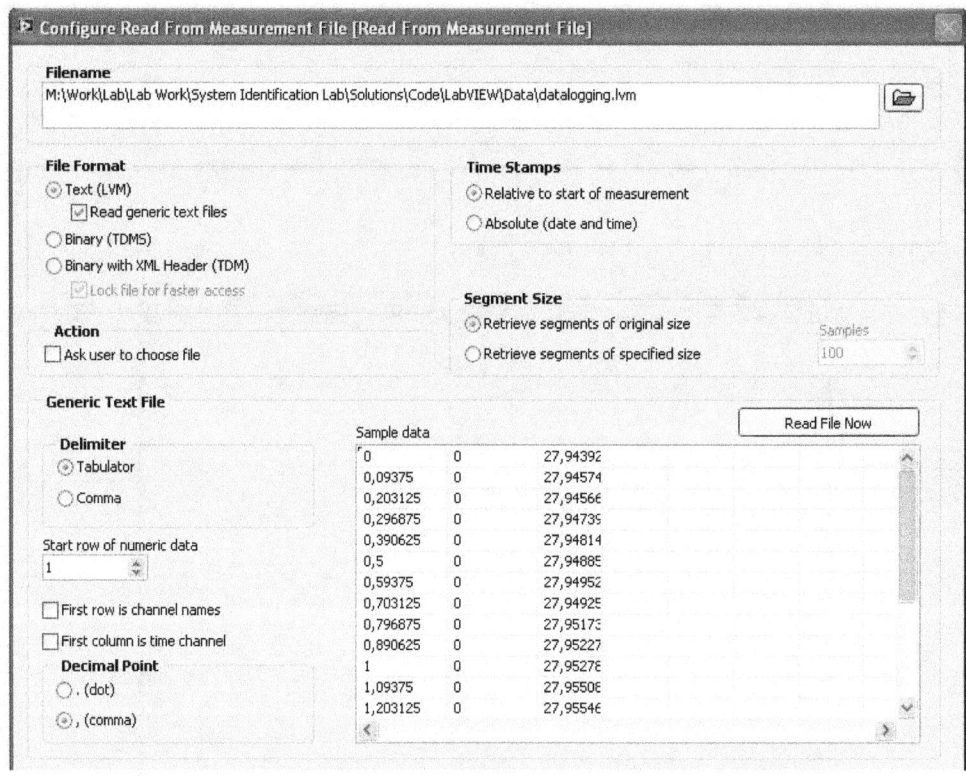

Recommended settings for the "**Read From Measurement File**" as seen in figure 6-6:

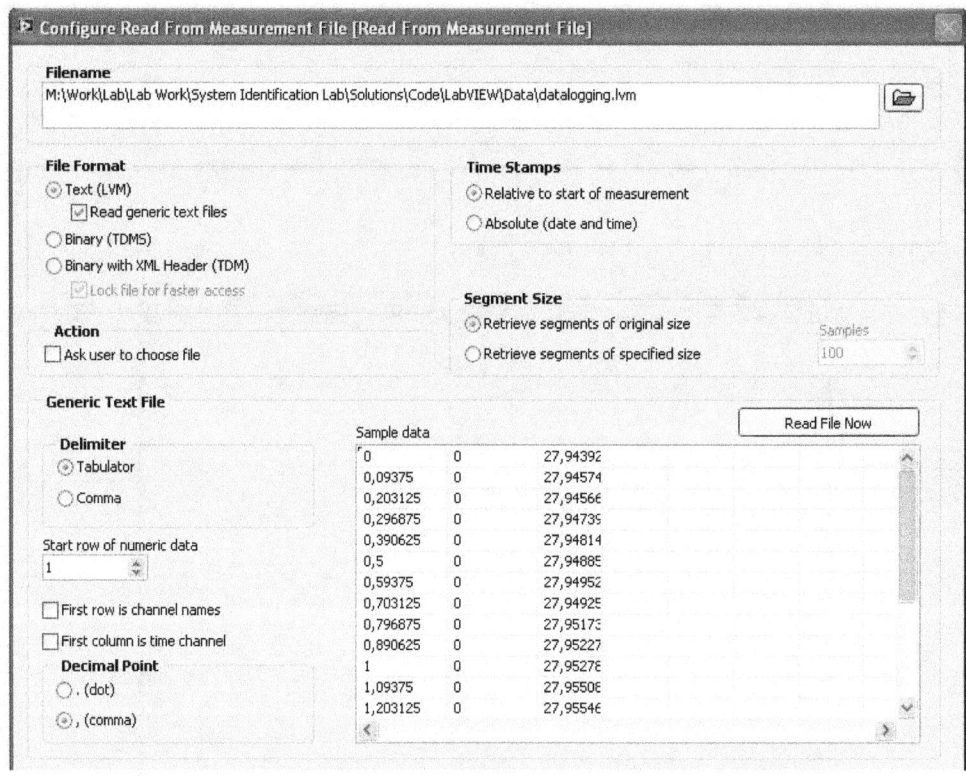

Figure 6-6

Example 5: Example of LabVIEW Program that read data from a Measurement File.

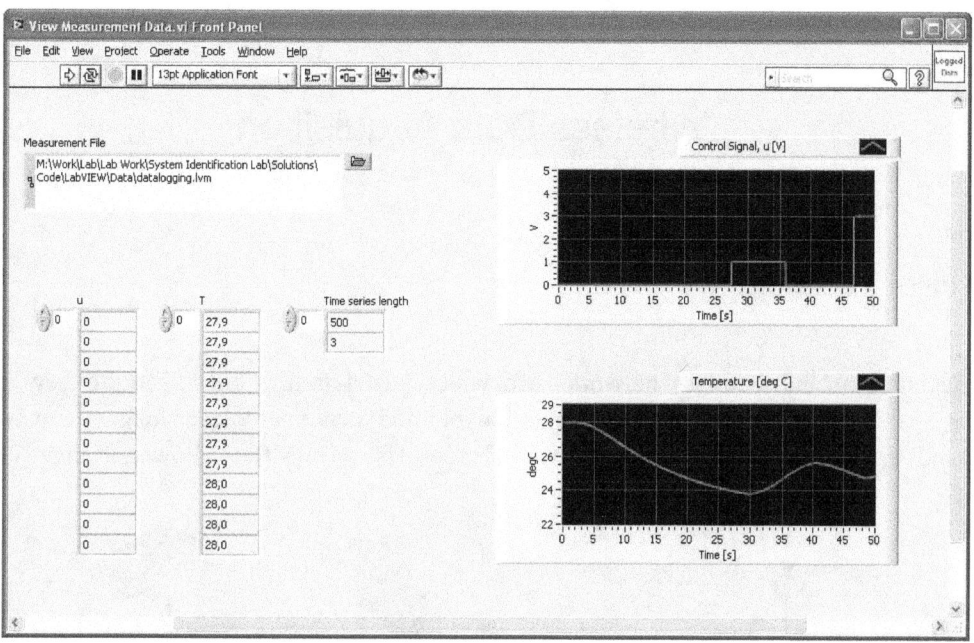

This application reads the data and plots it in 2 different graphs.

See its Block Diagram:

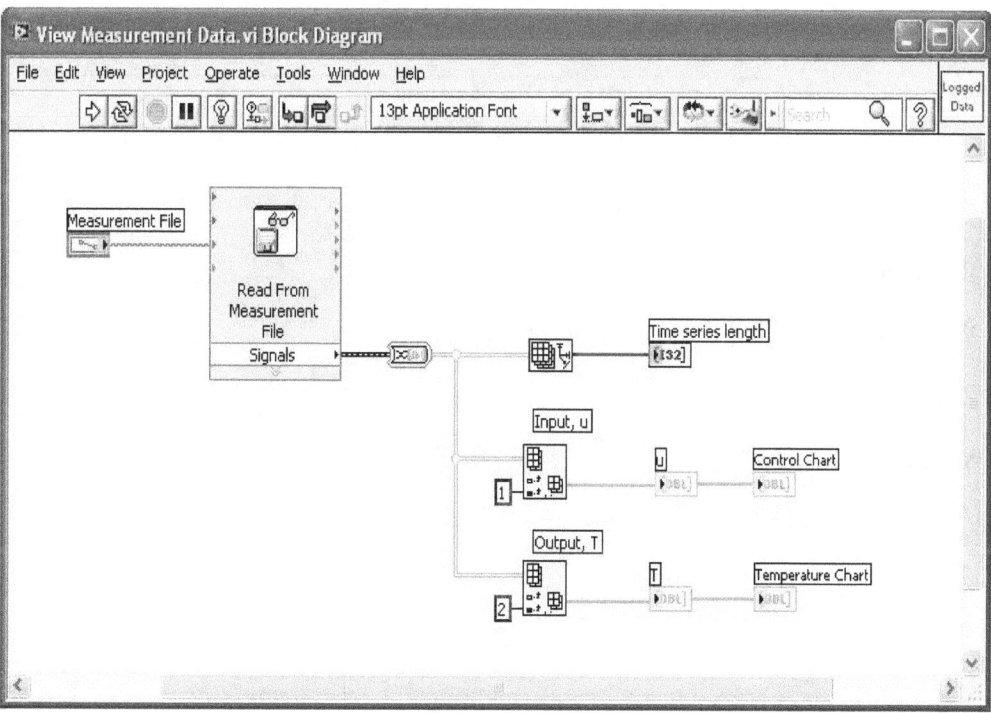

Chapter 7

Wireless Data Acquisition

7.1 Wireless Technology

Wireless technology and wireless networks are widely used today, but it's quite new in industrial automation systems. Wireless is a communication method that uses electromagnetic waves like (RF ,Infrared, and so on) instead of wire conductors. Trade-offs : (Data Rate, Power, Range , Security and Cost).

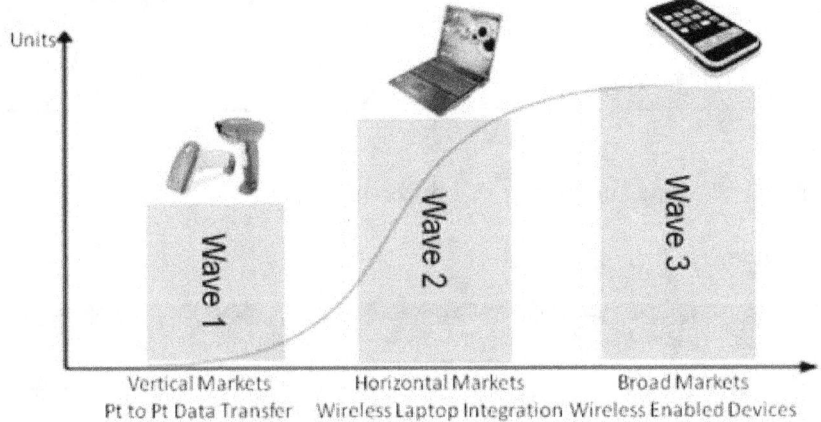

There are different technologies and wireless standards available:

- **Bluetooth**

- **Wireless USB**

- **ZigBee (IEEE 802.15.4)**

- **Wi-Fi (IEEE 802.11)**

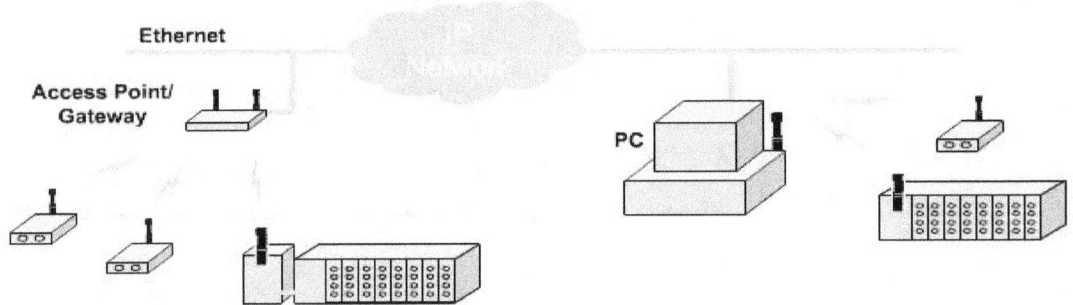

Different wireless technologies are shown in figure 7-1:

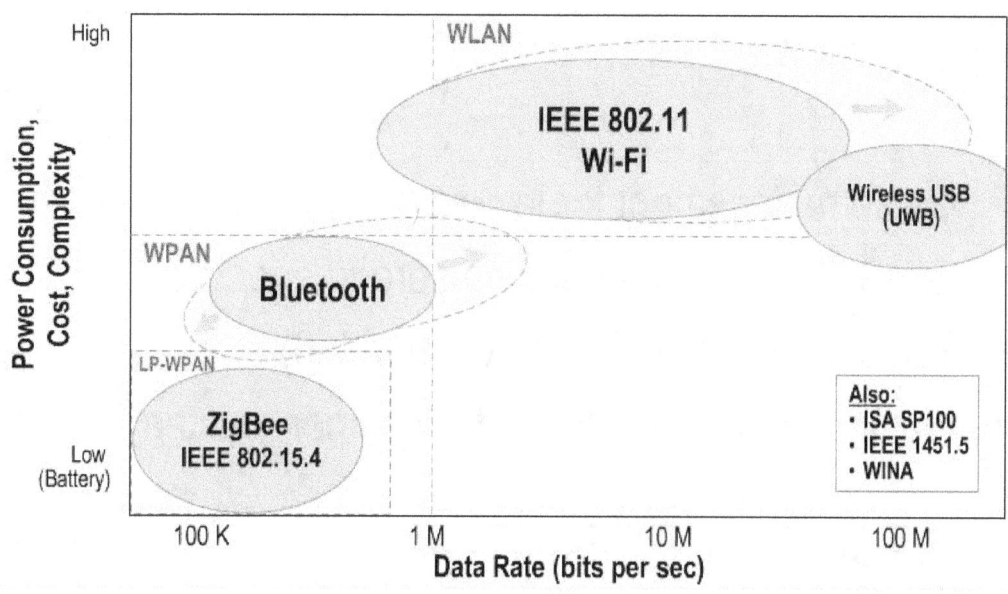

Figure 7-1

Figure 7-2 has given explicit comparison of the different wireless technologies, Wi-Fi and ZigBee are the primary wireless technologies for measurement and control systems.

	Wi-Fi 802.11 Family	Bluetooth 802.15.1	Wireless USB	ZigBee and 802.15.4
Applications	Enterprise, networking (Internet)	PC peripherals, cable replacement	PC peripherals, cable replacement, multimedia	Sensors, home/building automation, toys
Range	50 m	10 to 100 m	3 to 10 m	50 to 100 m
Data Rate	54 Mb/s (540 Mb/s)	750 kb/s	110 to 480 Mb/s	250 kb/s
Nodes per Network	>1,000	7	127	65,000
Battery Life	Hours	Days	Hours/days	Years
Setup/Usability	Better	Good	Better	Good → Best
Frequency	2.4 GHz, 5 GHz	2.4 GHz	3 to 10 GHz	900 MHz, 2.4 GHz
Security	Best	Good	Best/good	Better

Figure 7-2

A) Bluetooth

Bluetooth

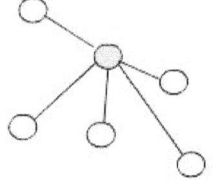

- Cable replacement for low-speed, low-cost peripherals
- 2.4 GHz frequency hopping spread spectrum (FHSS)
- 723 kb/s (up to 3 Mb/s in Version 2.0)
- Maximum of seven slave devices per master

Class	Max Power	Max Power	Range
1	100 mW	20 dBm	100 m
2	2.5 mW	4 dBm	10 m
3	1 mW	~0 dBm	1 m

Star Topology

B) Wireless USB

Wireless USB

- USB cable replacement over ultra-wideband (UWB)
- Throughput ~480 Mb/s at 3 m to 110 Mb/s at 10 m
- Competing standards have settled to one

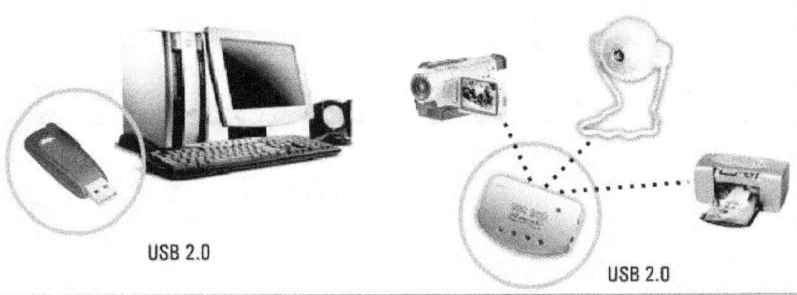

USB 2.0

USB 2.0

C) ZigBee (IEEE 802.15.4)

ZigBee (IEEE 802.15.4)

- Low-power wireless networking standard
- ZigBee builds on IEEE 802.15.4
- Targeting multiyear battery life
- Star, cluster tree, and mesh topologies

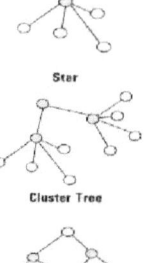

Band	Region	Data Rate	Range
868 MHz	Europe	20 kb/s	> 100 m
915 MHz	Americas	40 kb/s	> 100 m
2.4 GHz	Worldwide	250 kb/s	> 50 m

IEEE 802.15.4/ZigBee

- Popular for WSN devices
- IEEE 802.15.4 defines:
  - 868, 915 MHz, and 2.4 GHz radios
  - Up to 250 kb/s
  - Low-power communication
- ZigBee adds:
  - Device coordination
  - Network topologies
  - Interoperability with other wireless products

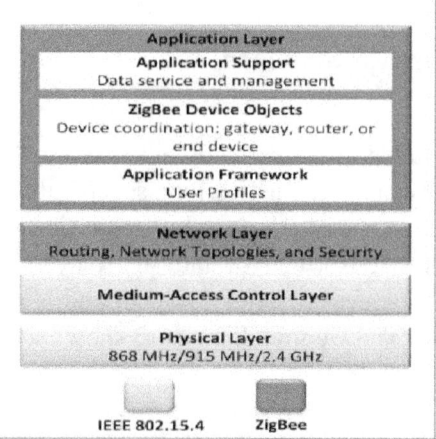

D) Wi-Fi (IEEE 802.11)

Wi-Fi (IEEE 802.11)

- Ubiquitous wireless networking standard
- Easy integration with existing IP network
- Evolving data rate, security, and other features

Version	Released	Frequency	Max Rate	Range
802.11	1997	2.4 GHz	2 Mb/s	~30 m
802.11b	1999	2.4 GHz	11 Mb/s	30 m
802.11a	1999	5 GHz	54 Mb/s	10 m
802.11g	2003	2.4 GHz	54 Mb/s	30 m
802.11n	2009?	2.4 GHz	~540 Mb/s	~50 m

7.2 Wireless Sensor Network (WSN)

A wireless sensor network (**WSN**) consists of spatially distributed autonomous sensors to cooperatively monitor physical or environmental conditions, such as temperature, sound, vibration, pressure, motion or pollutants.

The development of wireless sensor networks was motivated by military applications such as battlefield surveillance. They are now used in many industrial and civilian application areas, including industrial process monitoring and control, machine health monitoring, environment and habitat monitoring, healthcare applications, home automation, and traffic control. In addition to one or more sensors, each node in a sensor network is typically equipped with a radio transceiver or other wireless communications device, a small microcontroller, and an energy source, usually a battery.

The applications for WSNs are varied, typically involving some kind of monitoring, tracking, or controlling. Specific applications include habitat monitoring, object tracking, nuclear reactor control, fire detection, and traffic monitoring. In a typical application, a WSN is scattered in a region where it is meant to collect data through its sensor nodes.

7.2.1 Wireless Standard

Several standards are currently either ratified or under development for wireless sensor networks. **ZigBee** is a proprietary mesh-networking specification intended for uses such as embedded sensing, medical data collection, consumer devices like television remote controls, and home automation. **ZigBee** is promoted by a large consortium of industry players. Also, **Wireless HART** is an extension of the wireless technologies; hence HART Protocol and is specifically designed for Industrial applications like Process Monitoring and Control. Figure 7-3 shows such standards in wireless technology;

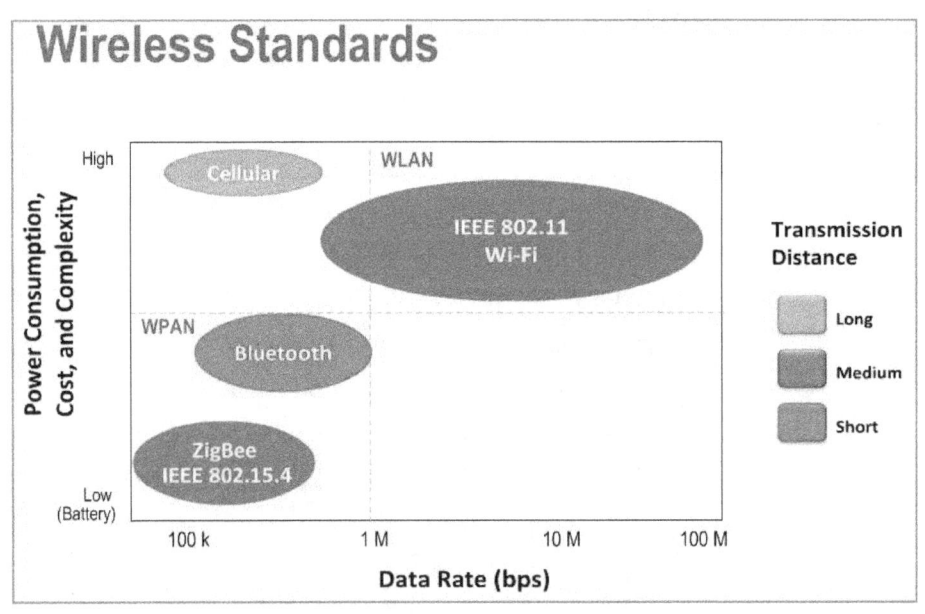

Figure 7-3

7.2.2 ZigBee (IEEE 802.15.4)

ZigBee is a low-cost, low-power, wireless mesh networking proprietary standard. The low cost allows the technology to be widely deployed in wireless control and monitoring applications, the low power-usage allows longer life with smaller batteries, and the mesh networking provides high reliability and larger range. Figure 7-4 shows the ZigBee (IEEE 802.15.4)

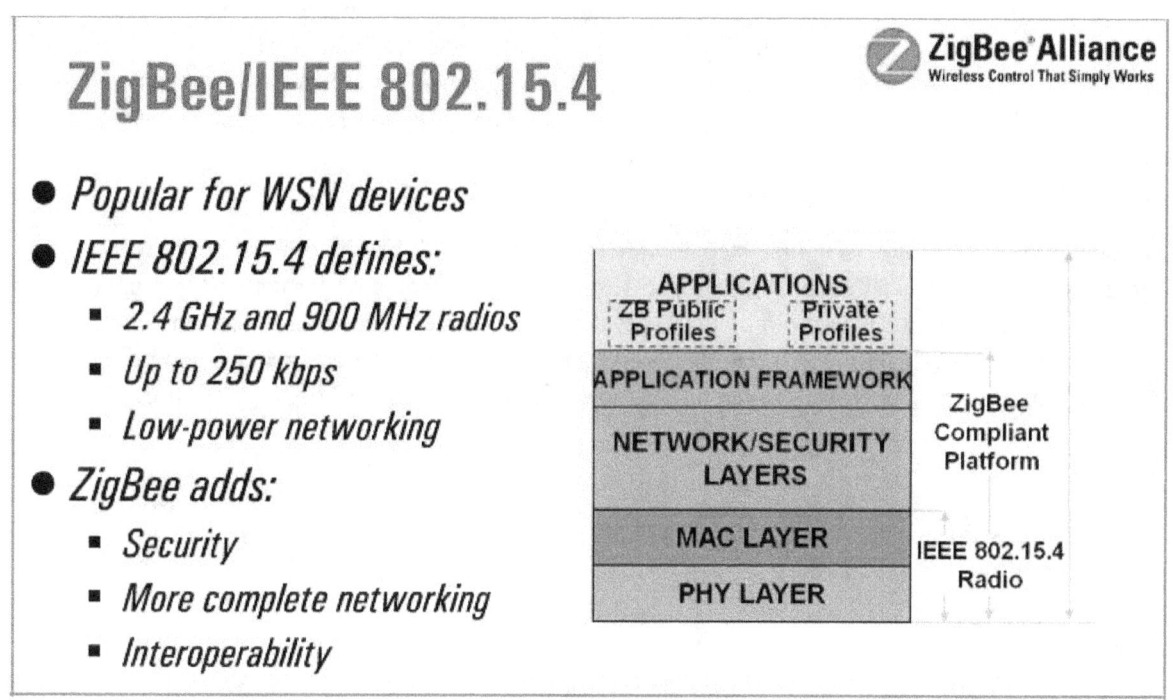

Figure 7-4

7.2.3 IEEE 802.11 & Wireless HART in LabVIEW

Wireless DAQ Products form National Instruments use the Wi-Fi IEEE 802.11 standard. NI WSN Measurement Systems use the ZigBee (IEEE 802.15.4) standard. **Read more about wireless technology and products from National Instruments.**

Wireless HART is an open-standard wireless networking technology developed by HART Communication Foundation. Developed as a multi-vendor, interoperable wireless standard, Wireless HART was defined specifically for the requirements of Process field device networks.

7.3 Wireless Mesh Network (WMN)

Mesh networking is a type of networking where each node in the network may act as an independent router, regardless of whether it is connected to another network or not. It allows for continuous connections and reconfiguration around broken or blocked paths by "hopping" from node to node until the destination is reached. A mesh network whose nodes are all connected to each other is a fully connected network. Mesh networks differ from other networks in that the component parts can all connect to each other via multiple hops.

A **wireless mesh network (WMN)** is a communications network made up of radio nodes organized in a mesh topology. Wireless mesh networks often consist of mesh clients, mesh routers and gateways. The mesh clients are often laptops, cell phones and other wireless devices while the mesh routers forward traffic to and from the gateways which may but need not connect to the Internet. The coverage area of the radio nodes working as a single network is sometimes called a mesh cloud. Access to this mesh cloud is dependent on the radio nodes working in harmony with each other to create a radio network. A mesh network is reliable and offers redundancy. When one node can no longer operate, the rest of the nodes can still communicate with each other, directly or through one or more intermediate nodes.

Wireless mesh architecture is a first step towards providing high-bandwidth network over a specific coverage area. Wireless mesh architecture's infrastructure is, in effect, a router network minus the cabling between nodes. It's built of peer radio devices that don't have to be cabled to a wired port like traditional WLAN access points (AP) do. Mesh architecture sustains signal strength by breaking long distances into a series of shorter hops. Below we've seen an example of a Wireless Mesh Topology (figure 7-5).

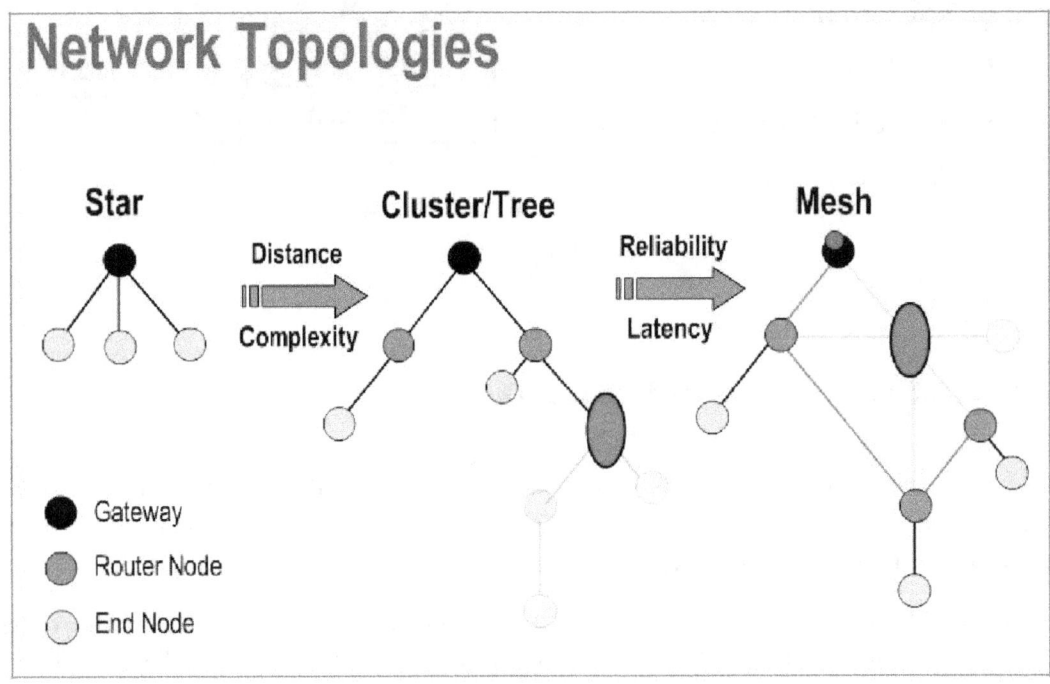

Figure 7-5

184

Wireless Mess Networks (WMN) is widely used in Wireless Sensor Networks (WSN). Below the following figure (7-6) has shown this feature.

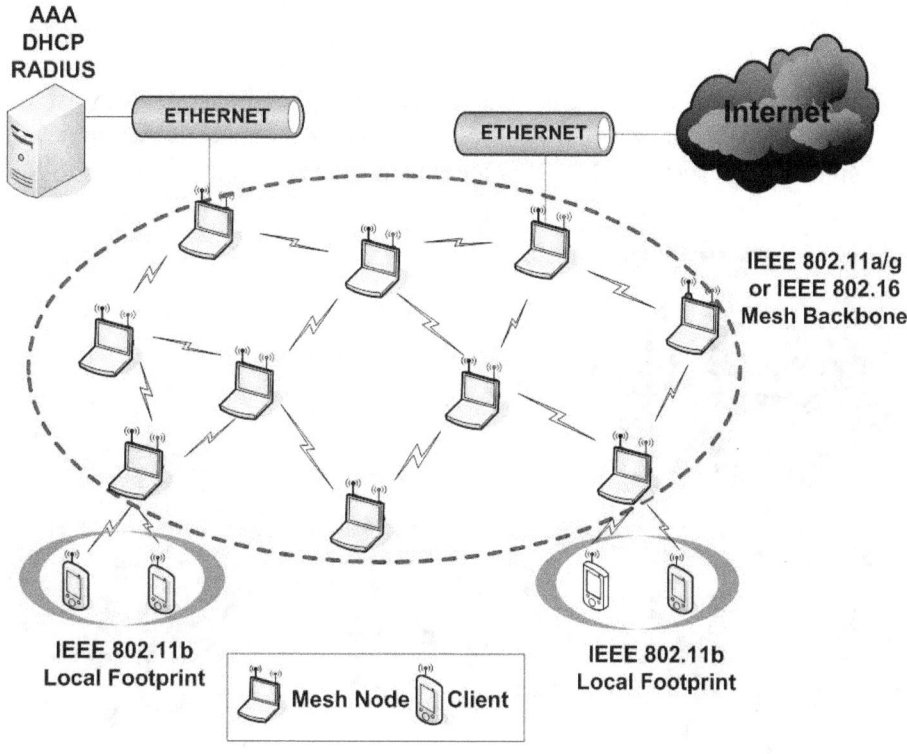

Figure 7-6

7.4 Wireless Data Acquisition in LabVIEW
Wi-Fi DAQ is:

- ➤ Simple – Direct sensors connectivity and graphical programming

- ➤ Secure – Highest commercially available data encryption and authentication

- ➤ Wireless – 802.11g

Wi-Fi data acquisition is an extension of PC-based data acquisition to measurement applications where cables are inconvenient or uneconomical. NI Wi-Fi data acquisition (DAQ) devices combine IEEE 802.11g wireless or Ethernet communication; direct sensor connectivity; and the flexibility of NI-DAQmx driver software for remote monitoring of electrical, physical, mechanical, and acoustical signals.

NI Wi-Fi DAQ devices can stream data on each channel at up to 250 kS/s. In addition, built-in NIST-approved 128-bit AES encryption and advanced network authentication methods offer the highest commercially available network security.

With the flexibility of NI **LabVIEW** graphical programming and the ubiquity of 802.11 network infrastructure, NI Wi-Fi DAQ makes it easy to incorporate wireless connectivity into new or existing PC-based measurement or control systems. The Data Flow in Wi-Fi DAQ is figure out as follows:

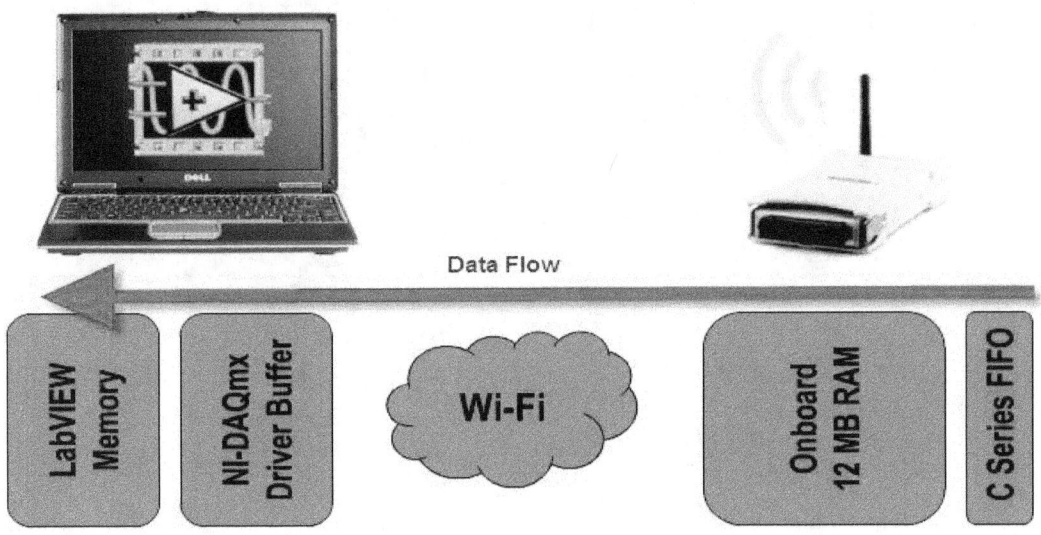

Unlike most wireless sensors or wireless sensor networks, wireless data acquisition devices are meant to stream data continuously back to a host PC or laptop. A wireless sensor node is typically a low-power, autonomous battery-operated device intended for long-term deployment in applications where measurements are needed only every few minutes, hours, or even days.

Wi-Fi data acquisition devices, on the other hand, behave in much the same way as a USB data acquisition device – a host PC collects data continuously (in real time) as the device acquires it. The data acquisition device may be battery-operated, but the focus is on the measurement versus the battery life. Also, Wi-Fi data acquisition devices use the near-ubiquitous wireless networking standard, IEEE 802.11 because of its higher bandwidth and broader applicability. Finally, because NI Wi-Fi data acquisition uses the same NI-DAQmx driver software as other NI data acquisition devices, you can develop your applications using NI LabVIEW, Lab Windows™/CVI; ANSI C/C++; or Microsoft C#, Visual Basic, or Visual Basic .NET.

7.5 NI Wireless DAQ Devices

Common components in Wi-Fi DAQ are as follows:

- DAQ Devices
- Wireless Access Points (WAP) or a Wireless Router
- Network Switches

7.5.1 NI Wi-Fi DAQ device connection

When setting up your NI Wi-Fi DAQ device, you may choose to connect to an existing enterprise network through a wireless access point or set up your own network with a wireless router. For existing IT infrastructure, NI Wi-Fi DAQ devices support WPA Enterprise and WPA2 Enterprise (IEEE 802.11i). If you set up your own network, you may use WEP, WPA Personal (WPA-PSK), or WPA2 Personal (WPA2-PSK) security. NI Wi-Fi DAQ devices also support ad hoc or peer-to-peer networks, which do not require any routers or access points.

However, ad hoc support is inconsistent across wireless network interface cards and is not secure. Communication through a wireless access point or wireless router is the preferred mode of operation.

7.5.2 NI Wi-Fi DAQ devices programming

All NI Wi-Fi and Ethernet DAQ devices use the same NI-DAQmx measurement services and driver software as other National Instruments PCI, PXI, and USB DAQ devices. NI-DAQmx measurement services software controls every aspect of your data acquisition system from configuration to programming. With NI-DAQmx software, you can quickly configure and acquire measurements using the DAQ Assistant and automatically generate code to get your application started quickly.

Also, NI-DAQmx have some features of virtual channels that automatically scale raw data into engineering units. It can be used the same driver API to program new NI Wi-Fi DAQ devices, incorporating wireless connectivity into your existing applications. Wi-Fi DAQ Devices from National Instruments are tabulated in the following table:

Module	Signal	Channels	Rate	Connectivity
Analog Input				
NI WLS-9205	±10V Programmable gain, 16-bit	32	250 kS/s	Spring terminal
NI WLS-9206	600 V Isolation, 16-bit	16	250 kS/s	Spring terminal
NI WLS-9211	Thermocouple, 24-bit	4	14 S/s	Screw terminal
NI WLS-9213	Thermocouple, 24-bit	16	75 S/s/ch	Spring terminal
NI WLS-9215	Simultaneous sampling, 16-bit	4	100 kS/s/ch	Screw terminal or BNC
NI WLS-9219	Universal (11 modes)	4	100 S/s/ch	Spring terminal
NI WLS-9234	IEPE (accelerometer and microphone), 24-bit	4	51.2 kS/s/ch	BNC
NI WLS-9237	Bridge completion, 24-bit	4	50 kS/s/ch	RJ50
Digital I/O				
NI WLS-9421	11 to 30 VDC sinking digital input	8	Software-timed	Screw terminal or D-Sub
NI WLS-9472	6 to 30 VDC sourcing digital output	8	Software-timed	Screw terminal or D-Sub
NI WLS-9481	60 VDC, 250 Vrms relay output	4	Software-timed	Screw terminal

The following wireless DAQ devices are available in many universities and research centers around the world:

- **NI WLS-9234** - Wireless 4-Channel **Accelerometer and Microphone Input**

- **NI WAP-3701 - Wireless access points** (WAPs) add wireless connectivity to all Ethernet-based NI programmable automation

- **NI UES-3880** - An eight-port unmanaged entry-level **switch** for networking between NI Ethernet-based controllers and devices using standard Ethernet protocols

- **PS-5** Power Supply

7.5.3 NI WLS-9234

The NI WLS-9234 is a four-channel IEEE 802.11 wireless or Ethernet C Series **dynamic signal acquisition module for making high-accuracy audio frequency measurements** from integrated electronic piezoelectric (IEPE) and non-IEPE sensors.

(4-Channel, 51.2 kS/s/ch, 24-Bit, ±5 V IEPE Input)

- IEEE 802.11b/g (Wi-Fi) wireless and Ethernet communications interfaces

- 51.2 kS/s per-channel maximum sampling rate; ±5 V input

- 24-bit resolution; 102 dB dynamic range; anti aliasing filters

- ➢ Software-selectable AC/DC coupling; AC-coupled (0.5 Hz)
- ➢ Software-selectable IEPE signal conditioning (0 or 2 mA)

The NI WLS-9234 is a four-channel IEEE 802.11 wireless or Ethernet C Series dynamic signal acquisition module for making high-accuracy **audio frequency measurements** from integrated electronic piezoelectric (IEPE) and non-IEPE sensors.

The WLS-9234 delivers 102 dB of dynamic range and incorporates software-selectable AC/DC coupling and IEPE signal conditioning for accelerometers and microphones. The four input channels simultaneously digitize signals at rates up to 51.2 kHz per channel with built-in anti aliasing filters that automatically adjust to your sampling rate.

The WLS-9234 is well-suited for noise and vibration analysis applications. The **NI Sound and Vibration Measurement Suite**, which specifically addresses these applications, has two components: the **NI Sound and Vibration Assistant** and **LabVIEW Analysis VIs** (functions) for power spectra, frequency response (FRF), fractional octave analysis, sound-level measurements, order spectra, order maps, order extraction, sensor calibration, human vibration filters, and torsional vibration.

WLS-9234 has 4 BNC connectors. The BNC connector is used for RF signal connections. The BNC connector is one of several radio frequency (RF) connectors on the market today.

7.5.4 NI WAP-3701

> ➤ **WAP** – Wireless Access Point

The NI WAP-3701 **wireless access points** (WAPs) add wireless connectivity to all Ethernet-based NI programmable automation controllers (PACs). A WAP-37x1 can connect wireless devices to a wired network and has an IP address so other wireless clients can access the device. A WAP-37x1 can also serve as a Dynamic Host Configuration Protocol (DHCP) server and configure other WAP-37x1 devices as wireless clients. A single WAP-37x1 typically supports up to 30 client devices, with the range depending on the wireless antennae selection. It also serves as a wireless bridge and allows two devices to talk to each other.

- EEE 802.11g/b wireless access point, client, and bridge

- Web-based management software included

- 64- and 128-bit wired equivalent privacy (WEP) security

- Redundant dual 24 VDC power inputs

- Metal enclosure, IP30 rated; Class I, Division 2 hazardous locations

- DIN-rail mounting support

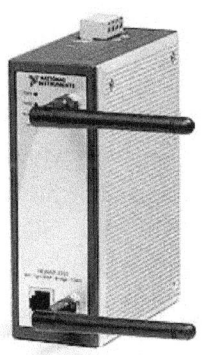

7.5.5 NI UES-3880

➢ **UES** – Unmanaged Ethernet Switch

The NI UES-3880 is an eight-port unmanaged entry-level **switch** for networking between NI Ethernet-based controllers and devices using standard Ethernet protocols. To network with NI programmable automation controllers (PACs), the UES-3880 provides industrial features including -40 to 70 °C operating temperature; Class I, Division 2 and ATEX Class 1, Zone 2 certifications for hazardous locations; and redundant dual 24 VDC power inputs.

7.5.6 PS-5 Power Supply

➢ **PS** – Power Supply

The PS-5 is a DIN-rail-mountable 24 VDC power supply that provides 5 A of current and is rated for operation from -25C to 60C. Two PS-5 supplies can be wired in parallel to provide up to 10 A. The PS-5 is recommended for industrial installations. The PS-4 is a DIN-rail-mountable 24 V power supply in Field Point packaging. The PS-3 and PS-2 are power supply "bricks" with a universal IEC power input. The PS-3 can also operate as a float charger for 12 VDC batteries. The PS-1 is a wall-mount power supply.

7.6 NI WSN Devices

A wireless sensor network (WSN) is a wireless network consisting of spatially distributed autonomous devices that use sensors to monitor physical or environmental conditions. These autonomous devices, or nodes, combine with routers and a gateway to create a typical WSN system. The distributed measurement nodes communicate wirelessly to a central gateway, which provides a connection to the wired world where you can collect, process, analyze, and present your measurement data. To extend distance and reliability in a WSN, you can use routers to gain an additional communication link between end nodes and the gateway.

The proprietary NI WSN protocol is based on IEEE 802.15.4 and ZigBee technology. The IEEE 802.15.4 communication standard defines the Physical and Medium Access Control layers in the networking model, providing communication in three frequency bands including the 2.4 GHz ISM band. ZigBee builds on the 802.15.4 standard with the network and application layers, offering features such as device coordination, reliability through mesh networking topologies, and the functionality to create user-defined profiles that allow for customization and flexibility within the protocol.

Question: What is the difference between NI Wi-Fi data acquisition (DAQ) and NI WSN devices?

NI Wi-Fi DAQ devices combine IEEE 802.11b/g (Wi-Fi) wireless or Ethernet communication, direct sensor connectivity, and the flexibility of NI-DAQmx software for a breadth of remote measurement and control options. Externally powered, NI Wi-Fi DAQ devices can stream continuous waveform data on each channel at more than 50 kS/s and offer the highest commercially available network security (WPA2 Enterprise).

NI WSN devices deliver low-power measurement nodes that operate for up to three years on 4 AA batteries and can be deployed for long-term, remote operation. The wireless measurement nodes communicate with a central gateway using a protocol based on IEEE 802.15.4 (ZigBee) to offer mesh routing capabilities that extend network distance and reliability. NI WSN systems support lower data rates to preserve power, are easily programmed using I/O variables, and currently accommodate thermocouple and ±10 V measurements, see figure 7-7.

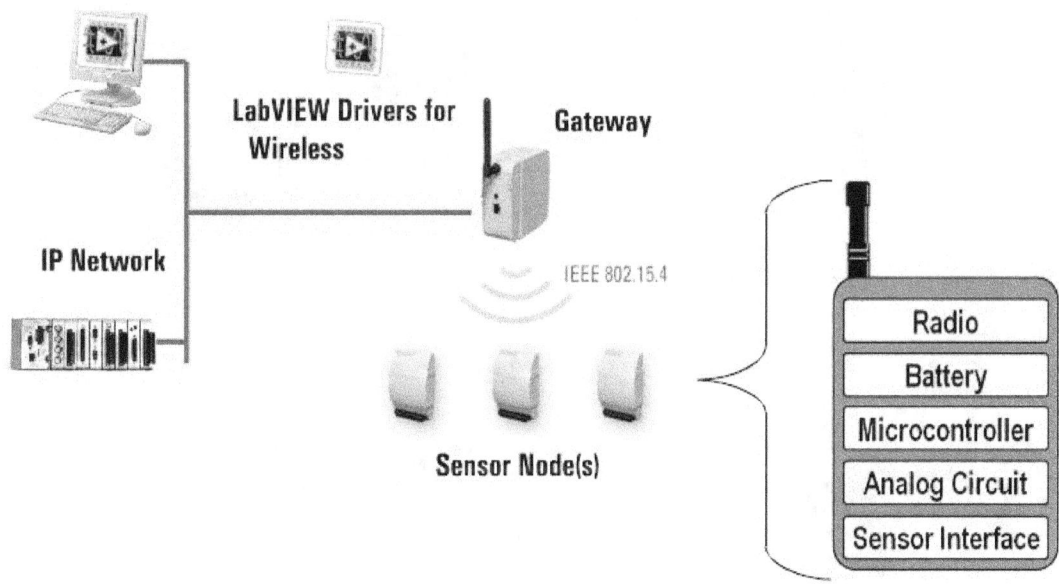

Figure 7-7

191

Many of the WSN systems today are based on ZigBee or IEEE 802.15.4 protocols due to their low-power consumption. ZigBee builds on the 802.15.4 layers to provide security, reliability through mesh networking topologies, and interoperability with other devices and standards as seen in figure 7-8.

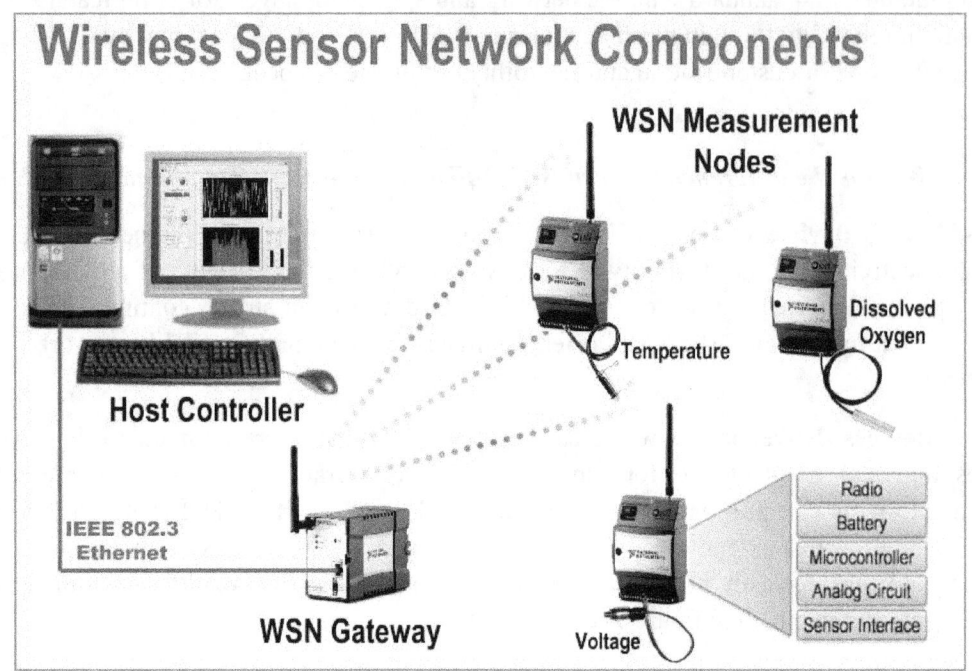

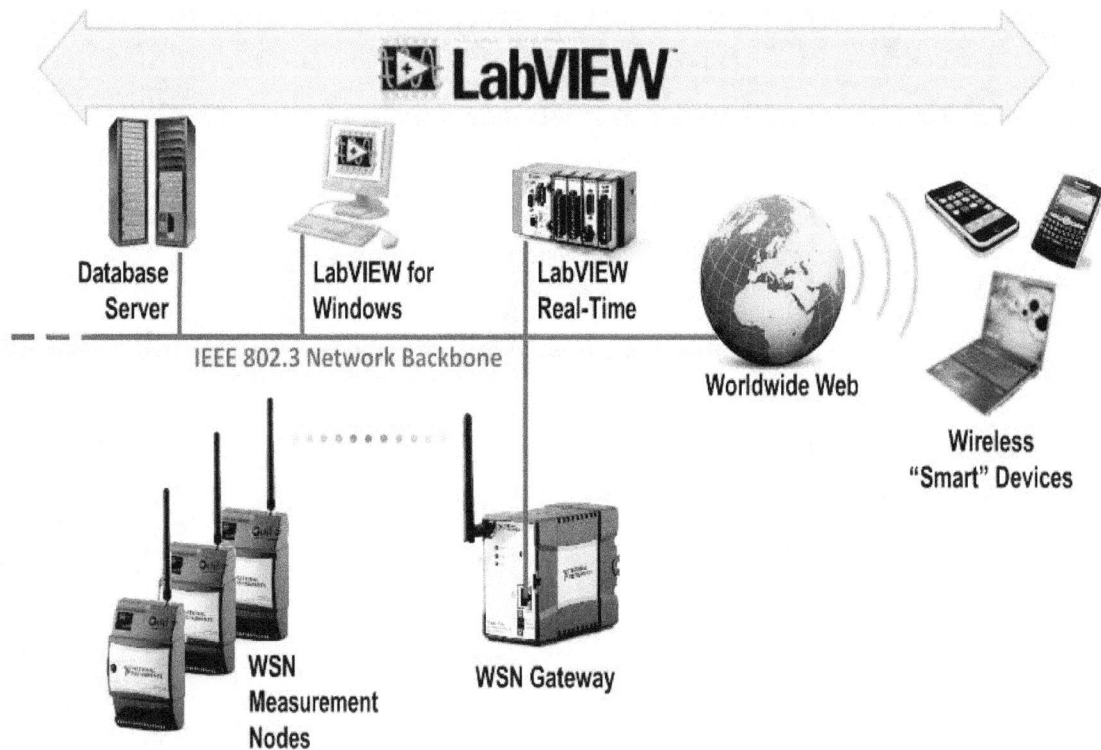

Figure 7-8

192

7.6.1 Devices

1. NI WSN-9791 - WSN Ethernet Gateway

The NI WSN-9791 Ethernet gateway coordinates communication between distributed measurement nodes and the host controller in your NI wireless sensor network (WSN). The gateway has a 2.4 GHz, IEEE 802.15.4 radio based on ZigBee technology to collect measurement data from the sensor network and a 10/100 Mbits/s Ethernet port to provide flexible connectivity to a Windows or LabVIEW Real-Time OS host controller.

2. NI WSN-3212 - Thermocouple Input Node

The NI WSN-3212 measurement node is a wireless device that provides four 24-bit thermocouple input channels and four bidirectional digital channels that you can configure on a per-channel basis for input, sinking output, or sourcing output.

3. NI WSN-3202 - Analog Input Node

The NI WSN-3202 measurement node is a wireless device that provides four ±10 V analog input channels and four bidirectional digital channels that you can configure on a per-channel basis for input, sinking output, or sourcing output.

7.6.2 LabVIEW Derivers

National Instruments offer the **NI-WSN software** has shown in figure 7-9.

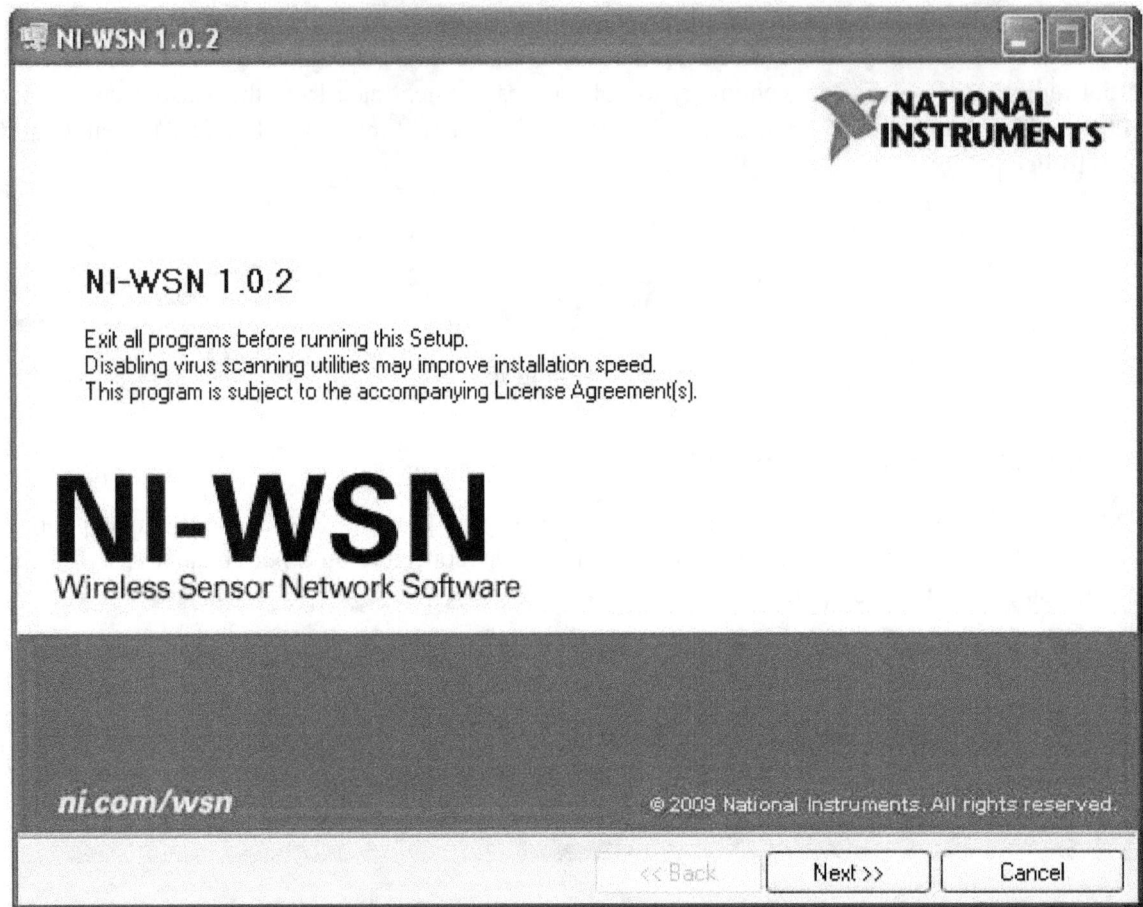

Figure 7-9

With NI-WSN software you can easily configure your network in the Measurement & Automation Explorer (MAX) utility. MAX provides an intuitive user interface to add and remove measurement nodes and configure wireless settings.

NI-WSN software provides seamless LabVIEW integration so that you can quickly and easily extract measurement data from your WSN. After adding the NI WSN Ethernet gateway to a LabVIEW Project, the nodes configured with the gateway in MAX automatically populate in the LabVIEW Project, giving you instant access to their I/O and properties.

Simply drag and drop I/O variables from the LabVIEW Project to a LabVIEW Block Diagram for data extraction, analysis, and presentation as shown in figure 7-10.

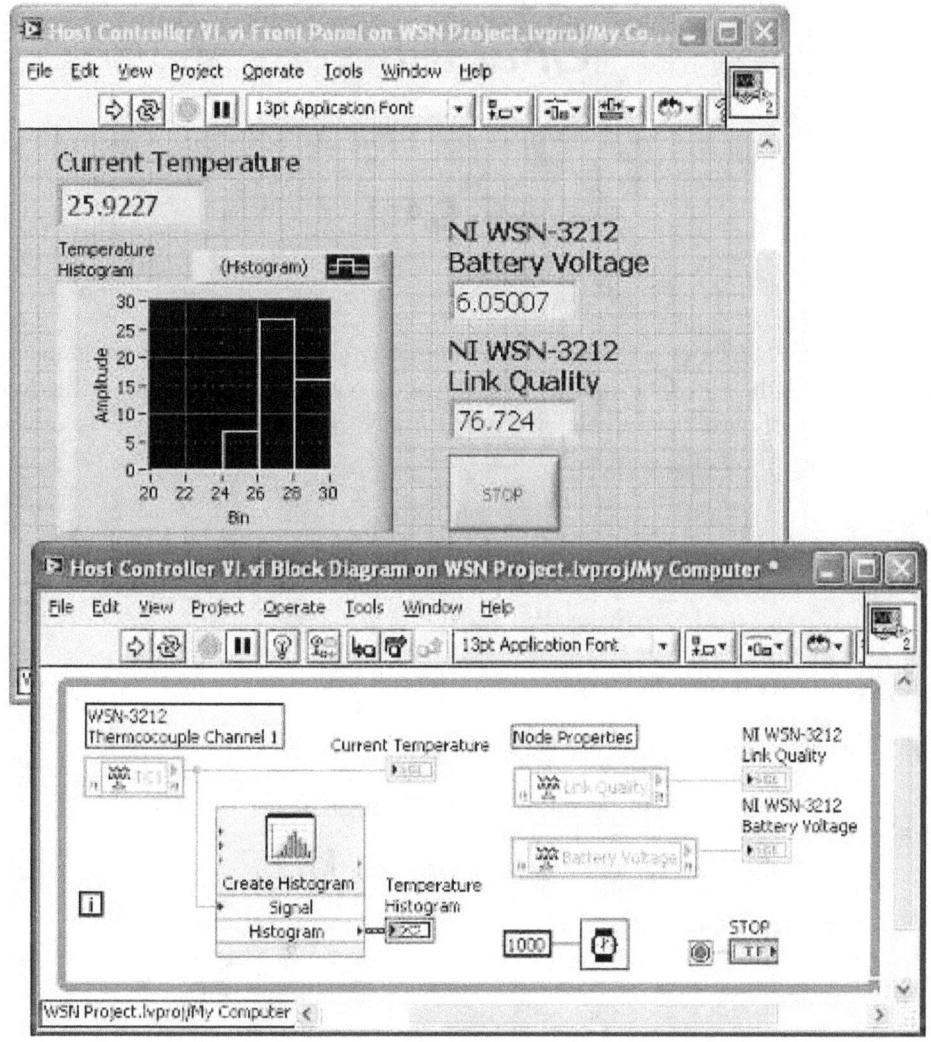

Figure 7-10

Also, National Instruments has offered WSN drivers for WSN equipments from other vendors.

Chapter 8

Wi-Fi DAQ

This chapter will go through how to transfer data acquired from a sensor to the computer wirelessly by using a Wi-Fi DAQ system. The data transmission relies on National Instruments technology, where NI WAP-3711 will broadcast wireless internet from any LAN. This documentation is created based on experiment by using the following equipment:

❑ **Hardware**

  o Temperature sensor PT-100

  o Transducer board

  o Wi-Fi equipment

  ▪ NI WAP-3711 (802.11g/b WAP/Bridge/Client)

  ▪ NI WLS-9163 (802.11g/b C Series Carrier)

  ▪ NI 9234 (DAQ)

❑ **Software**

  o Windows OS

  o LabVIEW

  o NI-DAQmx (MAX)

The data transferring procedure through Wi-Fi (Infrastructure) technology involves mainly five different steps;

Step 1: Physical devices, setup and connection. Devices: NI-WLS-9163, NI9234 (DAQ) and connection board shall be physically connected to each other.

Step 2: NI WAP-3711 configuration. Set up and configure NI WAP-3711 to manage the wireless data communication between the computer and NI WLS-9163 device.

Step 3: NI WLS-9163 configuration. Set up and configure NI WLS-9163 to transmit temperature values to NI WAP-3711.

Step 4: Configure your computer to get access to NI WAP-3711.

Step 5: Present the temperature value on the computer. Create a LabVIEW application to continuously retrieve and present the value on your computer. The acquired values are originally in Voltage format which needs to be converted into corresponding temperature.

8.1 Necessary Equipment

Software

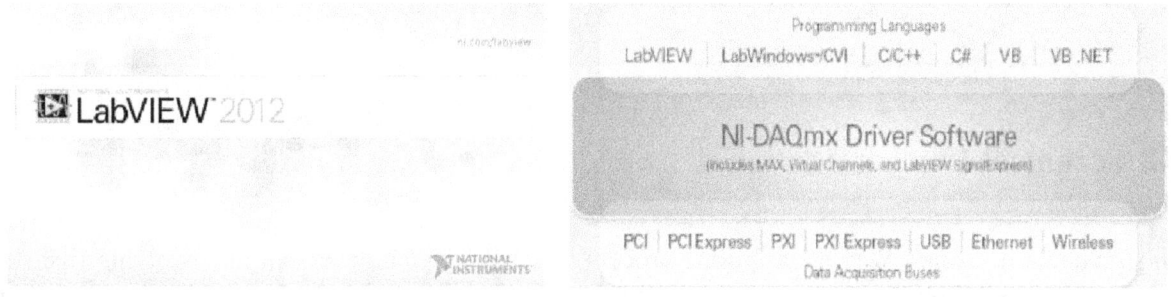

Hardware

NI WAP-3711

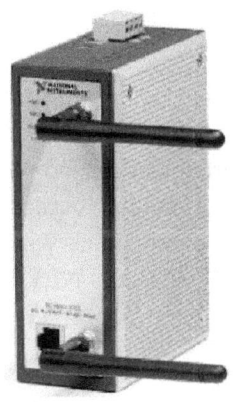

NI WLS-9163

Temperature sensor: PT-100

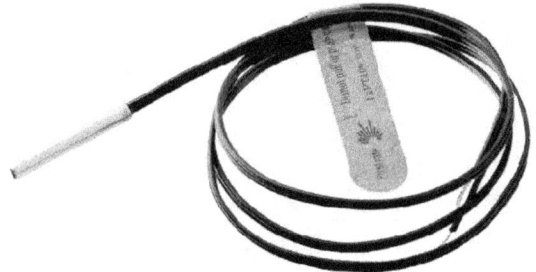

NI WLS-9234

Transducer board

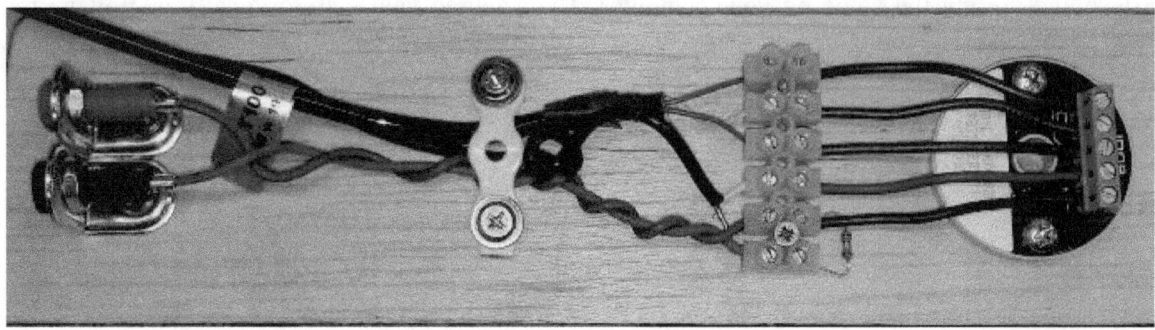

8.2 Technical Implementation

This section will explain the data acquiring process in details, which covers all the involving five steps stated above. By following the instructions, as the result temperature values obtained by the temperature sensor (PT-100) will be presented on the computer screen wirelessly. Wi-Fi (infrastructure) is the technology behind the wireless data transferring procedure. Here we use NI WAP-3711 to broad cast wireless network, which is based on the LAN. The wireless network has offered heavily encrypted and a lot of restrictions, by creating an own WAP data transmitted will be more flexible and give us more possibilities.

Step1: Physical Connection

PT-100 is a temperature sensor based on resistance (Ω) which various in corresponding to the changing in temperature. This element cannot be connected directly to the NI 9234 (DAQ) module because it requires input as Voltage. Due to this reason the two elements can only have communications with each other through the "transducer board". Additionally, both of the devices NI 9234 (DAQ) and NI WLS-9163 are needed to be attached to each other as one single physical component.

- Mount NI 9234 (DAQ) and NI WLS-9163 together as one physical unit.

    o Just simply slide the two modules together, make sure both of the modules are firmly connected with each other as shown in the Figure 8-1.

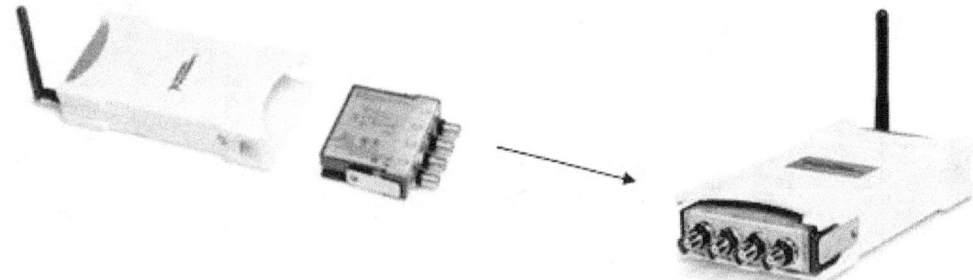

Figure 8-1 (NI 9234 and NI WLS-9163 attachment)

- PT-100 sensor and AC to DC switching power supply are already connected to the "transducer board". No further actions needed here.

- To read the temperature value from PT-100, the NI 9234 (DAQ) needs to be connected to the "transducer board". Study Figure 2-2 for details.

    o Same colors are connected to each other.

    o Attached the "coupling wire" to any of the pins on the NI 9234 (DAQ). AI0 pin used in this experiment.

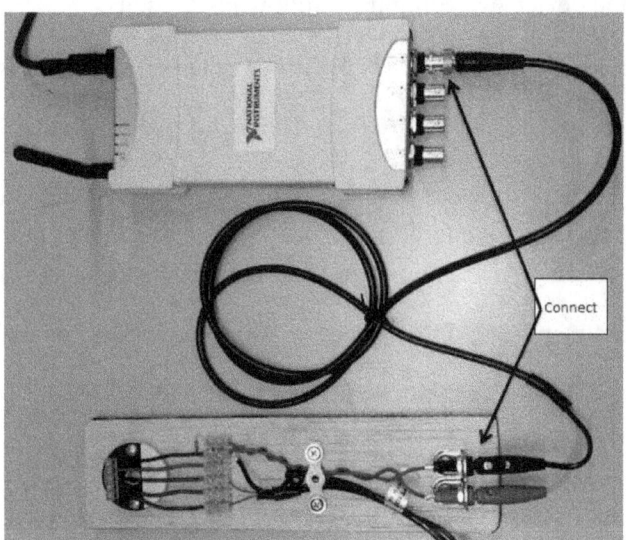

NI 9234 and transducer board connection

- Once the connection is established.

    o Simply plug the power supply to the wall outlet. The "Transducer board" will operate by 24 Voltage DC.

Step 2: NI WAP-3711 Configuration

The NI WAP-3711 device can be fully configures by using any web browsers. The ability for doing any changes on this device, users have to have password access. Once the users manage to log in, then they have full access to the device without any restrictions. The users are able to set new SSID name, new IP address, change security settings, change passwords and so on.

- In this particular case all we have to do is just simply establish connection between NI WAP-3711 and the LAN with an Ethernet cable. By using the default of NI WAP-3711 is correctly configured for this purpose.

  - Power up NI WAP-3711 and push the reset button for at least five seconds. To make sure the device is configured as default.

  - NI WAP should broadcast internet. Continue from step 3.

- Logging on to NI WAP-3711

  Notice that; This step is not necessary in this case, if so continue from step 3. However, if you have to configure the device just continue to follow instructions.

- Connect your computer to NI WAP-3711 by using Ethernet cable (straight-through or cross-over cable)

- The ability for different devices to have successful communications to each other, the requirement is that they all have to be in the same subclass (Subnet mask). Therefore we have to change the IP address on your computer so it is in the same subclass as NI WAP-3711.

  - Start menu → Control Panel → View network status and tasks.

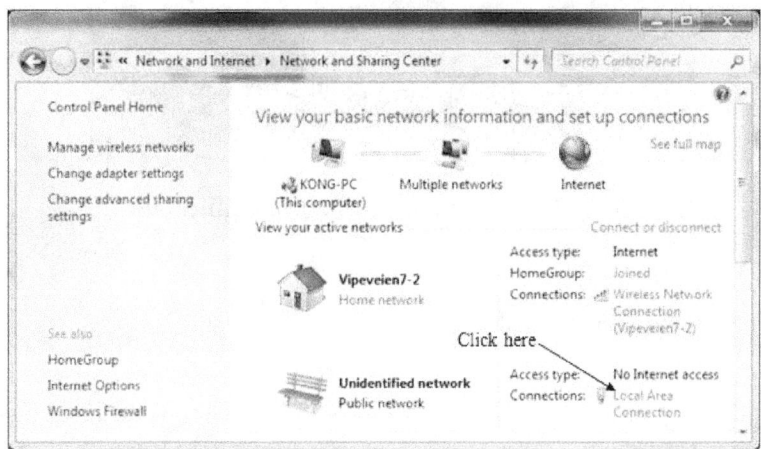

Local Area Connection

- Local Area Connection → Properties → Internet Protocol Version 4(TCP/IPv4) → Properties
- Select: "Use the following IP address" in the pop up window and type in the numbers exactly as illustrated in the Figure below → OK.

Note: the IP address: 102.168.127.X (where X can be any integer from 1 – 252 and 254)

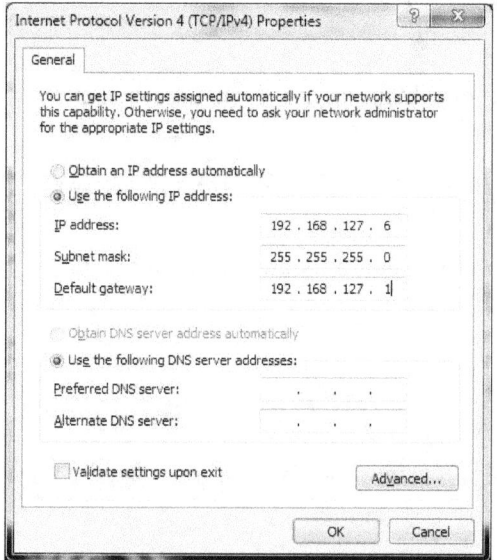

Assign address to Local Area Connection

- Open a web browser (Mozilla Firefox or Internet Explorer) from the address bar type: http://192.168.127.253 → Enter.

- Fill in User Name and Password.
- User Name: Admin
- Password: root

Logging into NI WAP-3711

- Now you should be logged on to NI WAP-3711 and have full access to the device.
- Configure NI WAP-3711

o **General** → Operational Mode
   Select AP/Bridge → Save
o **TCP/IP** → Addressing

   - Method of obtaining an IP address: Set
     Manually IP address: 192.168.127.253

   - Subnet mask: 255.255.255.0 Default gateway: 192.168.127.1 Host name: adv
     app

   - Make sure the numbers are identical as shown in the figure below → Save &
     Restart.

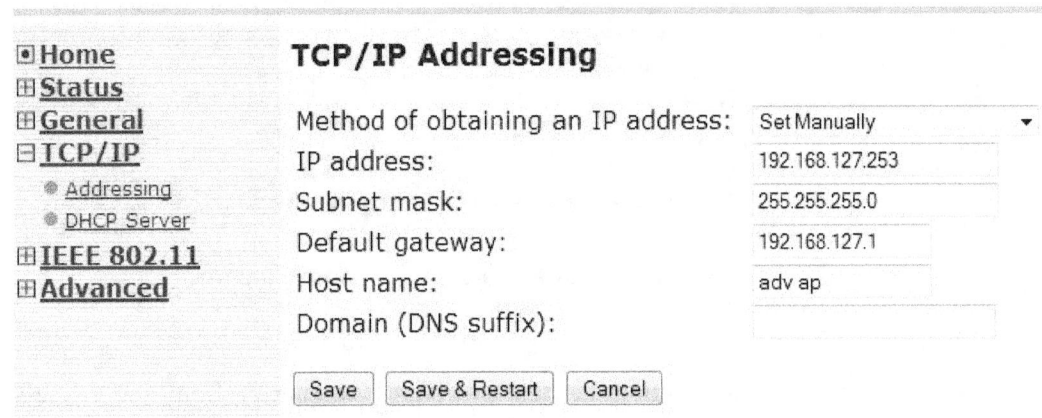

TCP/IP Addressing setting

- **TCP/IP** → DHCP Server Functionality: Enabled

  o Default gateway:
     192.168.127.1 Subnet mask:
     255.255.255.0 Primary DNS
     server: 192.168.127.1

  o Make sure your setting is identical to the Figure below.

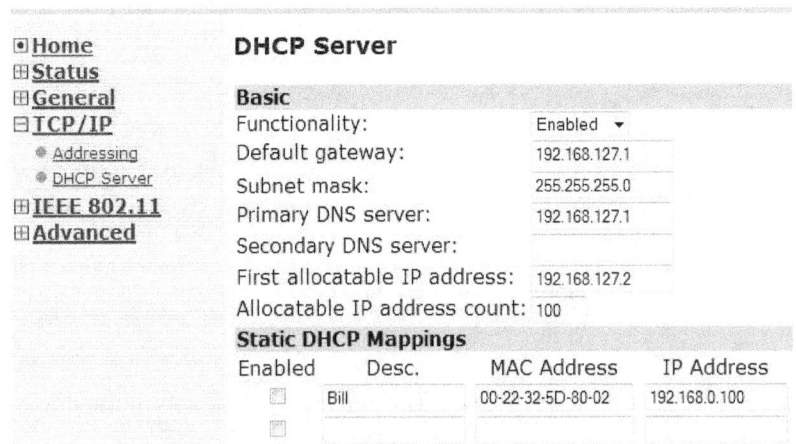

DHCP Server setting

- o **IEEE 802.11** → Communication

    - o Change the SSID name from NI to WAP (SSID can be given by any name) Network name (SSID): WAP

    - o Channel number: 3 (any integer 1-13)

    - o **Note:** Channels 1, 4 and 11 are very actively used. To avoid interference or other unexpected problems try to avoid these channels.

- o Save & Restart

    - o NI WAP-3711 is now fully configured.

Step 3: NI WLS-9163 Configuration

NI WLS-9163 is used to transmit the temperature data to NI WAP-3711 and can be configured by using Measurement & Automation (MAX). MAX is a software which allows users to fully access and configure settings of NI WLS-9163.

- Download and install Measurement & Automation (MAX).

    - o Open the link below and follow instructions given in the website to download and install NI-DAQmx 9.6 (MAX). **Notice that;** Choose latest version if available. http://joule.ni.com/nidu/cds/view/p/id/3423/lang/no

- Set Local Area Connection on your computer to obtain an IP address automatically

    - o Start menu → Control Panel → View network status and tasks.

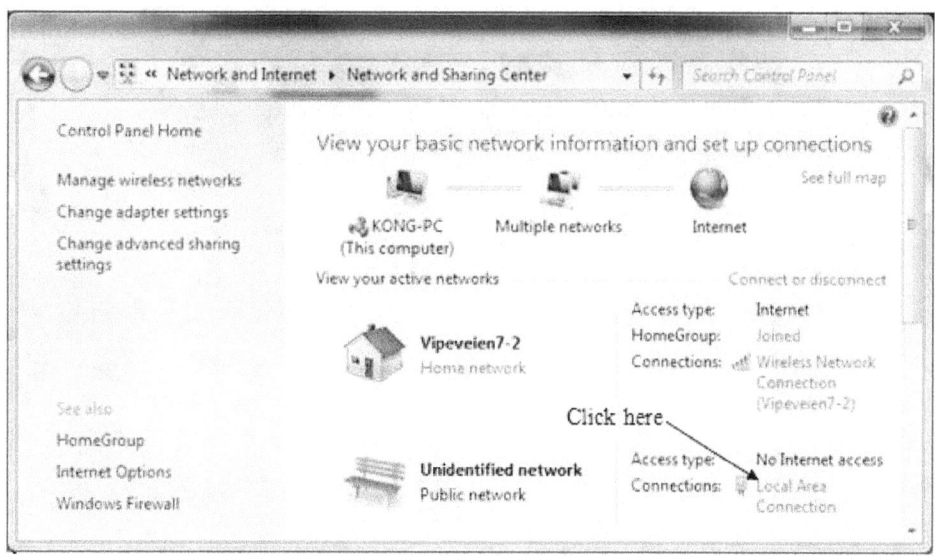

Local Area Connections

- Local Area Connection → Properties → Internet Protocol Version 4(TCP/IPv4) → Properties
- Select "Obtain an IP address automatically" in the pop up window → OK.
- Connect Ethernet cable from NI WLS-9163 to your computer. (Physical connection between your computer and NI WLS-9163 is established)
- Use MAX to detect NI WLS-9163 device and do the configuration
- Start Measurement and Automation (MAX) application by double click on the "NI MAX" icon on the desktop.
- Click the arrow (Devices and Interfaces) to view a list of all the DAQ devices. See the Figure below.
- Right click on Network Devices → click on Find Network NI-DAQmx devices. This will detect any current devices connected to your computer.

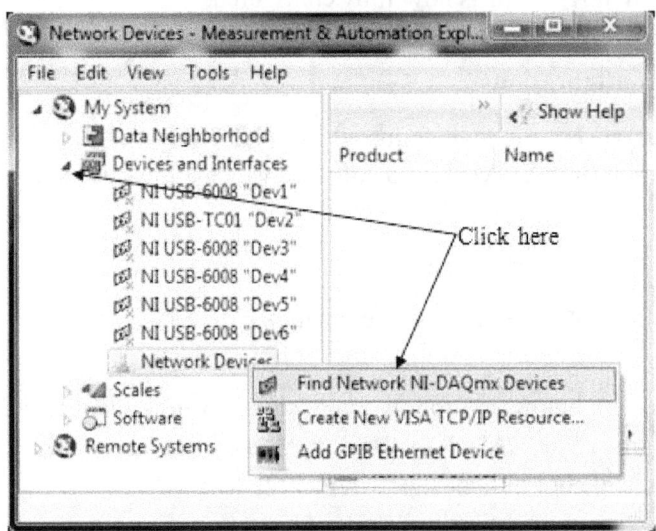

Detecting NI WLS-9163 by using MAX

Tick WLS-14215DD (or any NI-DAQmx device that is found) → Add Selected Devices.

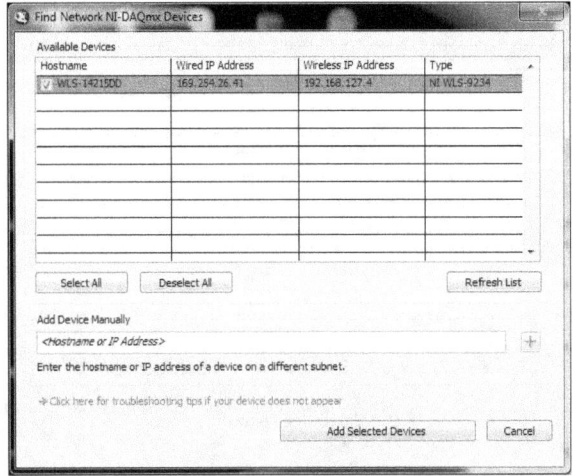

Add devices to MAX

204

- Detect and configure NI WLS-9163

  o Click NI WLS-9234 "WLS-14215DD" and create identical as shown on the Figure below.

  o Obtain IP Address Through:
  Static IP Address:
  128.39.35.240 Subnet Mask:
  255.255.254.0 Gateway:
  128.39.34.1

- DNS Server: 128.39.198.39 (or leave empty box).

    o **Notice that;** Country should be specified as United State. From experience attempts to set to Norway causes unsuccessful connection.

    o SSID: NI (By default NI WAP-3711 broad cast the SSID name as NI)

    o Authentication Type: Open (By default NI WAP-3711 is not encrypted by password, if it does: just type in the security type and password)

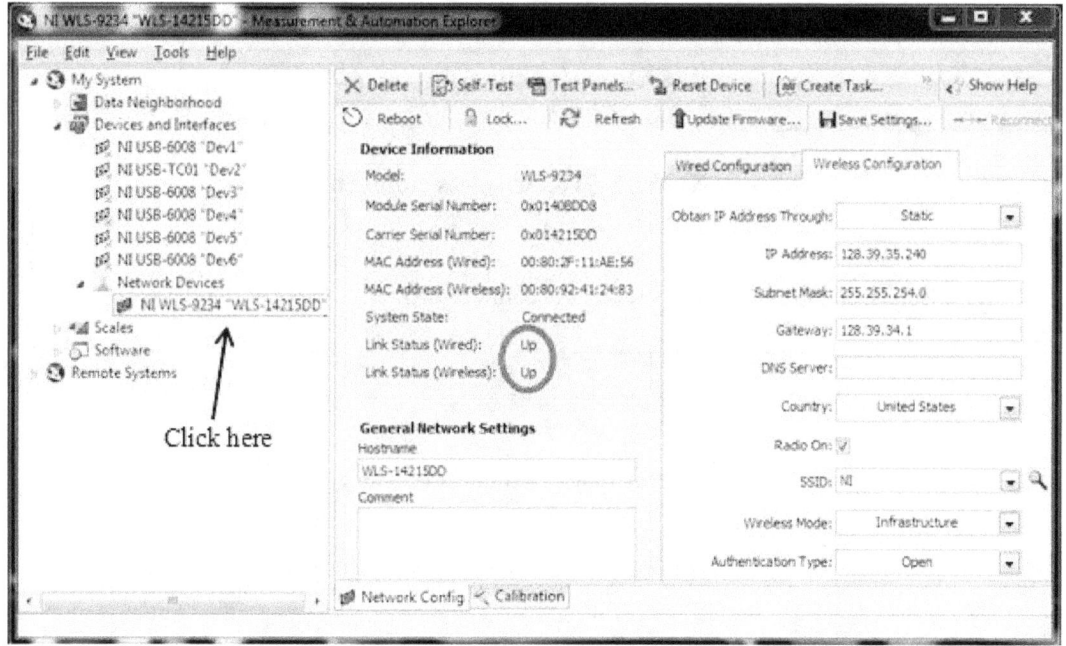

NI WLS-9163 configuration

o Click: Save Settings... → Refresh

o Press Refresh button several times until you obtain Wired and Wireless is shown **Up**.

o **Notice that;** If Wire and Wireless both are not shown as Up, just uptick Radio button On →
Save Settings → Tick on the Radio On button → Save Settings → Refresh.

o Remove the Ethernet cable from NI WLS-9163 → Click: Refresh

o **Notice that;** Wired down and Wireless Up (wireless communication between NI WLS-9163 and
NI WAP is successfully established)

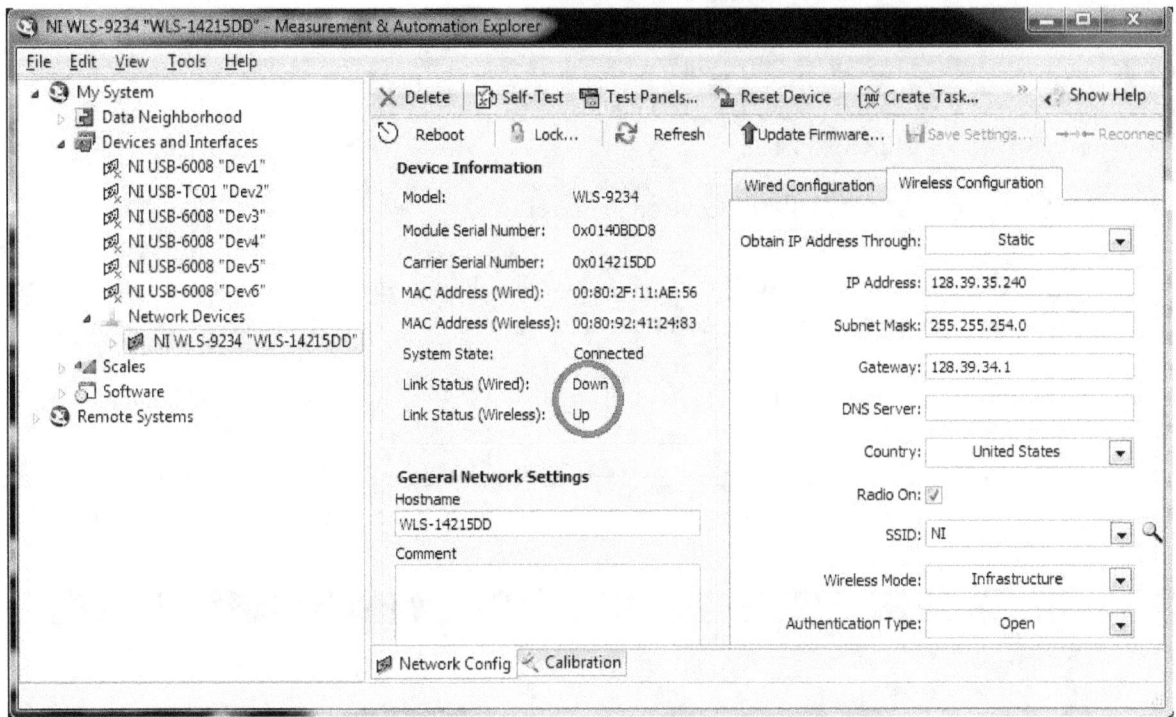

NI WLS-9163 wireless connection

o Click: Self-Test (To validate a successful connection).

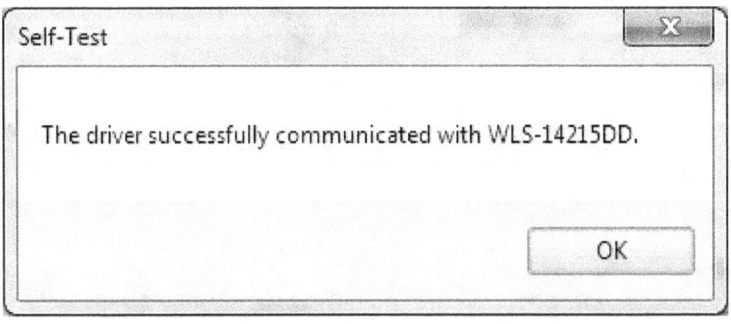

Successful communication confirmation

- Test to retrieve value wirelessly from NI WLS-9163
  - Click: Test Panels... Coupling: DC

  - Acquisition Mode: Continuous

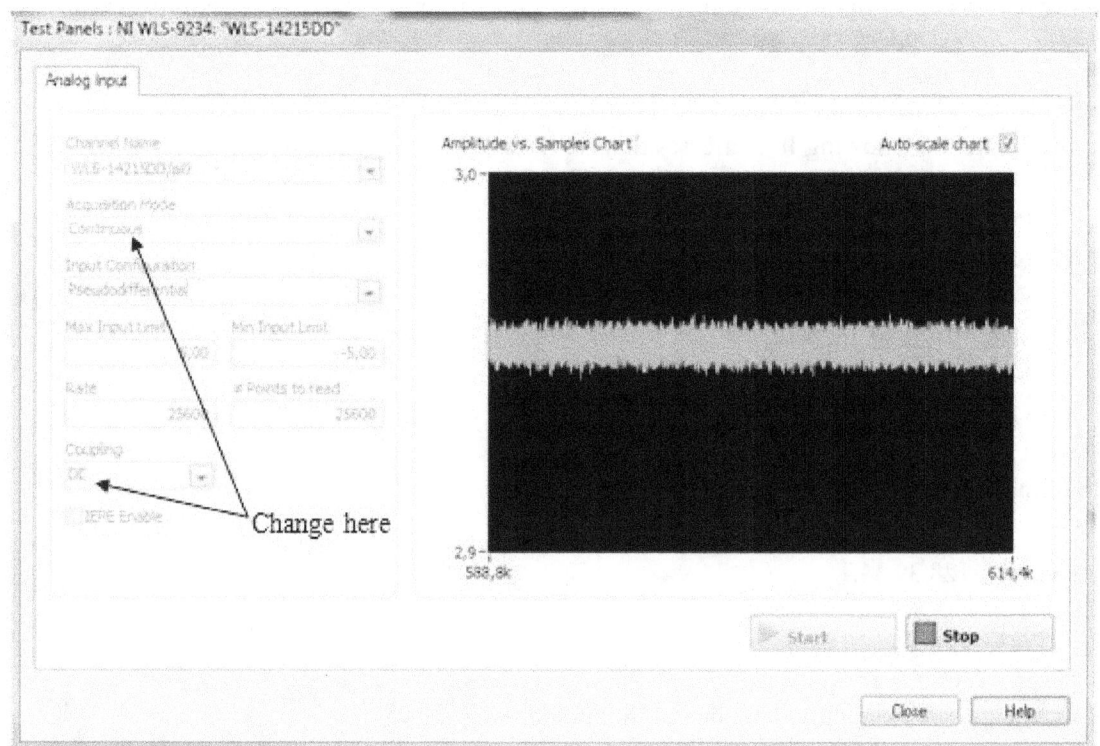

Result shown in Test Panel.

- All the obtaining values are presenting only in Voltage DC (1-5V).

- You are now successfully retrieving data from NI WLS-9163 and NI WAP-3711 to your computer completely wireless.

Step 4: Computer Configuration

At this moment your NI WAP-3711 should be able broadcast internet. Now we shall connect the computer to get access to NI WAP-3711. If the connection between your computer and this wireless network is successfully, it should have internet access. Use a web browser to validate a successful connection.

As mentioned from previous, wireless network heavily encrypted and causes easily complications. The LAN server will only allow a certain specific group of IP-addresses, in another word LAN server controls which IP-addresses shall have permission to the provided service. Logically as it is, every computer in the IT-labs carries a valid IP-number.

➢ **Demo :**

We need to find out two valid IP-addresses that are accepted by the LAN server. Once we know the allowable IP-numbers then we will assign our computer with one of the IP-addresses, while the other IP-address will be given to NI WLS-9163 device. It is obvious that every device has to have a valid IP-address that is accepted by the LAN server that been able to utilize the resources broadcasting through the Wireless Access Point.

Note: Let's take the following IP-addresses that are reserved for the laboratory context.

IP-address: 128.39.35.239
                     128.39.35.240
                     128.39.35.241
                     128.39.35.226

Subnet mask: 255.255.254.0

Gateway: 128.39.34.1

DNS: 128.39.198.39

- Assign your computer with one of the IP-addresses above

  o From your computer

    Click: Start menu → Control Panel → View network status and tasks.

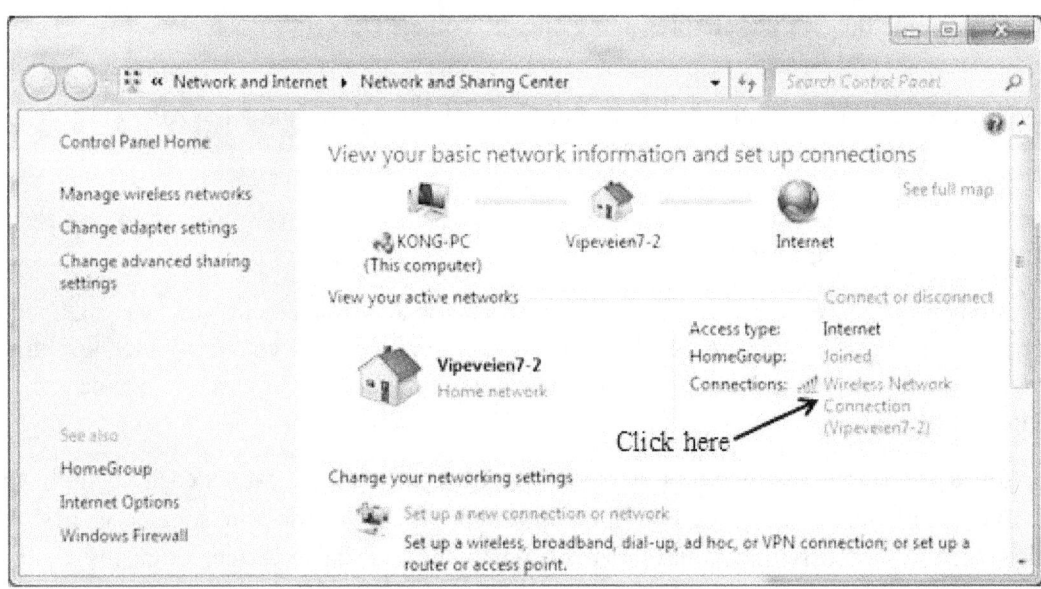

Wireless Network Connection

208

- Wireless Network Connection → Properties → Internet Protocol Version 4(TCP/IPv4) → Properties.

- Select "Use the following IP address" in the pop up window

- Type in the IP-addresses (parameters) above → OK, an example is shown in the following Figure.

    ▪ IP address: 128.39.35.239 Subnet mask: 255.255.254.0 Default gateway: 128.39.34.1
    ▪ Preferred DNS server: 128.39.198.39

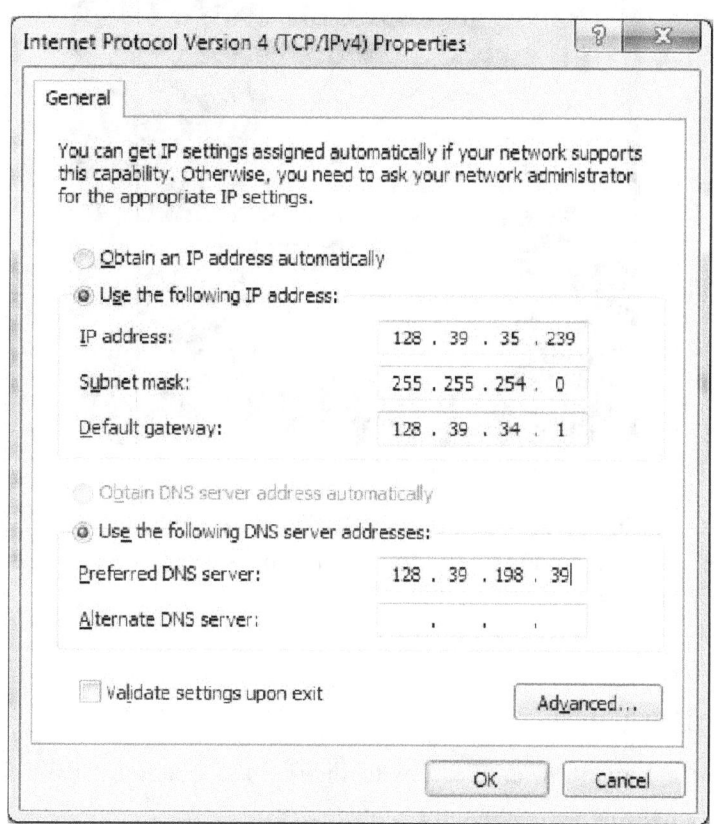

Assign Address to the Laptop

- Establish Wireless Network connection from your computer to WAP (in this case NI, which is the SSID name broadcasted by NI WAP-3711). The following figure shows Wireless network - Laptop connection.

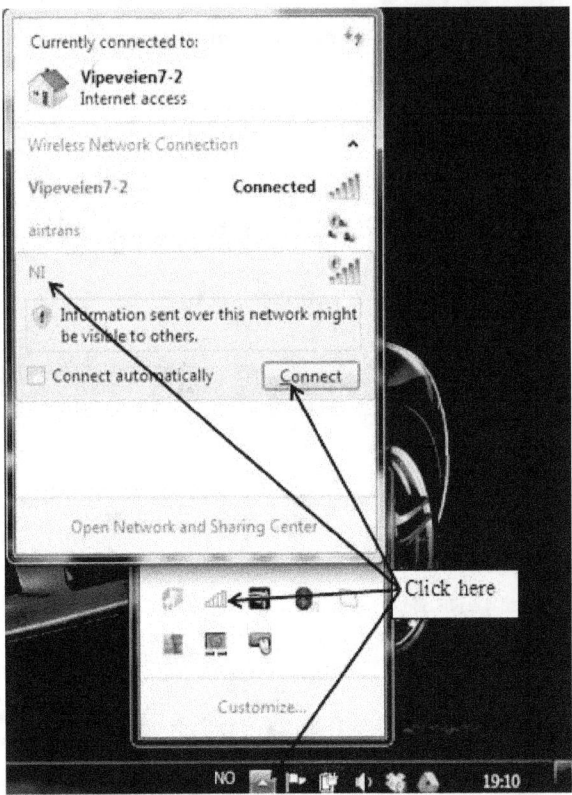

Currently, the computer should be connected to NI WAP-3711. To validate the successful connection, use a web browser and surf to some home pages. Now your computer has access to internet from your own Wireless Access Point your area.

Step 5: Temperature Presentation

- We will create a LabVIEW application to display the obtaining values in Voltage and then convert these values into corresponding temperatures.

- Start LabVIEW and create a New VI. Name this VI (application) and save.

- In the block diagram: right click to find DAQ Assistant. Follow instructions from the following Figure.

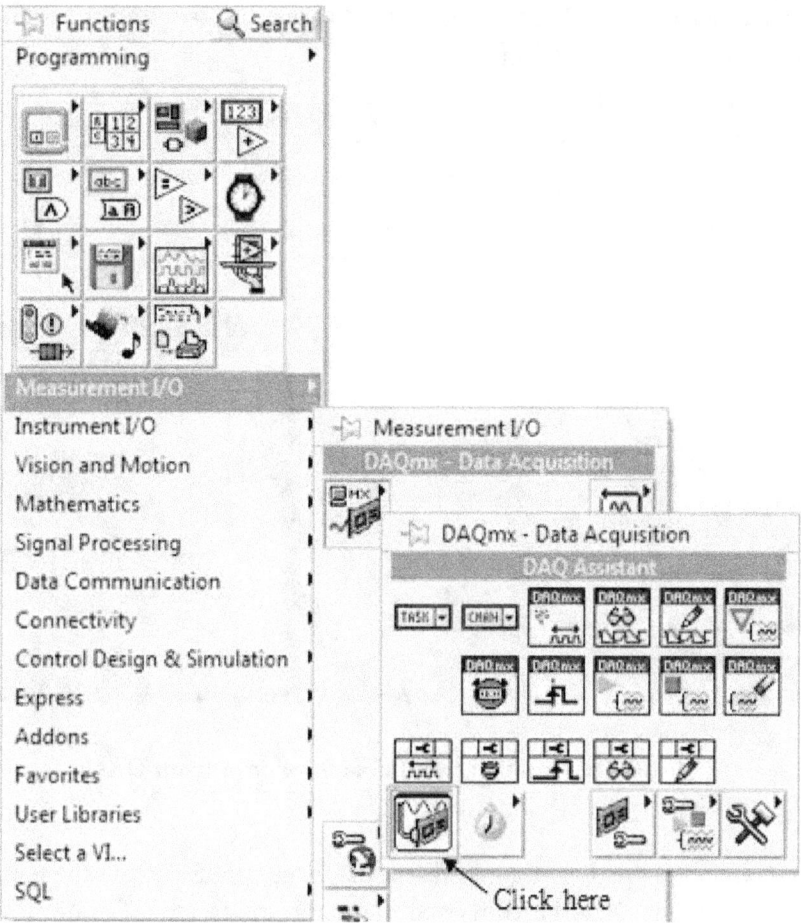

"DAQ Assistant" location

- Configure the DAQ Assistant to obtain the Voltage values.

    ▪ Click: Acquire Signals → Analog Input → Voltage → ai0 (depending on which of the pin of NI 9234 were used) → Finnish.

    ▪ From pop up Window: Click Device → Change to DC → OK

- Make sure your Block Diagram looks like the one illustrated in the following figure, the value is presenting in the Front Panel.

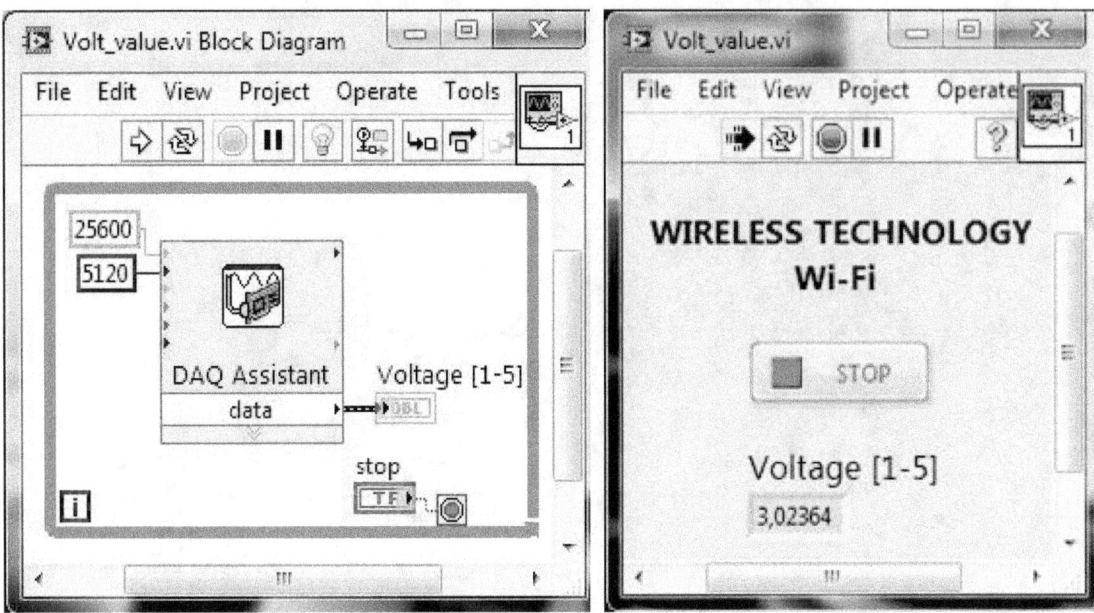

LabVIEW application to retrieve data from DAQ

Advance:

The front panel shown in the previous figure present the value captured from the temperature sensor PT-100 in Voltage. The value needs to be converted into temperature [°C]. Try to improve the LabVIEW application that shows both in Voltage and degree Celsius as illustrated in the following Figure.**Tips:**

- Use silver style as layout.
- Convert from Voltage to Temperature.
- Temperature [°C] = Voltage * a – b (define parameters a & b) 1 [V] = 0 [°C]

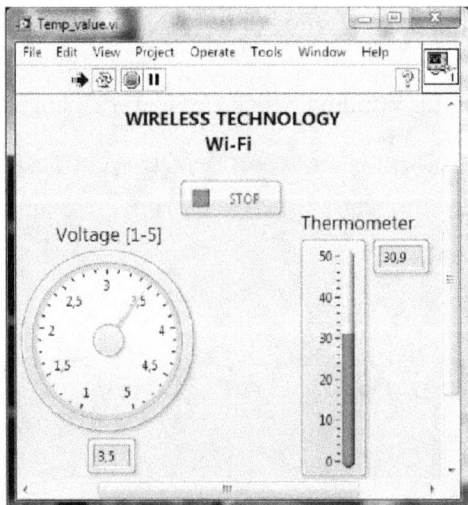

Raw and scaled data displayed on LabVIEW front panel

212

The block diagram of the improved front panel is shown in Figure below, where both of the parameters 12,5 are used to convert from Voltage to degree Celsius. **Note:** These parameters can only be used for this specific case, where 1-5 Volt are converted to 0-50 °C. For converting to a scale of 0-100 °C, just simply replace both of the parameters 12,5 with 25.

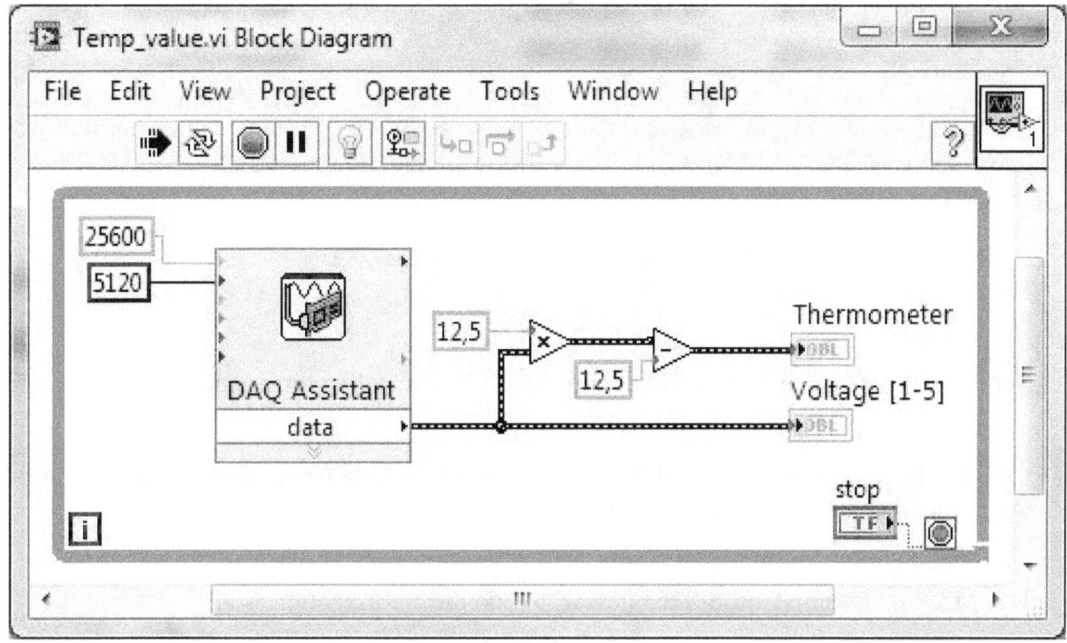

Voltage - Temperature scaled procedure in LabVIEW block diagram

If this have reached so far then one has to manage for presenting the environmental temperature by LabVIEW application. The raw value was captured by PT-100 sensor and transmitted completely wireless in form of Voltage DC, through Wi-Fi (infrastructure) technology.

8.3 Summarize procedure

This section will summarize what have been done to get a successful wireless data transferring by utilizing Wi-Fi (infrastructure) technology, which is implemented on Windows 7 professional OS. The data transmission relies on National Instruments technology, where NI WAP-3711 would be broadcast internet from any LAN. Five different steps are required to present the raw data into temperature degree (°C).

Step 1: Setup and connect all the physical devices together.

Step 2: Configure NI WAP-3711

In this particular case all we have to do is just simply establish connection between NI WAP-3711 and the LAN with an Ethernet cable.

213

o   Power up NI WAP-3711 and push the reset button for at least five seconds. To make sure the device is configured as default.

o   NI WAP should broadcast the internet.

Step 3: Configure NI WLS-9163

NI WLS-9163 is used to transmit the temperature value to NI WAP-3711 and can be configured by using Measurement & Automation (MAX). MAX is a software which allows users to fully access and configure settings of NI WLS-9163.

- Download and install Measurement & Automation (MAX).
  http://joule.ni.com/nidu/cds/view/p/id/3423/lang/no

- Set Local Area Connection on your computer to obtain an IP address automatically.

- Connect Ethernet cable from NI WLS-9163 to your computer. (Physical connection between your computer and NI WLS-9163 is established)

- Use MAX to detect NI WLS-9163 device and do the configuration.

  o   Obtain IP Address Through: Static IP Address: 128.39.35.240 Subnet Mask: 255.255.254.0 Gateway: 128.39.34.1

  o   DNS Server: 128.39.198.39 (or no input require).

  o   Country should be specified as United State.

  o   SSID: NI (By default NI WAP-3711 broad cast the SSID name as NI)

  o   Authentication Type: Open (By default NI WAP-3711 is not encrypted by password, if it does: just type in the security type and password)

  ▪   Click: Save Settings... → Refresh

  ▪   **Note:** Press Refresh button several times until you obtain Wired and Wireless is shown Up.

  ▪   If Wire and Wireless both are not shown as Up, just uptick Radio On → Save Settings → Tick on the Radio On button → Save Settings → Refresh.

  ▪   Remove the Ethernet cable from NI WLS-9163 → Click: Refresh

- ▪ Wired down and Wireless Up (wireless communication between NI WLS-9163 and NI WAP is successfully established)

  o Click: Self-Test (To validate a successful connection).

  o All the obtaining values are presenting only in Voltage DC (1-5V).

  o You are now successfully retrieving data wirelessly from NI WLS-9163 and NI WAP-3711 to your computer completely wireless, in addition your NI WLS-9163 is broadcasting the internet your LAN.

Step 4: Configure computer

- At this moment your NI WAP-3711 should be able to broadcast internet.

- Now we shall connect the computer to get access to NI WAP-3711.

  o If the connection between your computer and this wireless network is successfully, it should have internet access. Use a web browser to validate a successful connection.

- The LAN server will only allow a certain specific group of IP-addresses, in another word LAN server controls which IP-addresses shall have permission to the provided service. Logically as it is, every computer in the computer labs carries a valid IP-number.

- Find out two valid IP-addresses that are accepted by the LAN server. Once we know the allowable IP-numbers then we will assign our computer with one of the IP-addresses, while the other IP-address will be given to NI WLS-9163 device. It is obvious that every device has to have a valid IP-address that is accepted by the LAN server in able to utilize the resources broadcasting by the Wireless Access Point.

- The following IP-addresses are available for the laboratory context.
➢ 128.39.35.239

➢ 128.39.35.240

➢ 128.39.35.241

➢ 128.39.35.226
  Subnet mask: 255.255.254.0

  Default gateway: 128.39.34.1

  Preferred DNS server: 128.39.198.39

- Assign your computer with one of the IP-addresses stated above.

  o Since you already assigned 128.39.35.240 to NI WLS-9163, so you have to choose among the remaining addresses.

- Establish Wireless Network Connection from your computer to WAP (in this case NI, which is the SSID name broadcasted by NI WAP-3711).

- Now your computer should be connected to NI WAP-3711. To validate the successful connection, use a web browser and surf to some home pages. Now have access to internet from your own Wireless Access Point.

Step 5: Temperature presentation on screen

We created a LabVIEW application to display the obtaining values in Voltage and then convert these values into corresponding temperatures.

  o Start LabVIEW and create a New VI. Name this VI (application) and save.

  o In the block diagram: right click to find DAQ Assistant.

  o Configure the DAQ Assistant to obtain the Voltage values.

  o In the block diagram convert from Voltage to Temperature.

  o Focus on the layout in the front panel.

As the final result it may managed to transmit data completely wirelessly by Wi-Fi (infrastructure) technology and by using National Instruments equipments.

Chapter 9

Wireless DAQ using ZigBee

This chapter will go through how to transfer temperature data acquired from a sensor to the computer wirelessly by using ZigBee technology on Windows 7 operative system. This documentation is created based on experiment by using the following equipment:

- **Hardware**

    o Temperature sensor Pt-100

    o Connection board

    o ZigBee equipment

        ▪ NI WSN-9791 – WSN Ethernet Gateway
        ▪ NI WSN-3202 – Analog Input Node

- **Software**

    o Windows 7

        ▪ **Note:** Other OS versions besides "Professional" are experienced causing problems. (MAX is not able to detect NI WSN-9791)

    o NI-WSN-140 software

    o LabVIEW 2012

9.1 ZigBee Definitions

➢ **ZigBee** is a **specification** for a suite of high level communication protocols used to create **personal area networks** built from small, low-power **digital radios**. ZigBee is based on an **IEEE 802.15 standard**. Though low-powered, ZigBee devices can transmit data over long distances by passing data through intermediate devices to reach more distant ones, creating a **mesh network**; i.e., a network with no centralized control or high-power transmitter/receiver able to reach all of the networked devices. The decentralized nature of such **wireless ad hoc networks** make them suitable for applications where a central node can't be relied upon.

➢ **ZigBee** is used in applications that require only a low data rate, long battery life, and secure networking. It has a defined rate of 250 kbit/s, best suited for periodic or intermittent data or a single signal transmission from a sensor or input device. Applications include wireless light switches, electrical meters with in-home-displays, traffic management systems, and other consumer and industrial equipment that requires short-range wireless transfer of data at relatively low rates. The technology defined by the ZigBee specification is intended to be simpler and less expensive than other **WPANs**, such as **Bluetooth** or **Wi-Fi**.

➢ **ZigBee** networks are secured by 128 bit **symmetric encryption** keys. In home automation applications, transmission distances range from 10 to 100 meters **line-of-sight**, depending on power output and environmental characteristics.

➢ **ZigBee** is a low-cost, low-power, wireless mesh network standard. The low cost allows the technology to be widely deployed in wireless control and monitoring applications. Low power usage allows longer life with smaller batteries. Mesh networking provides high reliability and more extensive range. **ZigBee** chip vendors typically sell integrated radios and microcontrollers with between 60 KB and 256 KB flash memory.

➢ **ZigBee** operates in the industrial, scientific and medical (ISM) radio bands: 868 MHz in Europe, 915 MHz in the USA and Australia and 2.4 GHz in most jurisdictions worldwide. Data transmission rates vary from 20 kilobits/second in the 868 MHz frequency band to 250 kilobits/second in the 2.4 GHz frequency band.

The **ZigBee** network layer natively supports both star and tree typical networks, and generic mesh networks. Every network must have one coordinator device, tasked with its creation, the control of its parameters and basic maintenance. Within star networks, the coordinator must be the central node. Both trees and meshes allow the use of **ZigBee** routers to extend communication at the network level.

➢ **ZigBee** builds upon the physical layer and media access control defined in IEEE standard 802.15.4 (2003 version) for low-rate WPANs. The specification goes on to complete the standard by adding four main components: network layer, application layer, **ZigBee** device objects (ZDOs) and manufacturer-defined application objects which allow for customization and favour total integration.

Besides adding two high-level network layers to the underlying structure, the most significant improvement is the introduction of ZDOs. These are responsible for a number of tasks, which include keeping of device roles, management of requests to join a network, device discovery and security.

➢ **ZigBee** is not intended to support power line networking but to interface with it at least for smart metering and smart appliance purposes. Because **ZigBee** nodes can go from sleep to active mode in 30 ms or less, the latency can be low and devices can be responsive, particularly compared to Bluetooth wake-up delays, which are typically around three seconds. Because **ZigBee** nodes can sleep most of the time, average power consumption can be low, resulting in long battery life.

> **ZigBee** is built on top of the **physical layer** and **medium access control**(MAC layer) defined in the **IEEE standard 802.15.4** (2003 version) for low-rate **WPAN's**. The ZigBee specification then adds to the standard four main components: network layer, application layer, **ZigBee** device objects(ZDO's) and user-defined application objects which allows for customization and flexibility within the standard.

In addition to integrating two high-level network layers to the underlying structure, the most significant addition is the introduction of **ZigBee** Device Objects (ZDO's). ZDO's are responsible for multiple tasks, which include defining device roles, management of requests to join a network, device discovery and security. By nature, **ZigBee** is a "**mesh network**" architecture. In addition to the mesh topology the network layer supports two other types of topologies: **star** and **tree**. Every network must have one **ZigBee** Coordinator device, tasked with its creation, the control of its parameters and basic maintenance. Within star networks, the ZC is the central node. Both tree and mesh networks allow the use of ZigBee **Routers** to extend communication at the network layer.

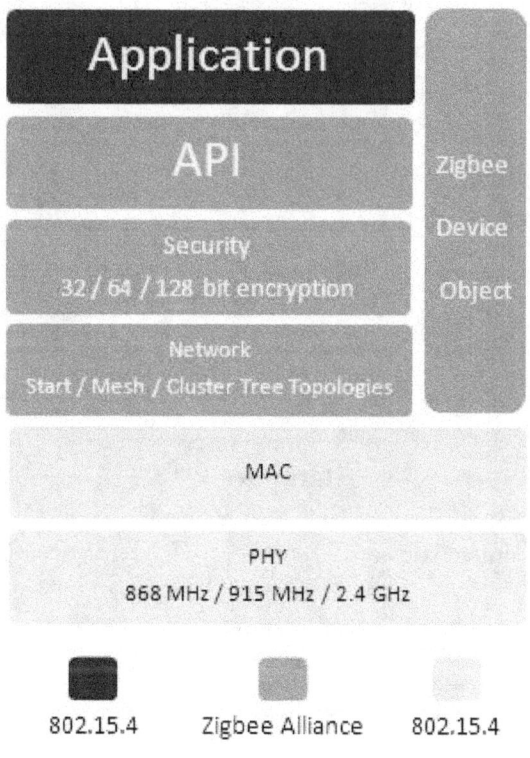

Figure 9-1 ZigBee Architecture

The data transferring procedure through ZigBee technology involves mainly three different steps;

219

Step 1: Connect the temperature sensor pt-100 to the node (NI WSN-3202).

Since pt-100 is a resistance based sensor and inputs to the node device are required to be in Voltage. Therefore, direct connections between the two devices are not possible and the devices have to be connected together via the "connection board".

Step 2: Install and configure NI-WSN-140 software.

Allows user to setup the Gateway, add the measurement node(s) and control the wireless network.

Step 3: Present the temperature value on the computer.

Create a LabVIEW application to continuously retrieve and present the value on the computer. The acquired values are originally in Voltage format which needs to be converted into corresponding temperature.

Necessary Equipment

Software

Hardware

NI WSN-9791-WSN Ethernet Gateway   NI WSN-3202 - Analog Input Node

Temperature sensor: PT-100

AC to DC switching power supply

Model: GPSUE-6

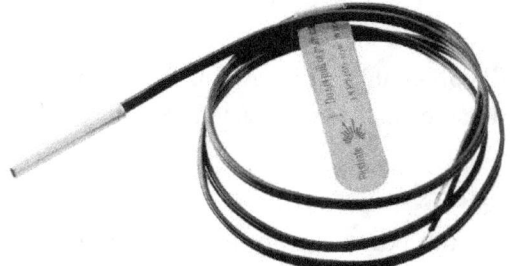

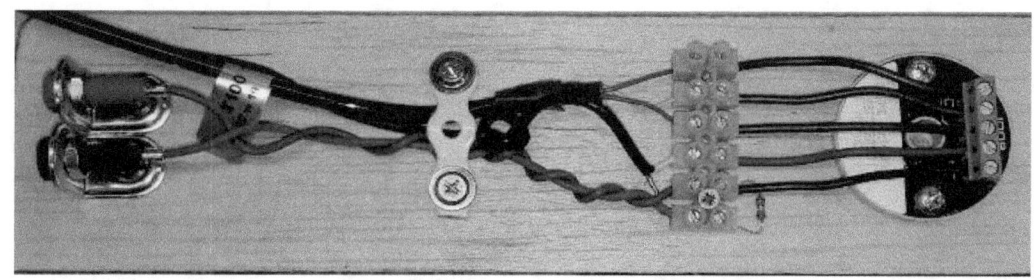

Transducer board

9.2 Technical Terminology

This section will explain the data acquiring process in details, which covers all the involving three steps stated above. By following the instructions, as the result temperature values obtained by the temperature sensor (Pt-100) will be presented on the computer screen wirelessly. ZigBee is the technology behind the wireless data transferring procedure.

- Range: 10 - 75 meters (depending on the environment)
- Radio Power Requirements: 1mW to 100mW
- Modulation Tech: DSSS and CSMA/CA

Step1: Physical Connection

Pt-100 is a temperature sensor based on a resistance (Ω) which various in corresponding to the changing in temperature, this element cannot be connected directly to the ZigBee node (NI WSN-3202) because the node only understands Voltage as input. Due to this reason the two elements can only have communications with each other through the "transducer board".

- Pt-100 and AC to DC switching power supply are already connected to the "Connection board".

- To read the temperature value from pt-100, the "NI WSN-3202 - Analog Input Node" needs to be connected to the "connection board", where red cable shall be connected to AI0 and blue cable to AI GND as shown in the Figure below.

- Once the connection is established, just simply plug the power supply to the wall outlet. The "Connection board" will operate by 24 Voltage DC.

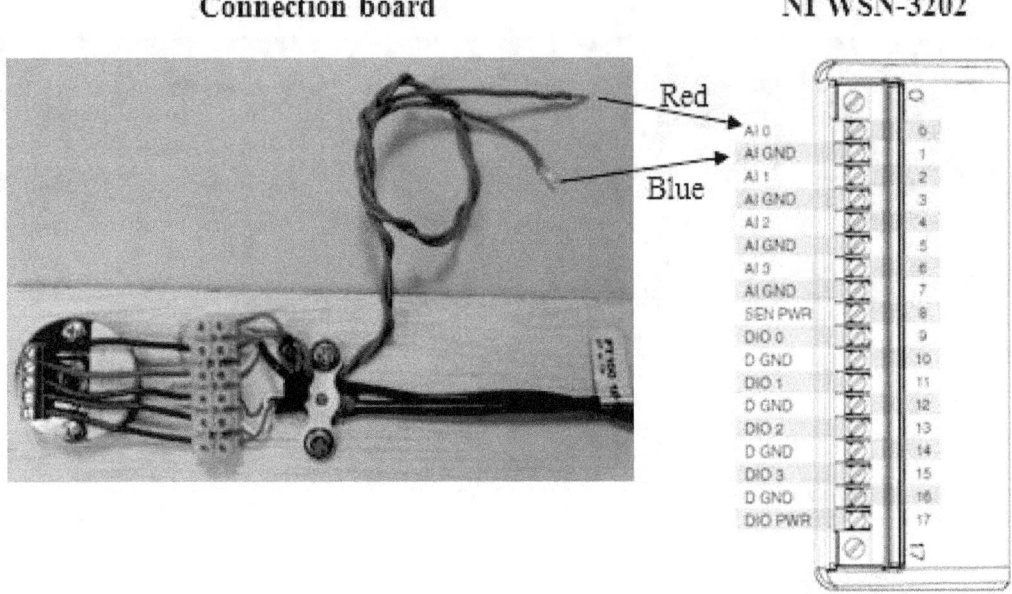

Figure 9-2 Connection board - NI WSN-3202 connection

Step 2: NI-WSN-140 Software

Ni-WSN-140 is computer software, through which the users are able to configure the network in the Measurement & Automation Explorer (MAX) utility. Max provides an intuitive user interface to add and remove measurement nodes and configure wireless settings. The settings have effects on the central control unit, NI WSN-9791 – WSN Ethernet Gateway.

1. Download & Install

- Open the link below to install the appropriate driver for NI WSN-971 – WSN Ethernet Gateway.

  o http://sine.ni.com/psp/app/doc/p/id/psp-917/lang/no

- From "NI Hardware Drivers" various options of drivers are available. Download the **latest version**.

    o Click "NI-WSN 1.4 - NI Wireless Sensor Networks" (Currently latest version)

    o Choose option 1 "NI Downloader: NIWSN140-downloader.exe (871.3 MB)" Support Operating system: Windows 7; Windows Vista; Windows XP 32-bit.

- The installing of NI-WSN is a straight forward procedure. Just simply follow instructions given of each steps and make sure to include "Real-Time Features" as shown in the following figure, and select the updates option.

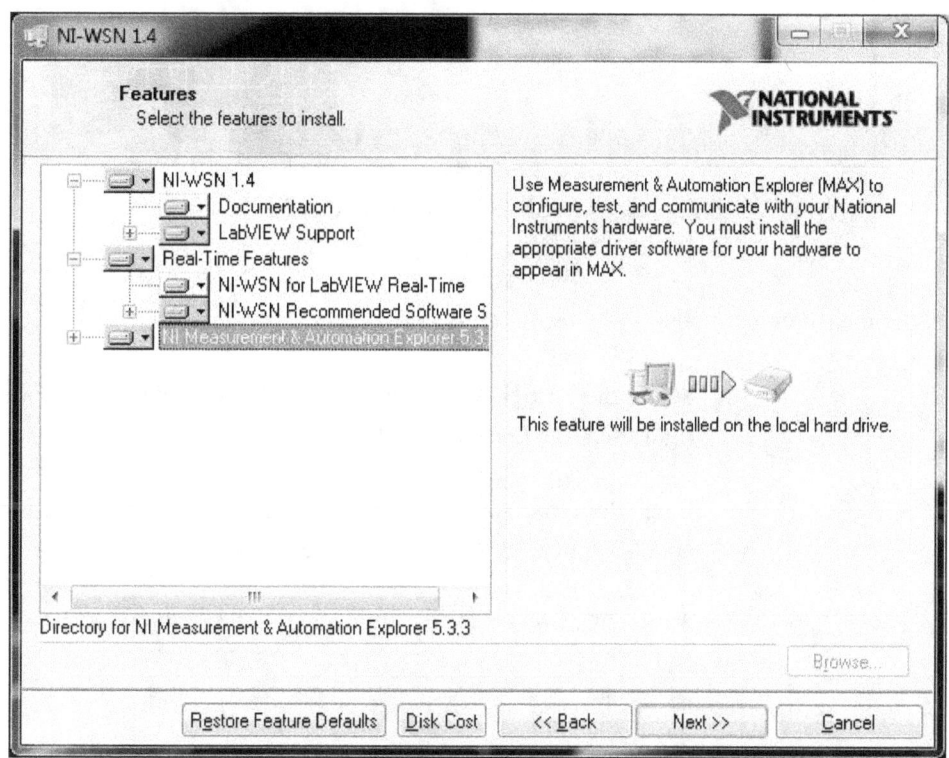

Figure 9-3 MAX Software installation

2. Configuration

The Central control unit, NI WSN-9791 – WSN Ethernet Gateway can be configured by using the Measurement and Automation Explorer (MAX).

- Launch MAX and expand "Remote Systems" to confirm that MAX has auto detected the WSN-9791 as shown on the Figure on the next page.

- If the gateway does not appear, consult the "Getting Started with NI Wireless Sensor Networks" tutorial for more information. Confirm or change the default settings on the "Network Settings" tab. Make sure to click "Apply" to save any changes that you made.

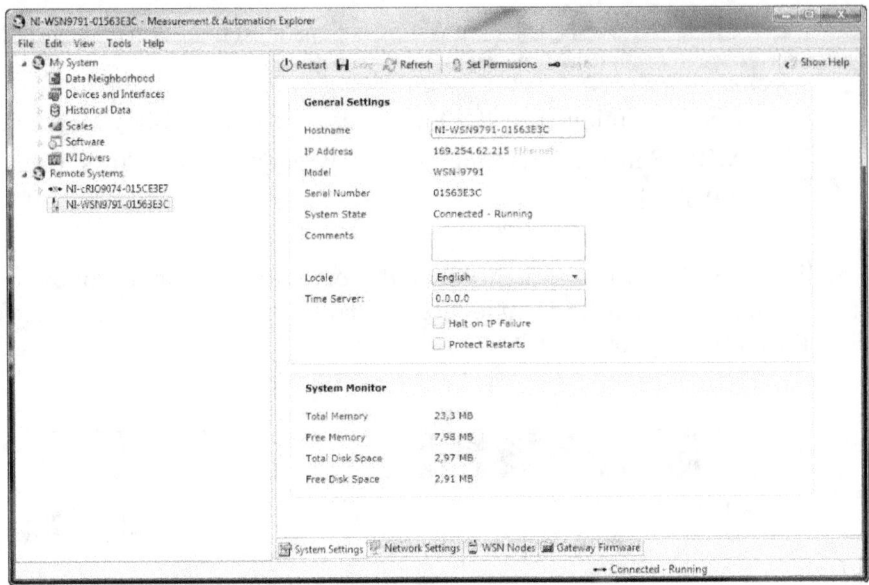

Figure 9-4 WSN-9791 detection by MAX

- Add the measurement node(s) to your wireless network

    o Select the "WSN Nodes" tab and click the "Add WSN Node" button or right-click the Gateway node.

    o Enter the type, serial number and ID number of the measurement node(s). Click "Apply" to save the changes.

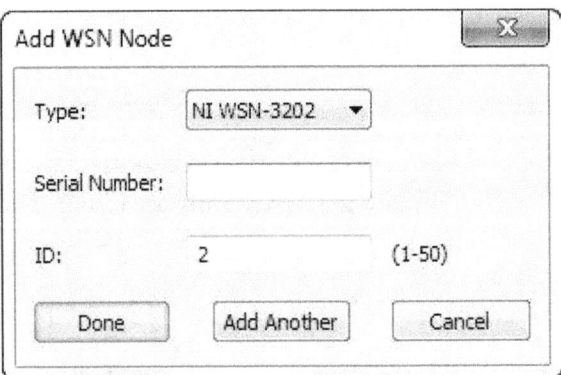

Figure 9-5 Assign an ID to a node

- Establish connection with the gateway

o To establish a connection with the gateway, press the "Signal Strength" button on each node for at least five seconds.

o Upon connection, Select Refresh all on the "WSN Nodes" tab to view the last communication time, battery state, link quality, and network mode of the measurement nodes.

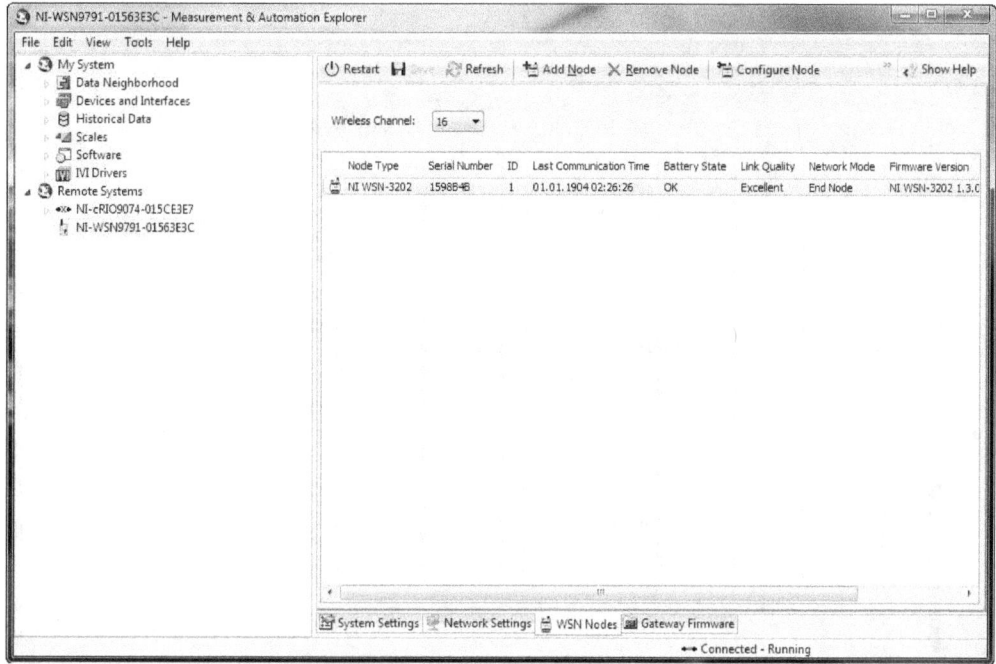

Figure 9-6 Information of the node

Step 3: Data Presentation

When the equipment is properly configured in MAX, LabVIEW can be used to present the temperature value on the computer screen.

- Start LabVIEW and create a New Project.

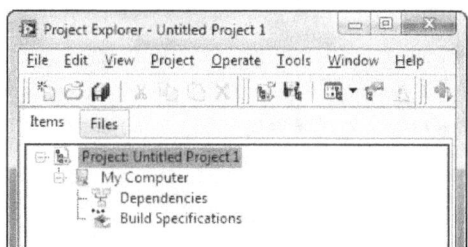

Create a new project in LabVIEW

- Add "WSN-9791" to the project by right-clicking on the project name and select "New » Targets and Devices". All modules will be listed with available I/O channels.

225

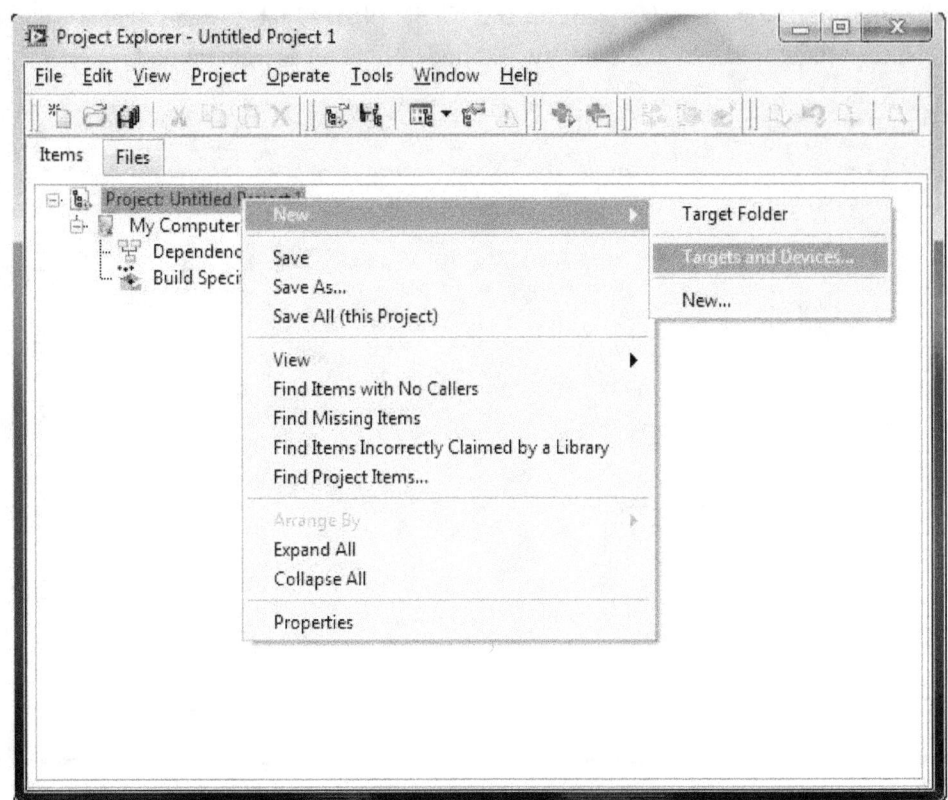

Add a node to the project

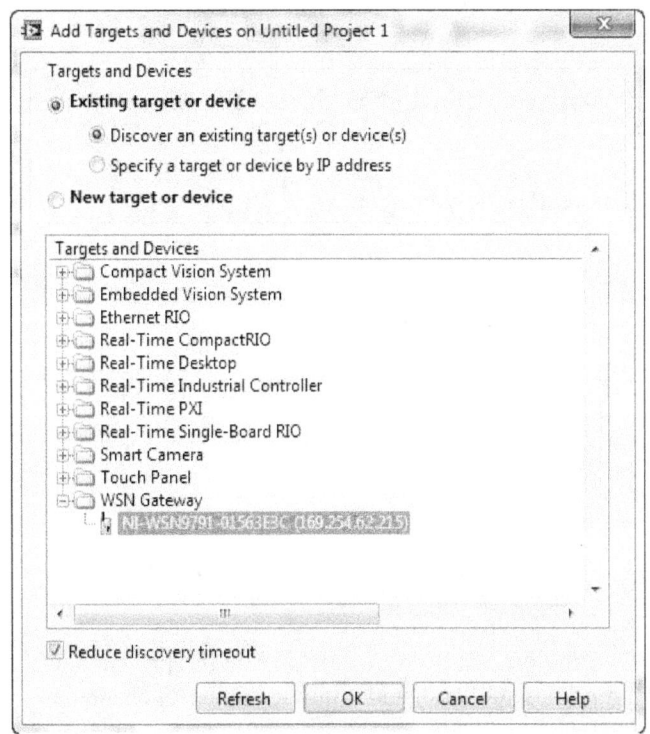

Device Selection

- In order to create a LabVIEW application, right-click on "My Computer" and select "New » VI" and **drag** the corresponding I/O variables from the Project Explorer to the LabVIEW block diagram, as shown on the Figure below.

- Build "VI" application and click Run to instantly acquire measurements from WSN.

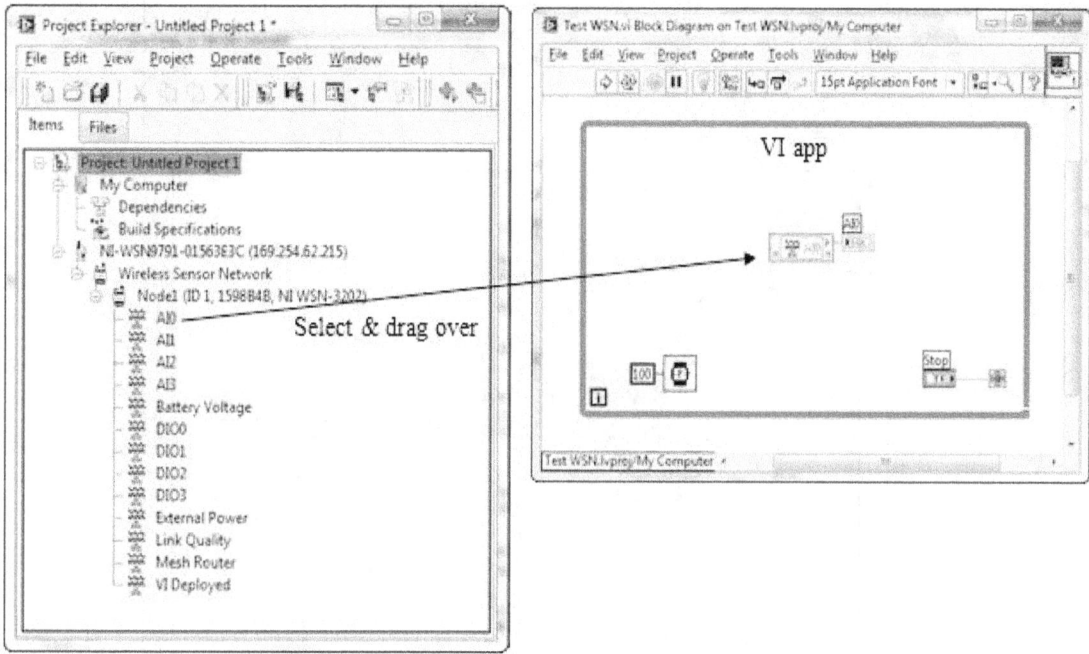

Figure 9-7 Select I/O port to be used in LabVIEW VI

Front panel of the LabVIEW program (VI app) reading from AI0 [Voltage 1-5]:

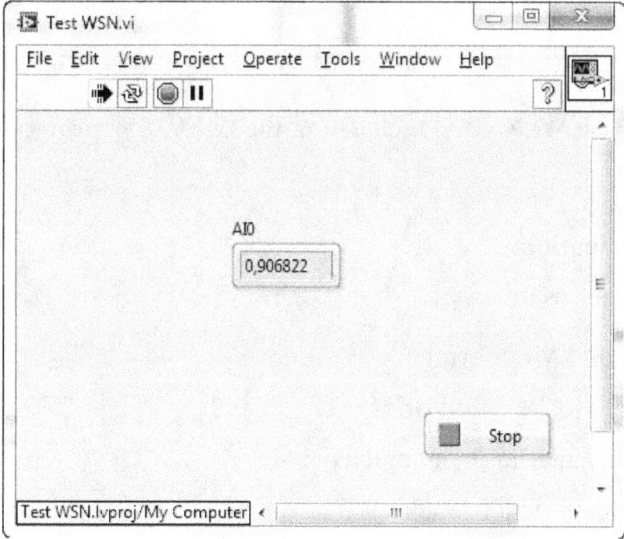

Successful data achievement

227

Advance: The front panel shown in the previous Figure present the value captured from the temperature sensor pt-100 in Voltage. The value needs to be converted into temperature [°C]. Try to improve the LabVIEW application that shows both in Voltage and degree Celsius as illustrated in the following Figures.

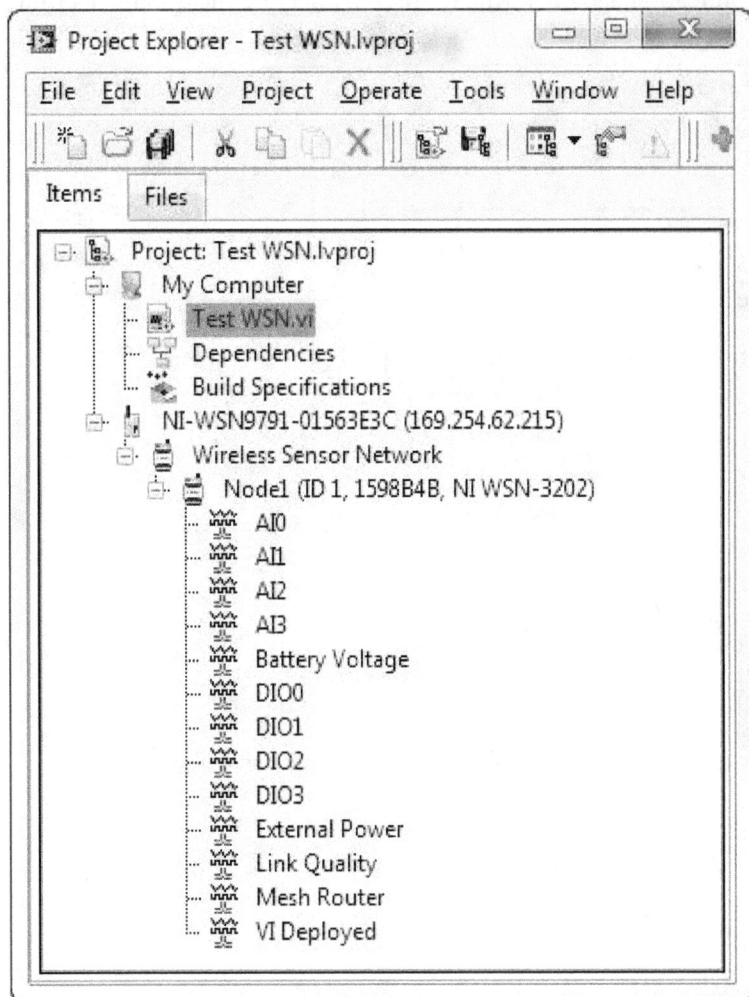

"Test WSN.vi" is included in the LabVIEW project

9.3 Some of ZigBee Applications

These applications are included as follows:

- Industrial Control and Monitoring

- Environmental and Health Monitoring

- Home Automation, Entertainment and Toys

- Security, Location and Asset Tracking

- Emergency and Disaster Response

- Military/Battlefield Applications

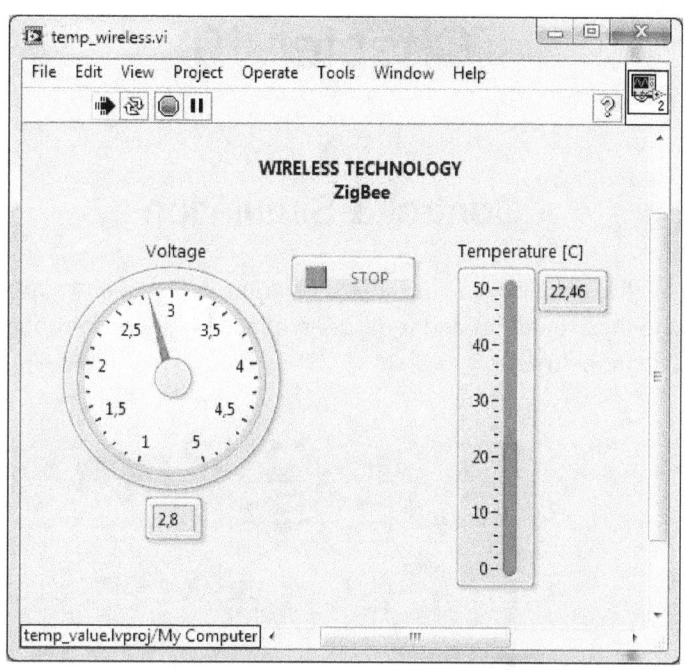

The Front panel displaying current Voltage and temperature data

The block diagram of the improved front panel is shown in the Figure below, where both of the parameters 12,5 are used to convert from Voltage to degree Celsius. **Note:** These parameters can only be used for this specific case, where 1-5 Volt are converted to 0-50 °C. For converting to a scale of 0-100 °C, just simply replace both of the parameters 12,5 with 25.

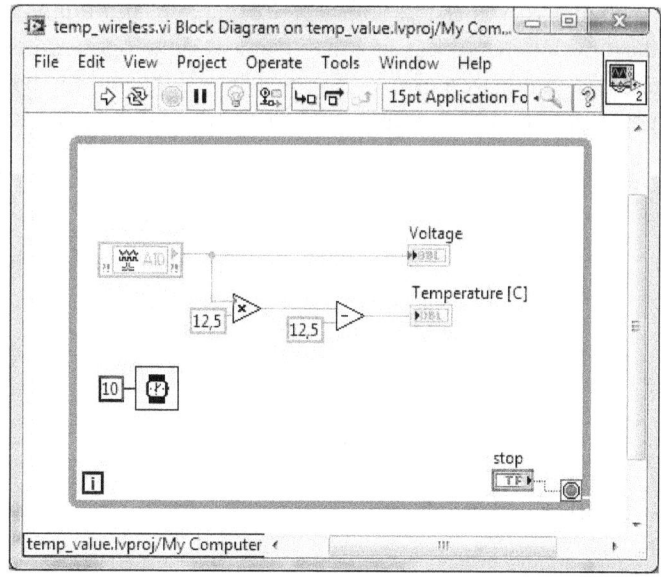

Voltage to temperature conversion

Chapter 10

Control & Simulation

This chapter will focus on the main aspects in LabVIEW control design and simulation modules and toolkits. All VIs related to these modules and toolkits are placed in the Control Design and Simulation palette as shown in the figure below:

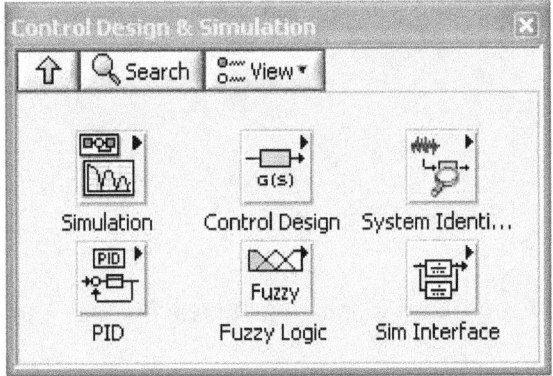

Control design is a process that involves developing mathematical models that describe a physical system, analyzing the models to learn about their dynamic characteristics, and creating a controller to achieve certain dynamic characteristics.

Simulation is a process that involves using software to recreate and analyze the behavior of dynamic systems, One can use the simulation process to lower product development costs by accelerating product development. Also, it uses the simulation process to provide insight into the behavior of dynamic systems you cannot replicate conveniently in the laboratory. Figure 10-1 has shown a closed-loop feedback control system:

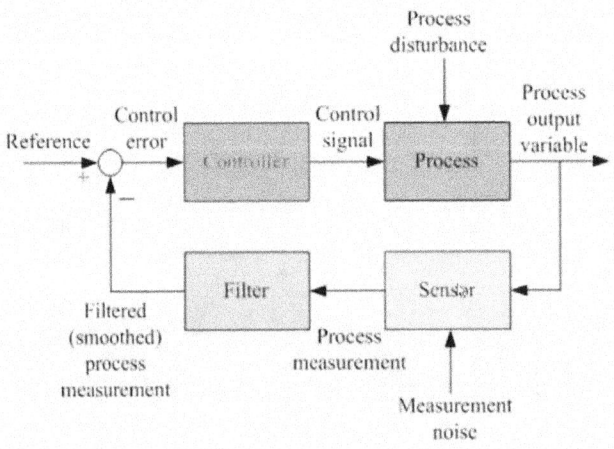

Figure 10-1

LabVIEW has several additional modules and Toolkits for Control and Simulation purposes, "**LabVIEW Control Design and Simulation Module**", "**LabVIEW PID and Fuzzy Logic Toolkit**", "**LabVIEW System Identification Toolkit**" and "LabVIEW Simulation Interface Toolkit". LabVIEW Math Script is also useful for Control Design and Simulation.

- LabVIEW Control Design and Simulation Module

- LabVIEW PID and Fuzzy Logic Toolkit

- LabVIEW System Identification Toolkit

- LabVIEW Simulation Interface Toolkit

10.1 LabVIEW Control Design and Simulation Module

With LabVIEW Control Design and Simulation Module you can construct plant and control models using transfer function, state-space, or zero-pole-gain. Analyze system performance with tools such as step response, pole-zero maps, and Bode plots. Simulate linear, nonlinear, and discrete systems with a wide option of solvers. With the NI LabVIEW Control Design and Simulation Module, you can analyze open-loop model behavior, design closed-loop controllers, simulate online and offline systems, and conduct physical implementations.

A) Simulation : The **Simulation** palette in LabVIEW:

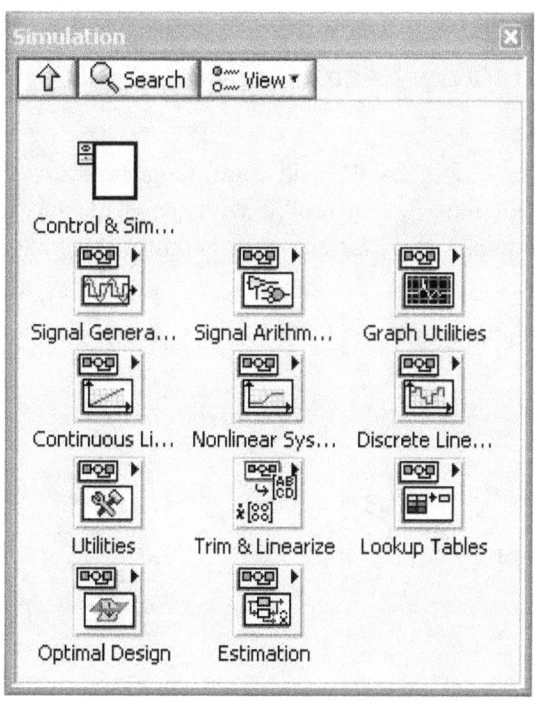

The main features in the Simulation palette are:

- **Control and Simulation Loop** - You must place all Simulation functions within a Control & Simulation Loop or in a simulation subsystem.

- **Continuous Linear Systems Functions** - Use the Continuous Linear Systems functions to represent continuous linear systems of differential equations on the simulation diagram.

- **Signal Arithmetic Functions** - Use the Signal Arithmetic functions to perform basic arithmetic operations on signals in a simulation system.

B) Control Design : The **Control Design** palette in LabVIEW:

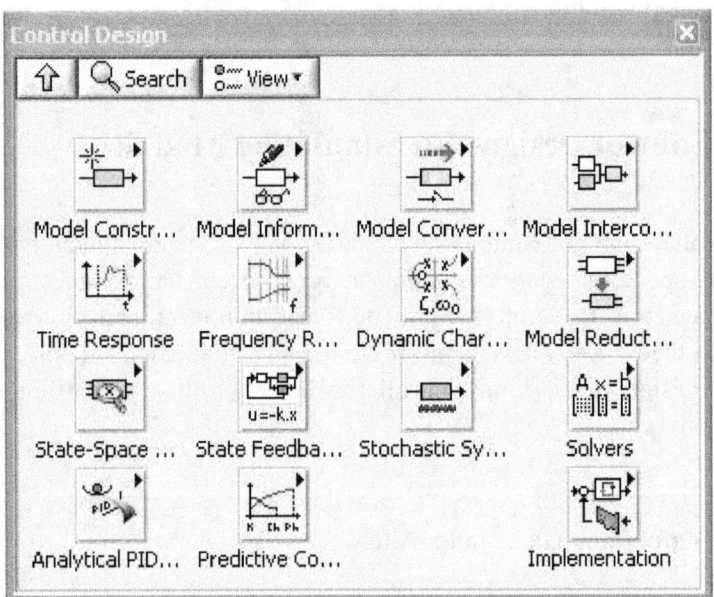

10.2 LabVIEW PID and Fuzzy Logic Toolkit

The NI LabVIEW PID and Fuzzy Logic Toolkit add control algorithms to LabVIEW. By combining the PID and fuzzy logic control functions in this toolkit with the math and logic functions in LabVIEW software, you can quickly develop programs for automated control. You may integrate these control tools with the power of data acquisition.

1. PID Control : The **PID** palette in LabVIEW:

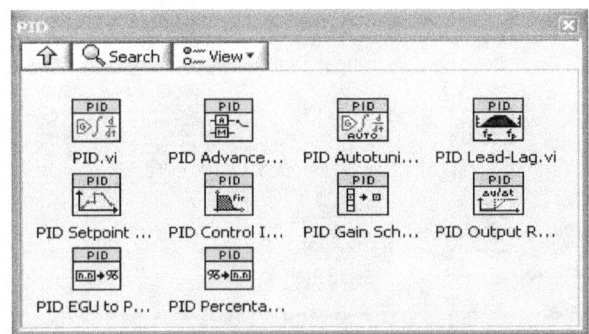

2. Fuzzy Logic : The **Fuzzy Logic** palette in LabVIEW:

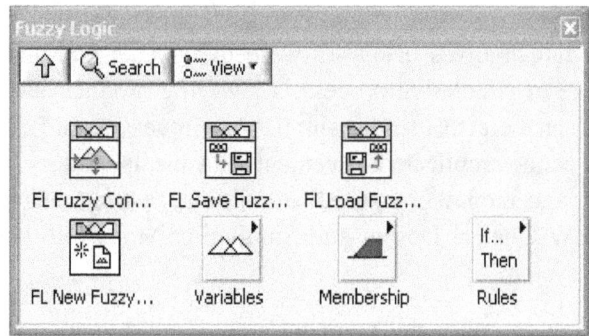

10.3 LabVIEW System Identification Toolkit

The "LabVIEW System Identification Toolkit" combines data acquisition tools with system identification algorithms for plant modeling. You can use the LabVIEW System Identification Toolkit to find empirical models from real plant stimulus-response information.

The **System Identification** palette in LabVIEW:

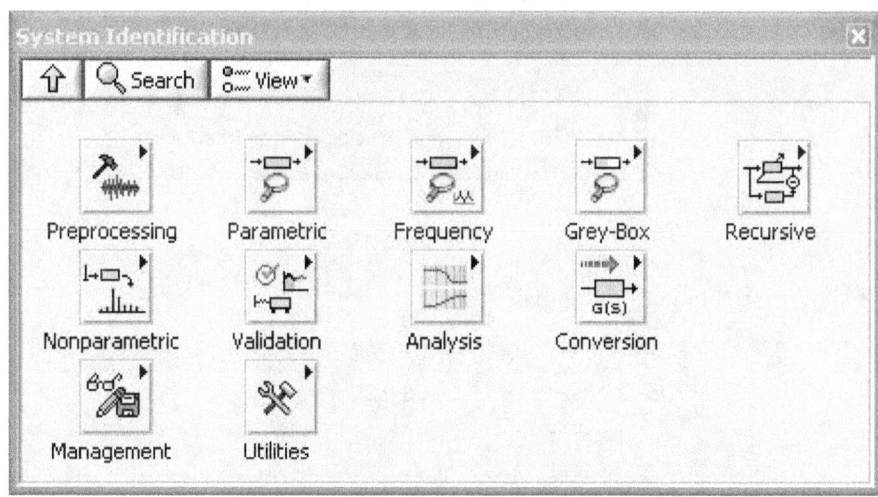

10.4 LabVIEW Simulation Interface Toolkit

The "LabVIEW Simulation Interface Toolkit" gives control system design and test engineers a link between the LabVIEW graphical development environment and Simulink software from The MathWorks. With the LabVIEW Simulation Interface Toolkit, you can easily build custom LabVIEW user interfaces to view and control your simulation model during run time. The "LabVIEW Simulation Interface Toolkit" will not be further investigated in this chapter.

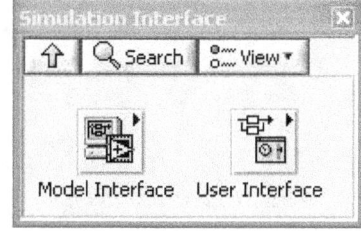

10.5 Simulation in LabVIEW

Simulation is a process that involves using software to recreate and analyze the behavior of dynamic systems, besides the usage of simulation process is lowering product development costs by accelerating product development. It can be also use the simulation process to provide insight into the behavior of dynamic systems which cannot replicate conveniently in the laboratory. For example, simulating a jet engine saves time, labor, and money compared to building, testing, and rebuilding an actual jet engine. One can use the LabVIEW Control Design and Simulation Module to simulate a dynamic system or a component of a dynamic system.

For example, you can simulate only the plant while using hardware for the controller, actuators, and sensors (Hardware-in-the-loop Simulation). A dynamic system model is a differential or difference equation that describes the behavior of the dynamic system. Use the Simulation VIs and functions to create simulation applications in LabVIEW. In the Control Design & Simulation palette we have the **Simulation** Sub palette, see the following figure 10-2:

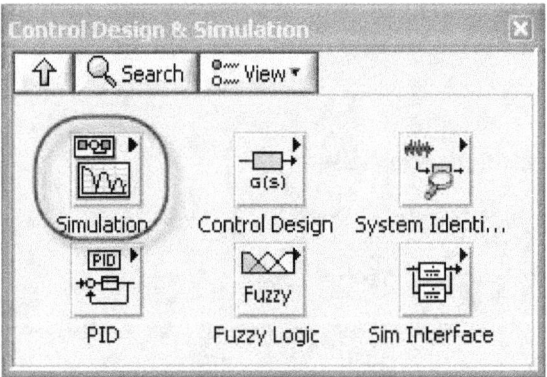

Figure 10-2

Figure 10-3 shows the Simulation Sub palette:

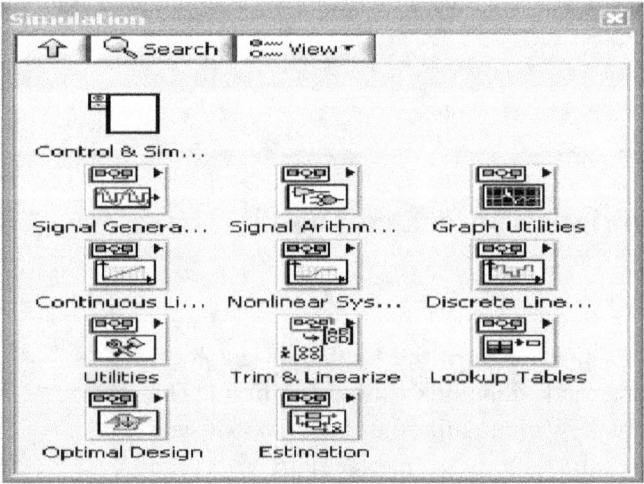

Figure 10-3

All the "Blocks" in the Simulation palette are not Sub VIs, i.e., we cannot double-click on them and open the Block Diagram because they have none. All the Blocks in the Simulation palette must be used <u>inside</u> the Control and Simulation Loop.

In the "**Simulation**" Sub palette we have the "Control and Simulation Loop" which is very useful in simulations:

It must place all Simulation functions within a Control & Simulation Loop or in a simulation subsystem. Also, it can be place simulation subsystems within a Control & Simulation Loop or another simulation subsystem, or it can be place simulation subsystems on a block diagram outside a Control & Simulation Loop or run the simulation subsystems as stand-alone VIs.

The Control & Simulation Loop has an Input Node (upper left corner) and an Output Node (upper right corner). Use the Input Node to configure simulation parameters programmatically. It can be configure these parameters interactively using the Configure Simulation Parameters dialog box. Access this dialog box by double-clicking the Input Node or by right-clicking the border and selecting Configure Simulation Parameters from the shortcut menu.

> **Configuration:**

When you place these blocks on the diagram you may double-click or right-click and then select

"Configuration…"

Example 1: Configuration Dialog box

For the "Transfer Function" (Simulation → Continuous Linear Systems) block we have the following Configuration window as seen the figure 10-4:

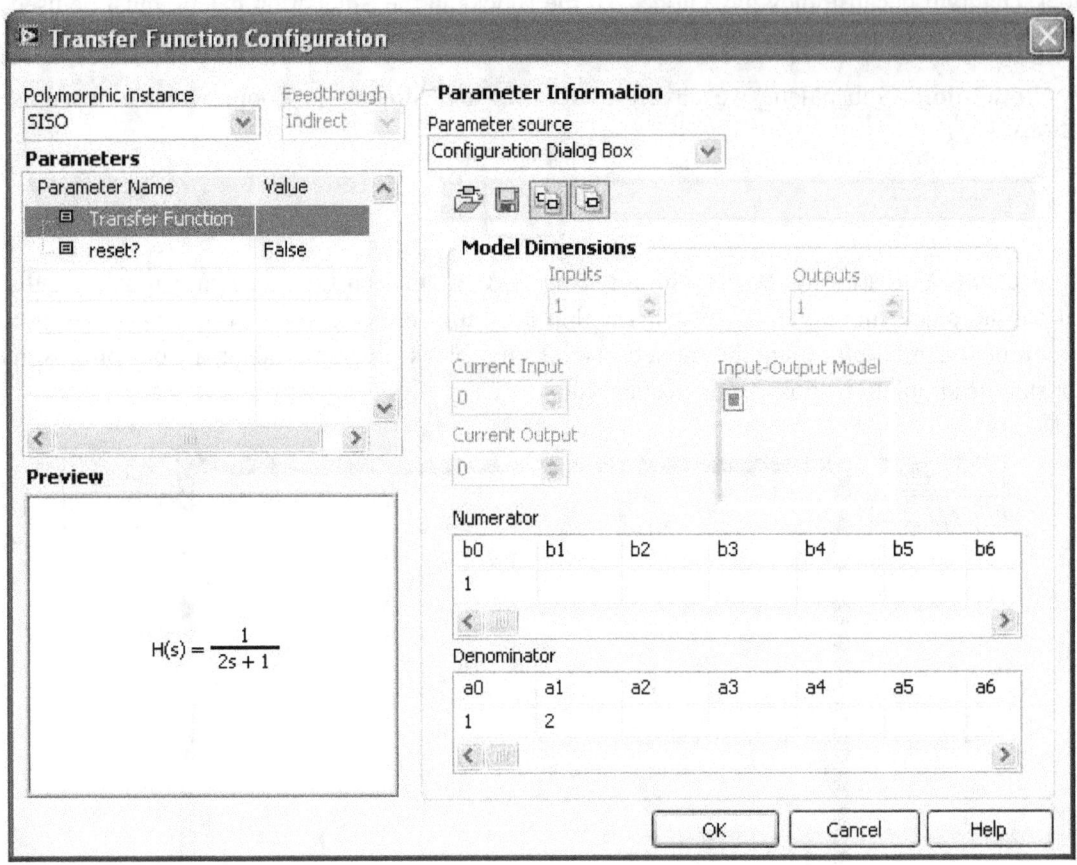

Figure 10-4

All the different blocks have their own different Configuration window.

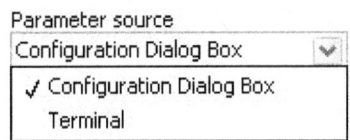

In the Parameter source you may select between:

- Configuration Dialog Box
- Terminal

If you select "Configuration Dialog Box" you enter the configuration in the Configuration window like we see above, while if you select "Terminal" that specific configuration is set from the Block Diagram like this:

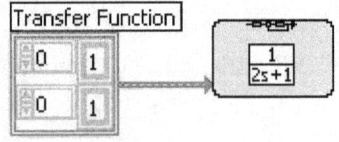

236

➢ **Icon Style:**

When you place the block on the block diagram you may select how that should appear. Right-click on the block/icon and select "Icon Style":

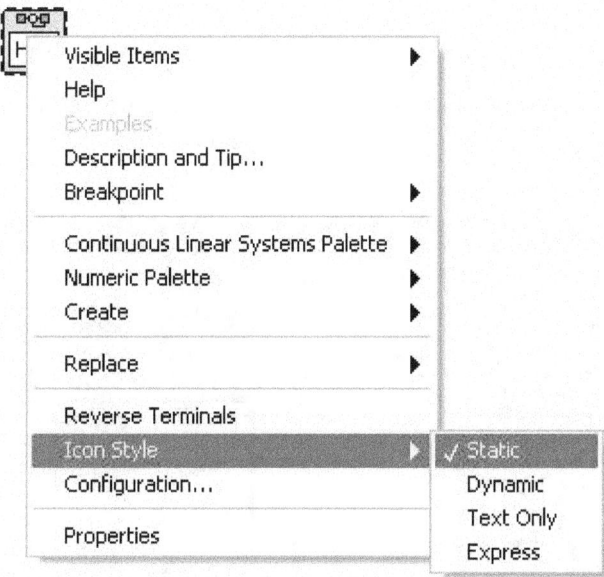

Example 2: Icon Style

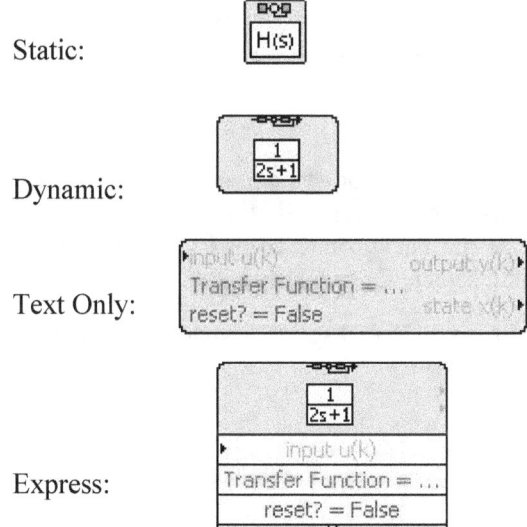 For the "Transfer Function" (Simulation → Continuous Linear Systems) block we have the following different icon styles:

It has seen for the Dynamic and Express styles that the appearance changes according to configuration parameters we set. Most of users may preferred the "static" icon style because it does not require lots of space on the diagram.

10.6 Simulation Sub system

It may create a **Simulation Subsystem** (File → New...) as shown in figure 10-5:

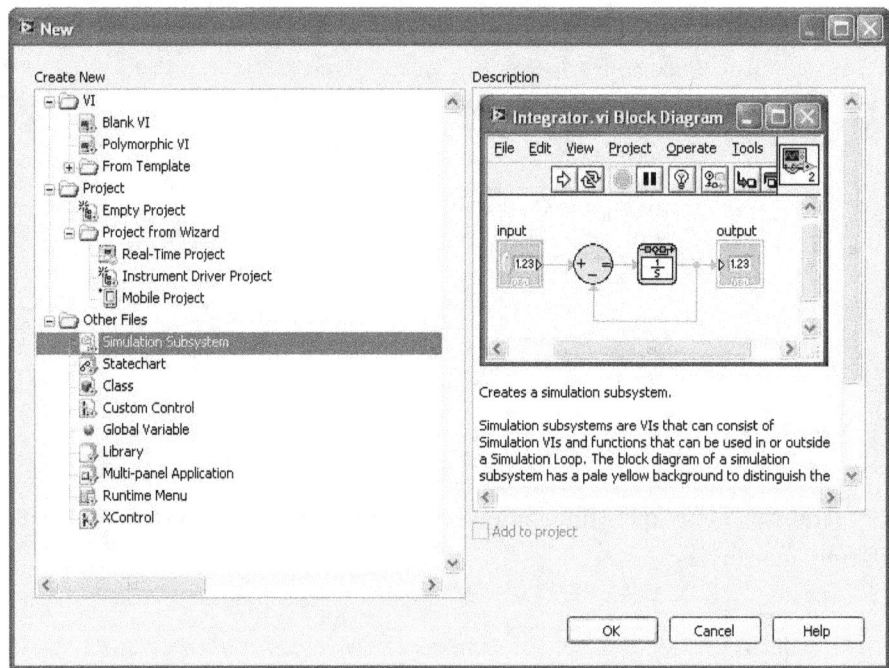

Figure 10-5

The Simulation Subsystem is very useful when dealing with larger simulation systems in order to create a more structured code. I recommend that you (always) use this feature. The Simulation Subsystem is almost equal to a normal LabVIEW Block Diagram but notice the background color is slightly darker. Note: In order to open the Simulation Subsystem, right-click and select "Open Subsystem".

The Simulation Subsystem may also be represented by different icons. If you select "dynamic" icon style, you will see a "miniature" version of the subsystem like the Figure 10-6:

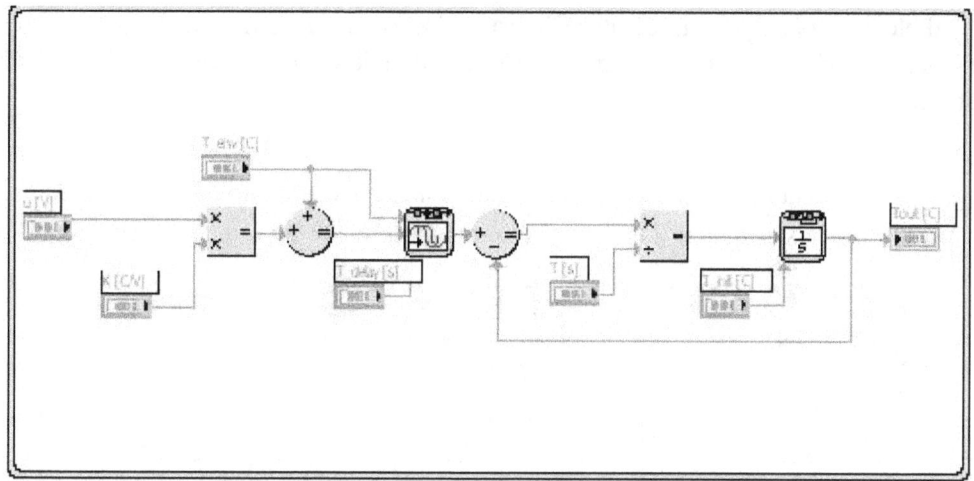

Figure 10-6

 It may drag in the corner in order to increase or decrease the dynamic icon. If you select "static" icon style you see the icon you created with the Icon Editor.

Like this icon :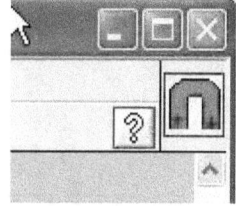

10.7 Continuous Linear Systems

In the "**Continuous Linear Systems**" Sub palette we want to create a simulation model:

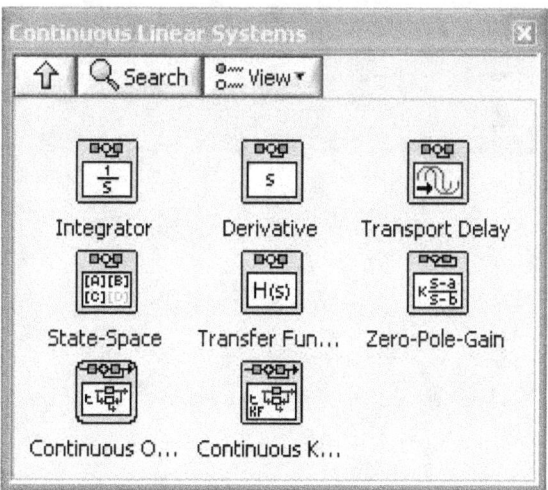

The most used blocks probably are Integrator, Transport Delay, State-Space and Transfer Function. When you place these blocks on the diagram you may double-click or right-click and then select "Configuration..."

Integrator - Integrates a continuous input signal using the ordinary differential equation (ODE) solver you specify for the simulation. The Configuration window for the Integrator block looks like this:

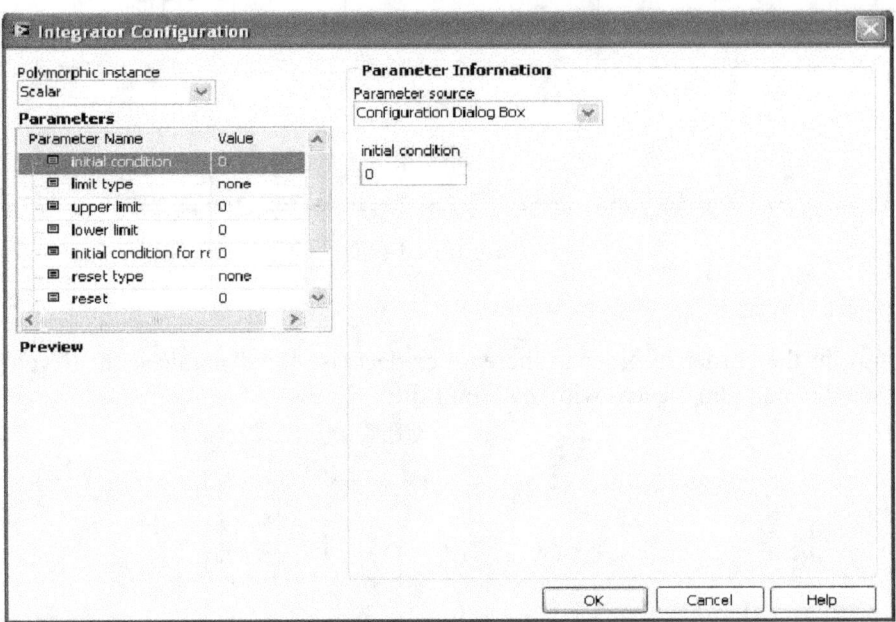

Transport Delay - Delays the input signal by the amount of time you specify.

The Configuration window for the Transport Delay block looks like the following figure:

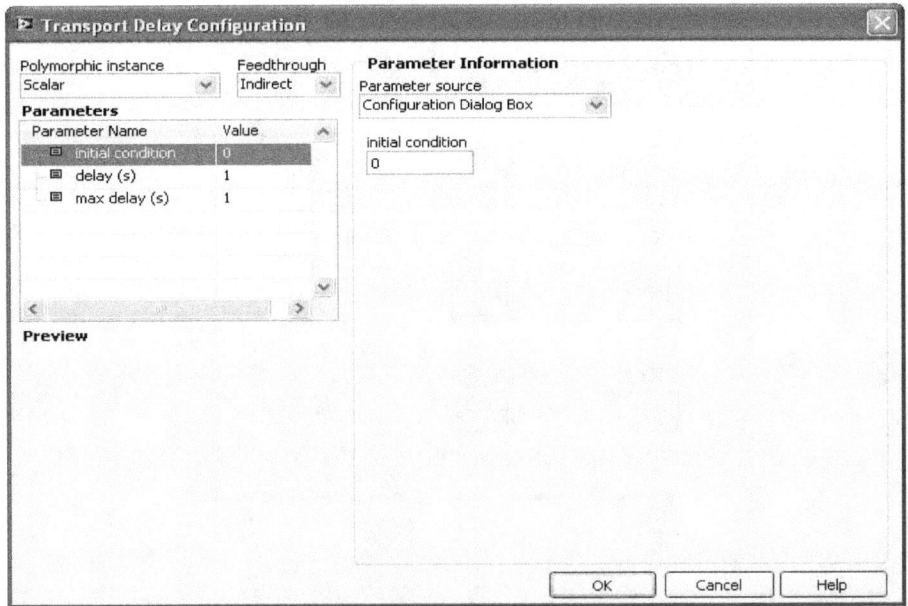

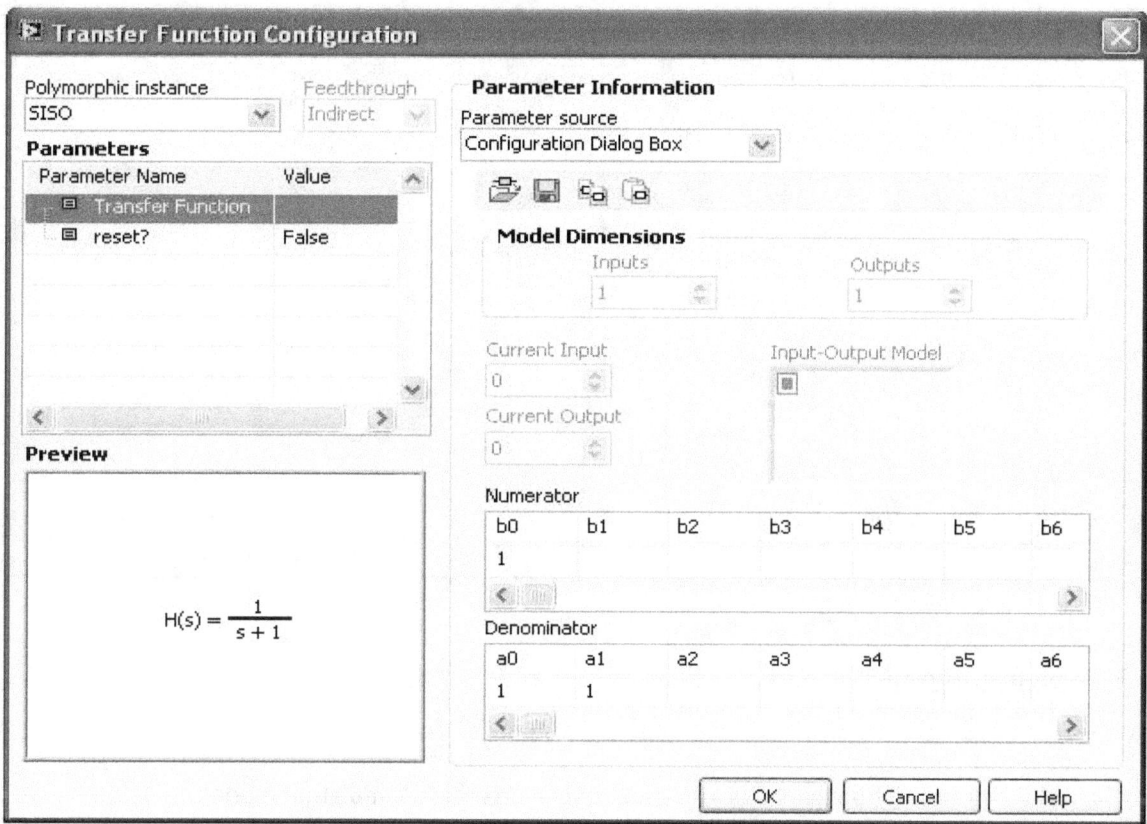 **Transfer Function** - Implements a system model in transfer function form. You define the system model by specifying the Numerator and Denominator of the transfer function equation.

The Configuration window for the Transfer Function block, see the figure below:

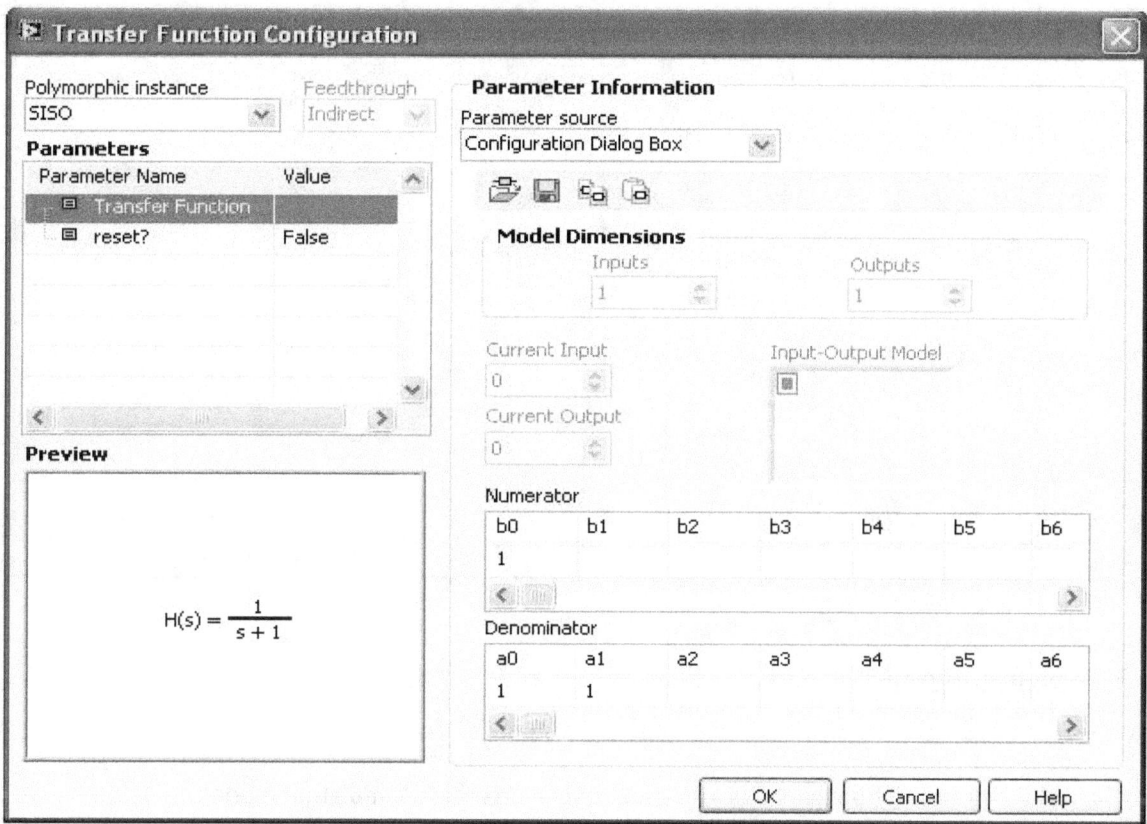

State-Space - Implements a system model in state-space form. You define the system model by specifying the input, output, state, and direct transmission matrices.

The Configuration window for the State-Space block looks like the figure below:

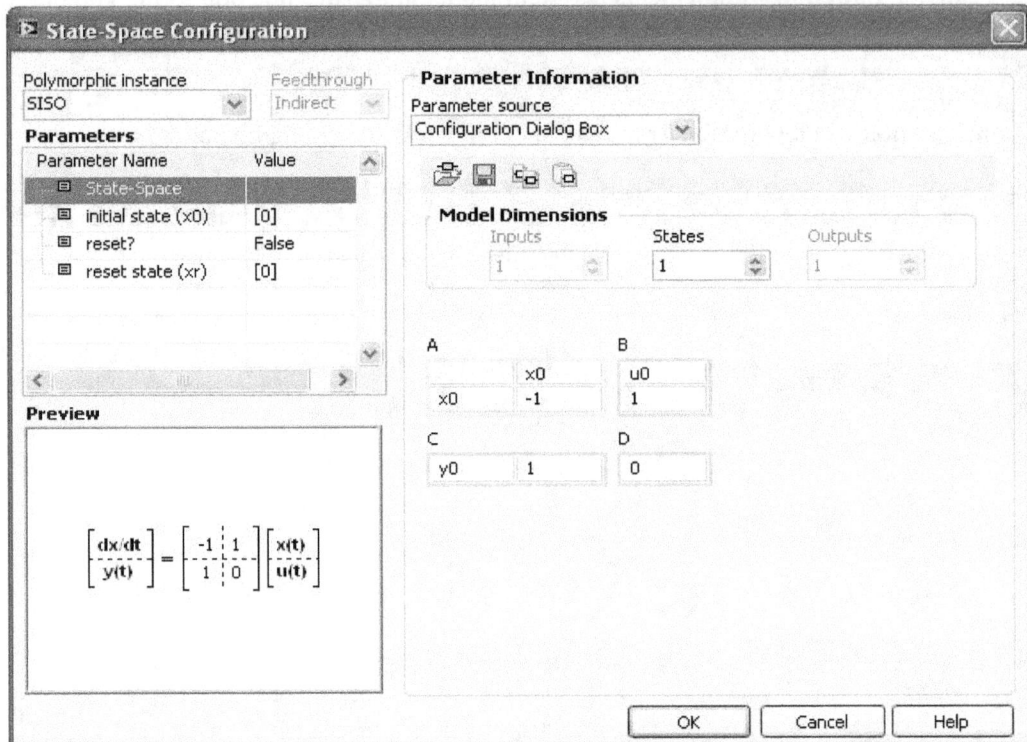

> **Signal Arithmetic:**

The "**Signal Arithmetic**" Sub palette is also useful when creating a simulation model:

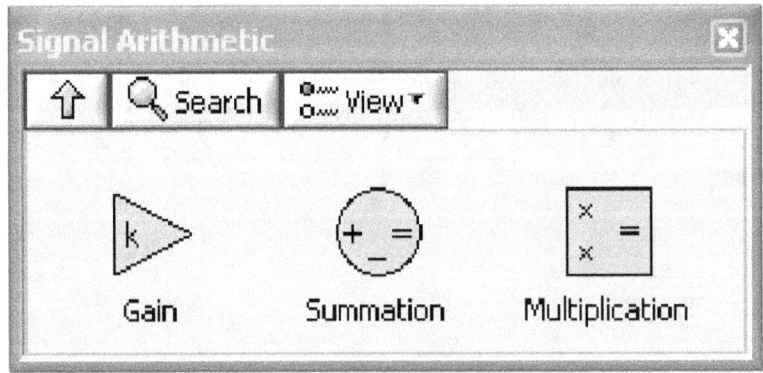

Example 3: Simulation Model

Below we've seen an example of a simulation model created in LabVIEW.

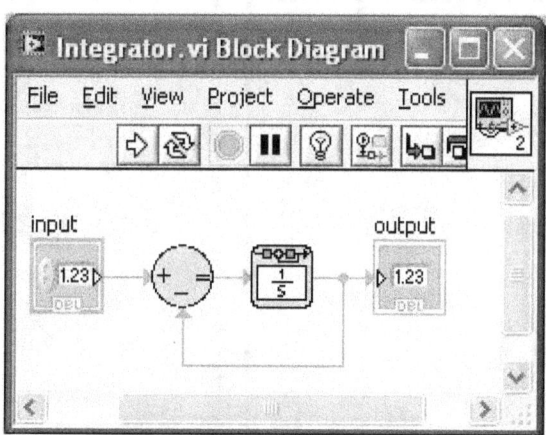

Example 4: Simulation

Below we've seen an example of a simulation model using the Control and Simulation Loop.

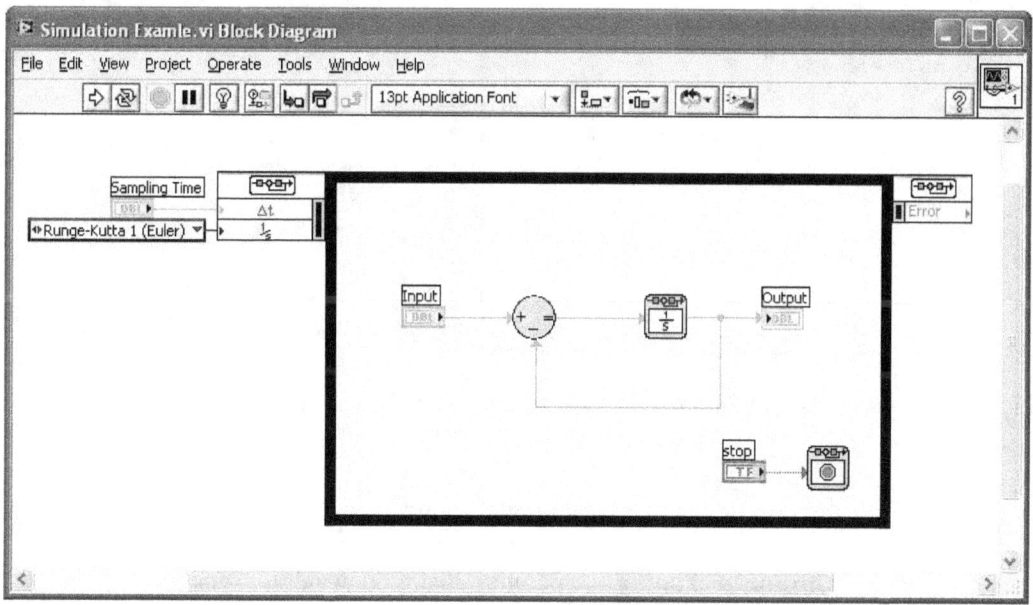

Notice the following:

Click on the border of the simulation loop and select

"Configure Simulation Parameters…"

The following window appears (Configure Simulation Parameters):

The two dialog windows "Configure Simulation Parameters" are shown side by side.

Left window (Simulation Parameters tab):

Simulation Time
Initial Time (s): 0
Final Time (s): Inf

Solver Method
ODE Solver: Runge-Kutta 1 (Euler)   □ Nan/Inf Check

Continuous Time Step and Tolerance
Step Size (s): 0,1
Minimum Step Size (s): 1E-10   Maximum Step Size (s): 1
Relative Tolerance: 0,001   Absolute Tolerance: 1E-7

Discrete Time Step
Discrete Step Size (s): 0,1   ☑ Auto Discrete Time

OK   Cancel   Help

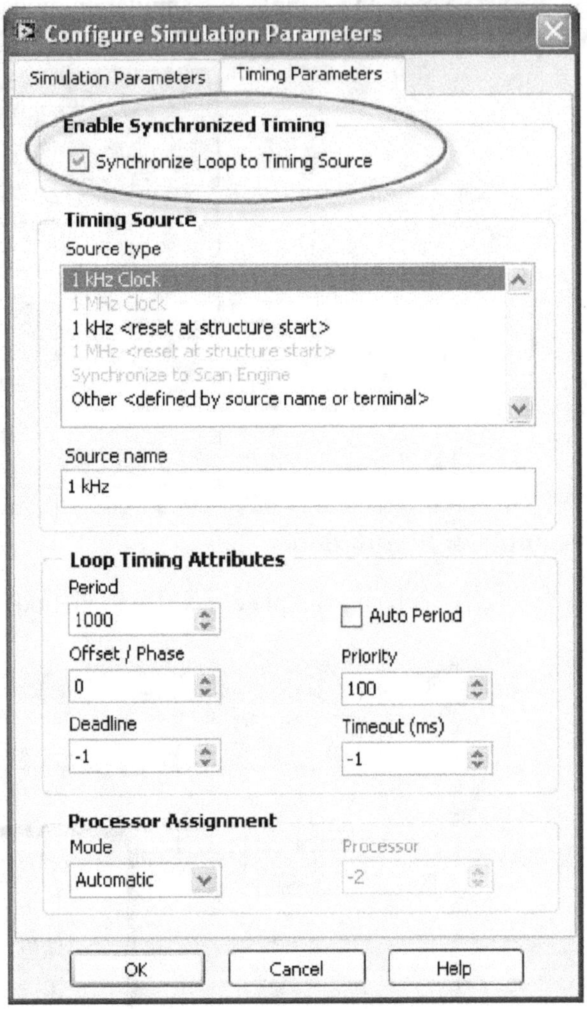

Right window (Timing Parameters tab):

Enable Synchronized Timing
☑ Synchronize Loop to Timing Source

Timing Source
Source type
1 kHz Clock
1 MHz Clock
1 kHz <reset at structure start>
1 MHz <reset at structure start>
Synchronize to Scan Engine
Other <defined by source name or terminal>

Source name
1 kHz

Loop Timing Attributes
Period: 1000   □ Auto Period
Offset / Phase: 0   Priority: 100
Deadline: -1   Timeout (ms): -1

Processor Assignment
Mode: Automatic   Processor: -2

OK   Cancel   Help

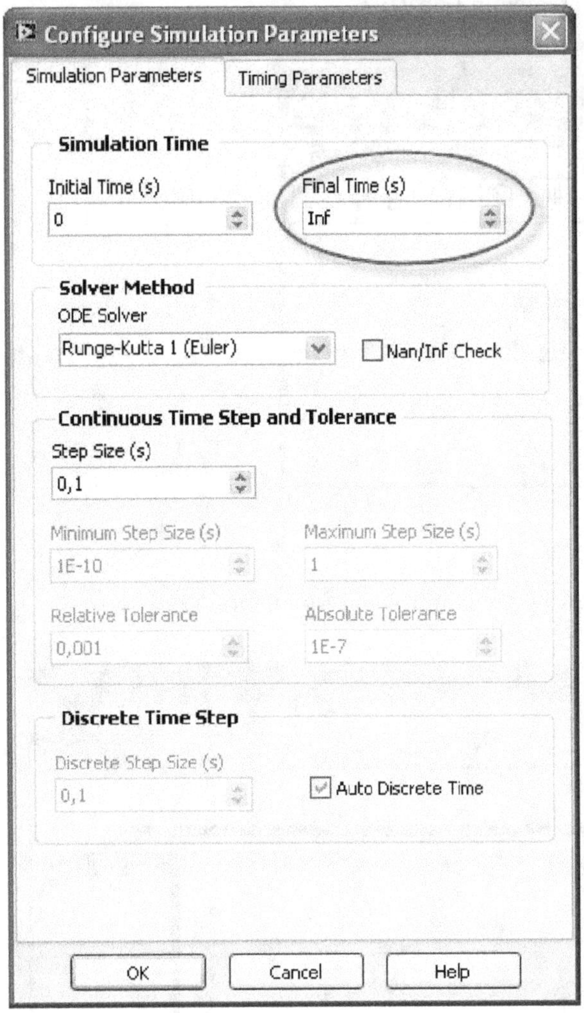

In this window you set some Parameters regarding the simulation, some important are:

- **Final Time (s)** – set how long the simulation should last. For an infinite time set "**Inf**".

- **Enable Synchronized Timing** - Specifies that you want to synchronize the timing of the Control & Simulation Loop to a timing source. To enable synchronization, place a checkmark in this checkbox and then choose a timing source from the Source type list box.

It may also set some of these Parameters in the Block Diagram:

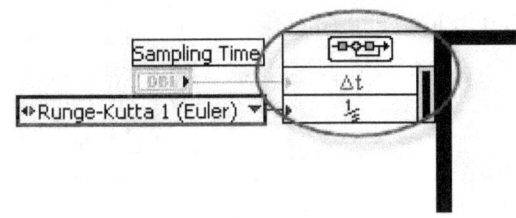

By using the mouse to increase the numbers of Parameters and right-click and select "Select Input".

Exercise: *Simulation of a spring-mass damper system*

In this exercise you will construct a simulation diagram that represents the behavior of a dynamic system, and it will simulate a spring-mass damper system. Where t is : the simulation time, F(t) is an external force applied to the system, c is the damping constant of the spring, k is the stiffness of the spring, m is a mass, and x(t) is the position of the mass. The following Figure shows this dynamic system:

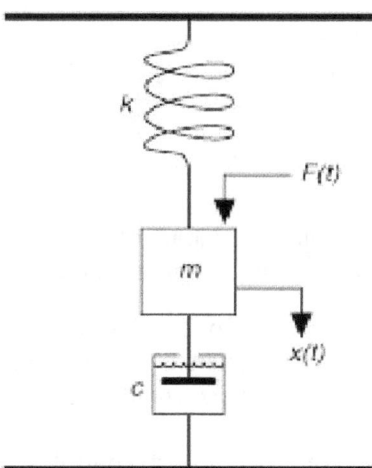

The goal is to view the position x(t) of the mass m with respect to time t. one can calculate the position by integrating the velocity of the mass, and it can be calculate the velocity by integrating the acceleration of the mass. If you know the force and mass, you can calculate this acceleration by using Newton's Second Law of Motion, given by the following equation:

Force = Mass × Acceleration

Therefore; Acceleration = Force / Mass

Substituting terms from the differential equation above yields the following equation: You will construct a simulation diagram that iterates the following steps over a period of time.

Creating the Simulation Diagram

You create a simulation diagram by placing a Control & Simulation Loop on the LabVIEW block diagram.

- Launch LabVIEW and select File» New VI to create a new, blank VI.

- Select Window» Show Block Diagram to view the block diagram.

245

- It can be press the <Ctrl-E> keys to view the block diagram.

- If you are not already viewing the Functions palette, select View» Functions Palette to display this palette.
- Select Control Design & Simulation» Simulation to view the Simulation palette.
- Click the Control & Simulation Loop icon.

- Move the cursor over the block diagram. Click to place the top left corner of the loop, drag the cursor diagonally to establish the size of the loop, and click again to place the loop on the block diagram.

The simulation diagram is the area enclosed by the Control & Simulation Loop. Notice the simulation diagram has a pale yellow background to distinguish it from the rest of the block diagram. You can resize the Control & Simulation Loop by dragging its borders.

Configuring Simulation Parameters

The Control & Simulation Loop contains the parameters that define how the simulation executes. Complete the following steps to view and configure these simulation parameters.

- Double-click the Input Node, attached to the left side of the Control & Simulation Loop, to display the Configure Simulation Parameters dialog box. You also can right-click the loop border and select Configure Simulation Parameters from the shortcut menu.

- Ensure the value of the **Final Time (s)** numeric control is 10, which specifies that this tutorial simulates ten seconds of time.

- Click the ODE Solver pull-down menu to view the list of ODE solvers the Control Design and Simulation Module includes. If the term (variable) appears next to an ODE solver, that solver has a variable step size. The other ODE solvers have a fixed step size. Ensure a checkmark is beside the default ODE solver **Runge-Kutta 23 (variable)**.

- Because this ODE solver is a variable step-size solver, you can specify the **Minimum Step Size**

    o and **Maximum Step Size (s)** this ODE solver can take. Enter 0.01 in the Maximum Step Size

    o numeric control to limit the size of the time step this ODE solver can take.

- Click the Timing Parameters tab to access parameters that control how often the simulation executes.

- Ensure the Synchronize Loop to Timing Source checkbox does not contain a checkmark. This option specifies that the simulation executes without any timing restrictions. Use this option when you want the simulation to run as fast as possible. If you are running this simulation in real-time, you can place a checkmark in this checkbox and configure how often the simulation executes.

- Click the OK button to save changes and return to the simulation diagram.

Building the Simulation

The next step is to build the simulation by placing Simulation functions on the simulation diagram and wiring these functions together. Note that you can place most Simulation functions only on the simulation diagram, that is, you cannot place Simulation functions on a LabVIEW block diagram. Complete the following steps to build the simulation of this dynamic system.

Placing Functions on the Simulation Diagram

- Open the Simulation palette.

- Select the **Signal Arithmetic** palette and place a **Multiplication** function on the simulation diagram. You will use this function to divide the force by the mass to calculate the acceleration.

- Double-click the Multiplication function to display the Multiplication Configuration dialog box. You can double-click most Simulation functions to view and change the parameters of that function.

- The function currently displays two symbols on the left side of the dialog box. This setting specifies that both incoming signals are multiplied together. Click the bottom symbol to change it to a symbol. This Multiplication function now divides the top signal by the bottom signal.

- Click the OK button to save changes and return to the simulation diagram.

- Right-click the Multiplication function and select Visible Items» Label from the shortcut menu. Double-click the Multiplication label and enter Calculate Acceleration as the new label.

- Return to the Simulation palette and select the **Continuous Linear Systems** palette.

- Place an Integrator function on the simulation diagram. You will use this function to calculate velocity by integrating acceleration.

- Label this Integrator function Calculate Velocity.

- Press the <Ctrl> key and click and drag the Integrator function to another location on the simulation diagram. This action creates a copy of the Integrator function, which you will use to calculate position by integrating velocity. Label this new Integrator function Calculate Position.

- Select the Graph Utilities palette and place two **SimTime Waveform** functions on the simulation diagram. You will use these functions to view the results of the simulation over time.

- Each SimTime Waveform function has an associated Waveform Chart. Label the first waveform chart Velocity and the second waveform chart Position.

- Arrange the functions to look like the following simulation diagram.

- Save this VI by selecting File» Save. Save this VI to a convenient location as "Spring-Mass Damper Example.vi".

The Block Diagram should now look like the figure below:

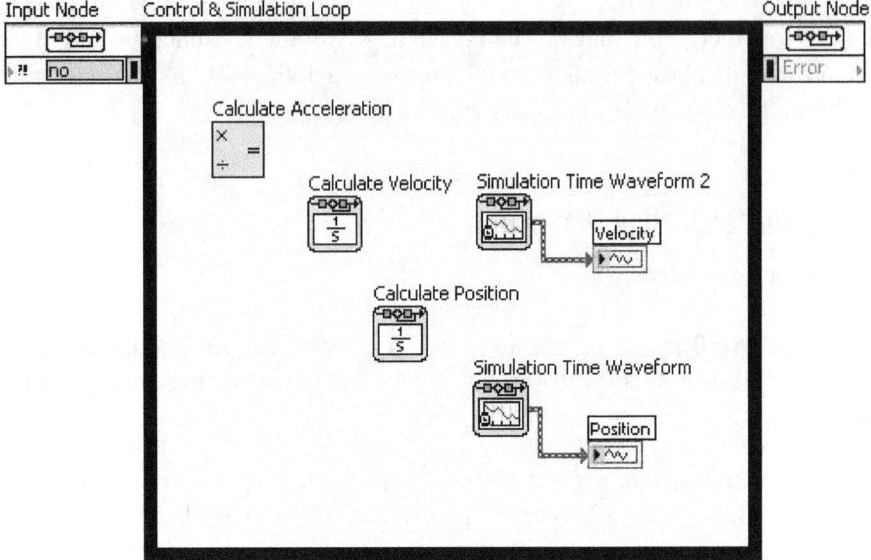

Wiring the Simulation Functions Together

The next step is wiring the functions together to represent the flow of data from one function to another.

Note: Wires on the simulation diagram include arrows that show the direction of the dataflow, whereas wires on a LabVIEW block diagram do not show these arrows. Complete the following steps to wire these functions together.

- Right-click the Operand1 input of the Calculate Acceleration function and select Create» Control from the shortcut menu to add a numeric control to the front panel window.

- Label this control Force.

- Double-click this control on the simulation diagram. LabVIEW displays the front panel and highlights the Force control.

- Display the block diagram and create a control for the Operand2 input of the Calculate Acceleration function. Label this new control Mass.

- Wire the Result output of the Calculate Acceleration function to the input of the Calculate Velocity function.

- Wire the output of the Calculate Velocity function to the input of the Calculate Position function.

- Right-click the wire you just created and select Create Wire Branch from the shortcut menu. Wire this branch to the Value input of the SimTime Waveform function that has the Velocity waveform chart.

- Wire the output of the Calculate Position function to the Value input of the SimTime Waveform function that has the Position waveform chart.

248

The Block Diagram should now look like the figure below:

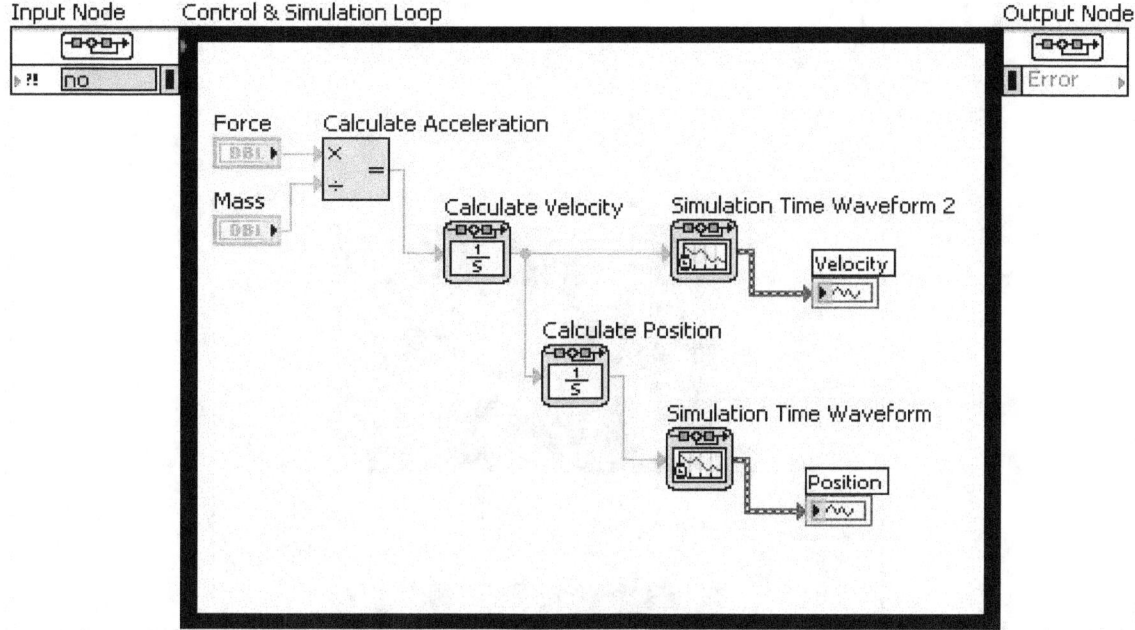

Running the Simulation

You now can run this simulation to test that the data is flowing properly through the Simulation functions.

Complete the following steps to run this simulation.

- Select Window» Show Front Panel, or press <Ctrl-E>, to view the front panel of this simulation. The front panel displays the following objects: a control labeled Force, a control labeled Mass, a waveform chart labeled Velocity, and a waveform chart labeled Position.

- If necessary, rearrange these controls and indicators so that all objects are visible.

- Enter -9.8 in the Force numeric control. This value represents the force of gravity, 9.8 meters per second squared, acting on the dynamic system.

- Enter 1 in the Mass numeric control. This value represents a mass of one kilogram.

- Click the Run button, or press the <Ctrl-R> keys, to run the VI.

The Front Panel should look like the figure 10-7:

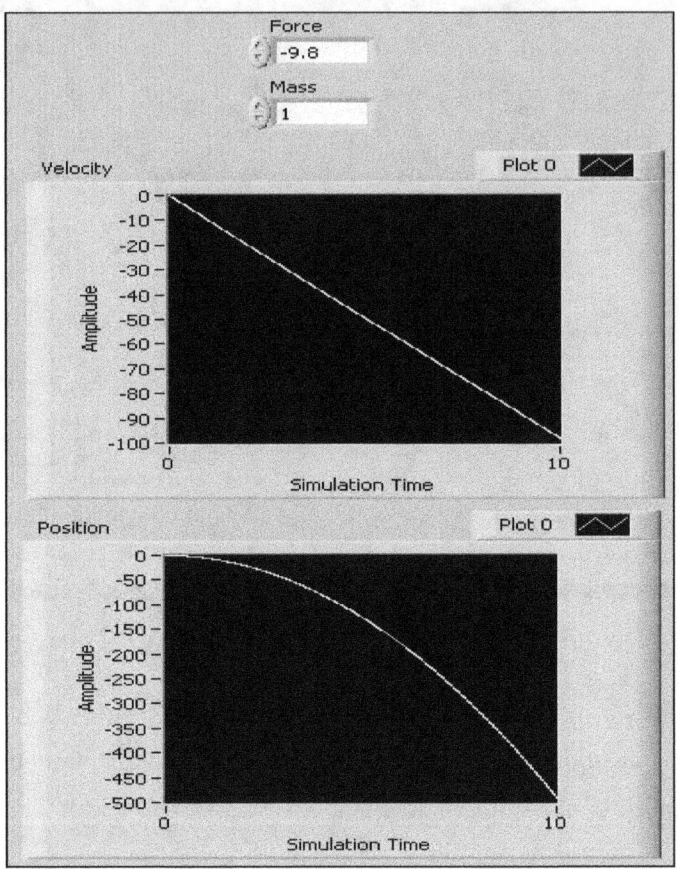

Figure 10-7

In the Figure above notice that the force of gravity causes the mass position and velocity to constantly decrease. However, in the real world, a mass attached to a spring oscillates up and down. This simulated spring does not oscillate because the simulation diagram does not represent damping or stiffness. This represents these factors to have a complete simulation of the dynamic system.

Representing Damping and Stiffness

Representing damping and stiffness involves feeding back the velocity and position, each multiplied by a different constant, to the input of the Calculate Acceleration function. Recall the following differential equation this VI simulates.

In the previous representation, multiply the stiffness constant k by the mass position x(t) then subtract these quantities from the external force applied to the mass. Complete the following steps to represent damping and stiffness in this dynamic system model.

250

- View the simulation diagram.

- Select the Signal Arithmetic palette and place a **Summation** function on the simulation diagram. Move this function to the left of the Force and Mass controls.

- Double-click the Summation function to configure its operation. By default, the Summation function displays the following three input terminals. This configuration subtracts one input signal from another.

- Click the symbol twice to change this terminal to the another symbol. This Summation function now subtracts the top and bottom input signals from the left input signal.

- Click the OK button to save changes and return to the simulation diagram.

- Select the Signal Arithmetic palette and place a Gain function on the simulation diagram. Move this function above the existing simulation diagram code but still within the Control & Simulation Loop.

- The input of the Gain function is on the left side of the function, and the output is on the right side. You can reverse the direction of these terminals to indicate feedback better. Right-click the Gain function and select Reverse Terminals from the shortcut menu. The Gain function now points toward the left side of the simulation diagram.

- Label this Gain function Damping.

- Press the <Ctrl> key and drag the Gain function to create a separate copy. Move this copy below the existing simulation diagram code but still within the Control & Simulation Loop. Label this function Stiffness.

- Right-click the wire connecting the Force control to the Calculate Acceleration function and select Delete Wire Branch from the shortcut menu. Move the Force control to the left of the Summation function, and wire this control to the Operand 2 input of the Summation function.

- Create wires 1–5 as indicated in the Figure below. The simulation diagram now fully represents the equation that defines the behavior of the dynamic system.

- Press <Ctrl-S> to save the VI.

The Block Diagram should now look like the figure 10-8:

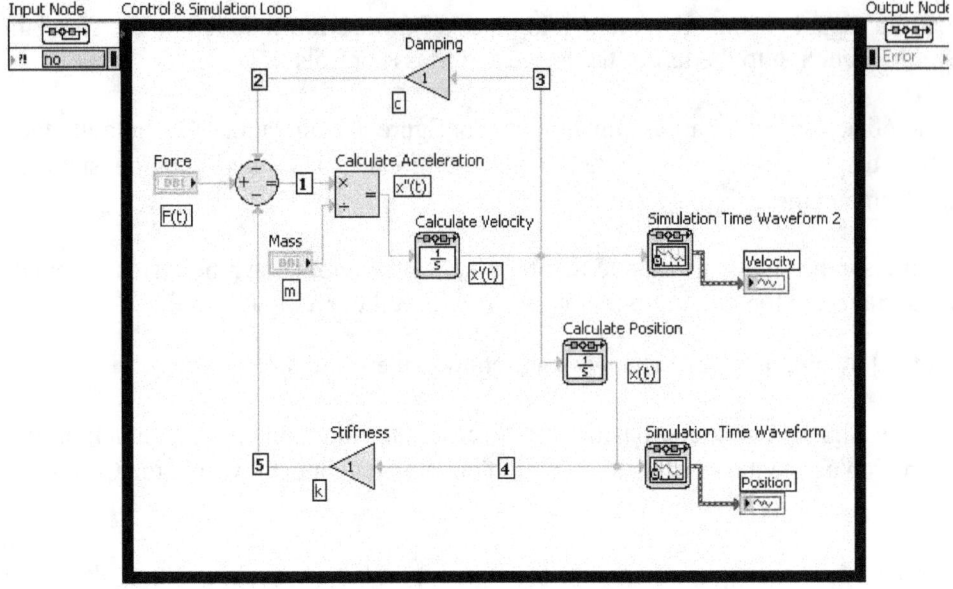

Figure 10-8

Configuring the Stiffness of the Spring

Before you run the simulation again, you must configure the stiffness of the simulated spring.

Complete the following steps to configure this Simulation function.

1. Double-click the Stiffness function to display the Gain Configuration dialog box.

2. Enter 100 in the gain numeric control. This value represents a stiffness of 100 Newton per meter.

3. Click OK to return to the simulation diagram. Notice that the Stiffness function displays 100.

4. Display the front panel and ensure the Force control is set to -9.8 and the Mass control is set to 1.

5. Run the simulation. The Velocity and Position charts display the behavior of the mass as the spring oscillates. Notice the new behavior compared to the last time you ran the simulation. This time, the velocity and position do not constantly decrease. Both values oscillate, which is how a spring behaves in the real world.

6. Change the value of the Mass control to 10 and run the simulation again. Notice the different behavior in the Velocity and Position charts. The 10 kg mass forces the spring to oscillate less frequently and within a smaller velocity/position range.

The Front Panel should look like the figure 10-9:

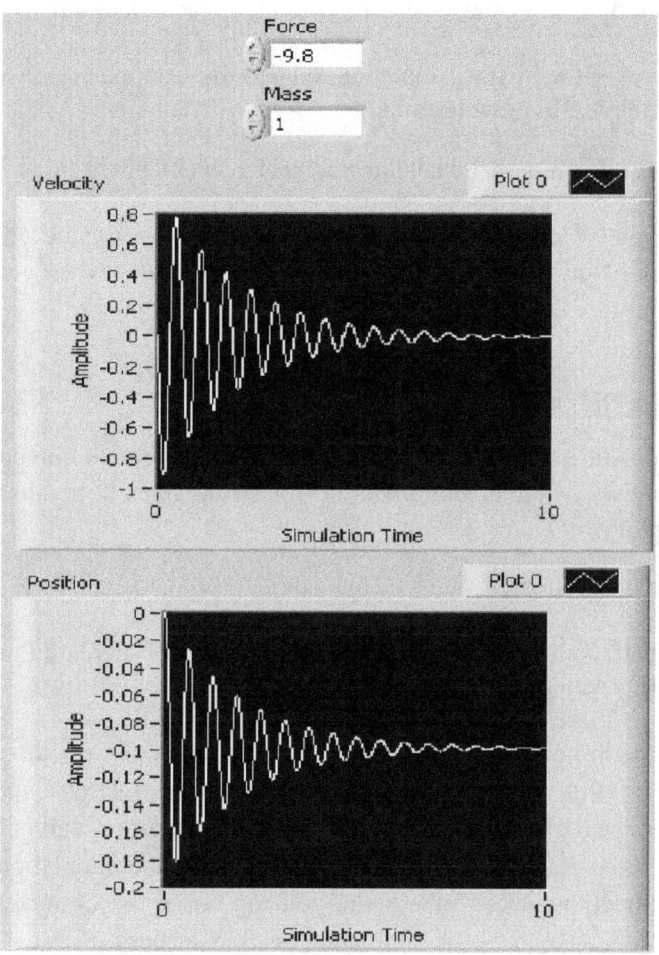

Figure 10-9

Configuring Simulation Functions Programmatically:

The previous section provided information about configuring Simulation functions using the configuration dialog box. Instead of using the configuration dialog box, you can improve the interactivity of a simulation by creating front panel controls that configure a Simulation function programmatically. Complete the following steps to configure the Stiffness function programmatically.

- If you are not already viewing the Context Help window, press <Ctrl-H> to display this window.

- Display the block diagram and move the cursor over the Stiffness function. Notice this function has only one input terminal.

- Display the Gain Configuration dialog box of the Stiffness function.

- Select Terminal from the Parameter source pull-down menu. This action disables the gain numeric control.

- Click the OK button to save changes and return to the block diagram.

- Move the cursor over the Stiffness function. Notice the Context Help window displays the Gain function with the new gain input terminal.

- Create a control for this input, and label the control gain (k).

- View the front panel. Notice the new control gain (k). Enter a value of 100 for this control and run the simulation. Notice the behavior is exactly the same as when you used the configuration dialog box to configure the Stiffness function.

Modularizing Simulation Diagram Code

You can create simulation subsystems to divide simulation diagrams into components that are modular, reusable, and independently verifiable. Complete the following steps to create a simulation subsystem from this simulation diagram.

- View the simulation diagram.

- Select the Calculate Acceleration, Calculate Velocity, and Calculate Position functions by pressing the <Shift> key and clicking each function.

- Select Edit» Create Simulation Subsystem. LabVIEW replaces these three functions with a single function that represents the simulation subsystem, which is circled in the Figure below. The inputs and outputs of the simulation subsystem include the inputs and outputs of all the functions you selected. Also, notice the amount of blank space on the simulation diagram. Because you combined three functions into a subsystem, you can resize the Control & Simulation Loop and reposition the functions to make the simulation diagram easier to view.

- Press <Ctrl-S> to save the simulation diagram. LabVIEW prompts you to save the simulation subsystem you just created. Click the Yes button and save this simulation subsystem as "Newton.vi". You now have a simulation subsystem that obtains the position of a mass by using Newton's Second Law of Motion.

Note: You can resize the simulation subsystem to better display its simulation diagram. Also, it can be double-click the simulation subsystem to display the configuration dialog box of that simulation subsystem.

The simulation subsystem should look like the figure 10-10:

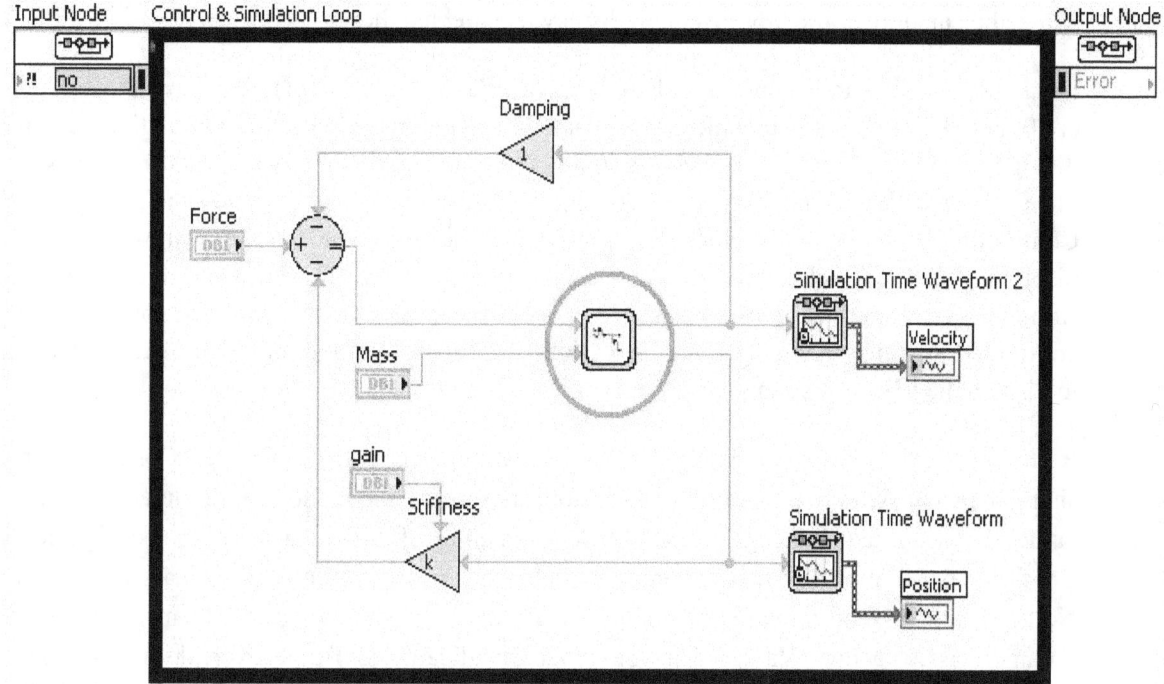

Figure 10-10

Editing the Simulation Subsystem

Edit the simulation subsystem "Newton.vi" by right-clicking this subsystem and selecting Open Subsystem from the shortcut menu. View the simulation diagram. Notice this simulation subsystem does not contain a Control & Simulation Loop, but the entire background is pale yellow to indicate a simulation diagram. If you place this simulation subsystem in a Control & Simulation Loop, the simulation subsystem inherits all simulation parameters from the Control & Simulation Loop.

If you run this subsystem as a stand-alone VI, you can configure the simulation parameters by selecting Operate» Configure Simulation Parameters. Any parameters you configure using this method do not take effect when the subsystem is within another Control & Simulation Loop. If you place this simulation subsystem on a block diagram outside a Control & Simulation Loop, you can configure the simulation parameters by double-clicking the simulation subsystem to display the configuration dialog box of that simulation subsystem.

Configuring Simulation Parameters Programmatically:

Earlier in this exercise, you used the Configure Simulation Parameters dialog box to configure the parameters of "Spring-Mass Damper Example.vi". You also can configure simulation parameters programmatically by using the Input Node of the Control & Simulation Loop. Complete the following steps to configure simulation parameters programmatically.

255

- View the simulation diagram of "Spring-Mass Damper Example.vi".

- Move the cursor over the Input Node to display resizing handles.

- Drag the bottom handle down to display all available Node inputs. You use these inputs to configure the simulation parameters without displaying the Configure Simulation Parameters dialog box. You also can right-click the Input Node and select Show All Inputs from the shortcut menu. Notice the gray boxes next to each input. These boxes display values you configure in the Configure Simulation Parameters dialog box. For example, the third gray box from the top displays 10.0000, which is the value of the Final Time numeric control that you configured. The fifth gray box from the top displays RK 23. This box specifies the current ODE solver, which you configured as Runge-Kutta 23 (variable). Move the cursor over the left edge of each Node input to display the label of that input.

- Right-click the input terminal of the ODE Solver input and select Create» Constant from the shortcut menu. A block diagram constant appears outside the Control & Simulation Loop. The value of this constant is Runge-Kutta 1 (Euler), which is different than what you configured in the Configure Simulation Parameters dialog box. However, the gray box disappears from the Input Node, indicating that the value of this parameter does not come from the Configure Simulation Parameters dialog box. Values that you programmatically configure override any settings you made in the Configure Simulation Parameters dialog box.

The Input Node should now looks like the following figure 10-11:

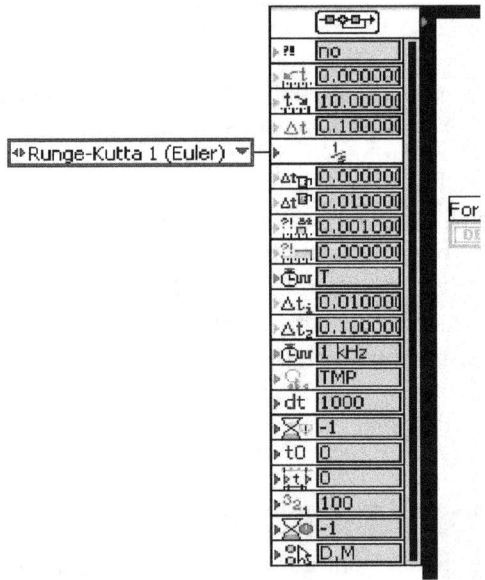

Figure 10-11

This exercise introduced the following concepts:

- The simulation diagram reflects the dynamic system model you want to simulate. This dynamic system model is a differential or difference equation that represents a dynamic system.

- The Control & Simulation Loop contains the parameters that define the behavior of the simulation. The Control & Simulation Loop also defines the visual boundary of the simulation diagram. Double-click the Input Node of the Control & Simulation Loop to access configurable parameters, and can be expand the Input Node to access these parameters.

- The Simulation palette contains the VIs and functions you use to build a simulation. One can double-click most Simulation functions to display a dialog box that configures that function. You also can create input terminals for function inputs.

- It can be create simulation subsystems to modularize, encapsulate, validate, and re-use portions of the simulation diagram.

10.8 PID Control

Currently, the Proportional-Integral-Derivative (PID) algorithm is the most common control algorithm used in industry. Often, people use PID to control processes that include heating and cooling systems, fluid level monitoring, flow control, and pressure control. In PID control, you must specify a process variable and a set point. The process variable is the system parameter you want to control, such as temperature, pressure, or flow rate, and the set point is the desired value for the parameter you are controlling. A PID controller determines a controller output value, such as the heater power or valve position. The controller applies the controller output value to the system, which in turn drives the process variable toward the set point value.

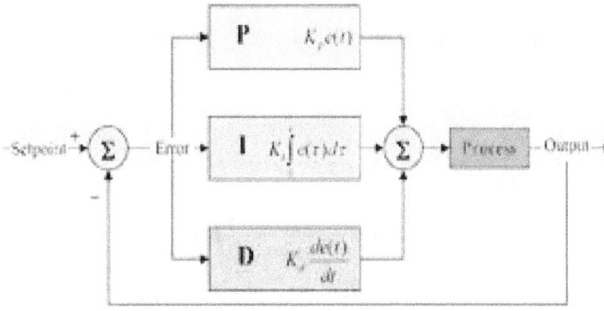

The PID controller compares the set point (SP) to the process variable (PV) to obtain the error (e):

$$e = SP - PV \dots\dots\dots\dots\dots\dots\dots\dots\dots\dots\dots\dots\dots(1)$$

Then the PID controller calculates the controller action, u(t), where Kc is controller gain:

$$u(t) = K_c\left(e + \frac{1}{T_i}\int_0^t edt + T_d\frac{de}{dt}\right) \quad\text{...........................(2)}$$

Ti is the integral time in minutes, also called the reset time, and Td is the derivative time in minutes, also called the rate time. The following formula represents the proportional action:

$$u_p(t) = K_c e \quad\text{..(3)}$$

The following formula represents the integral action:

$$u_I(t) = \frac{K_c}{T_i}\int_0^t edt \quad\text{...............................(4)}$$

The following formula represents the derivative action:

$$u_D(t) = K_c T_d\frac{de}{dt} \quad\text{................................(5)}$$

In the "**PID**" Sub palette we have the functions/Sub VIs for PID Control. I recommend that you use the "**PID Advanced.vi**".

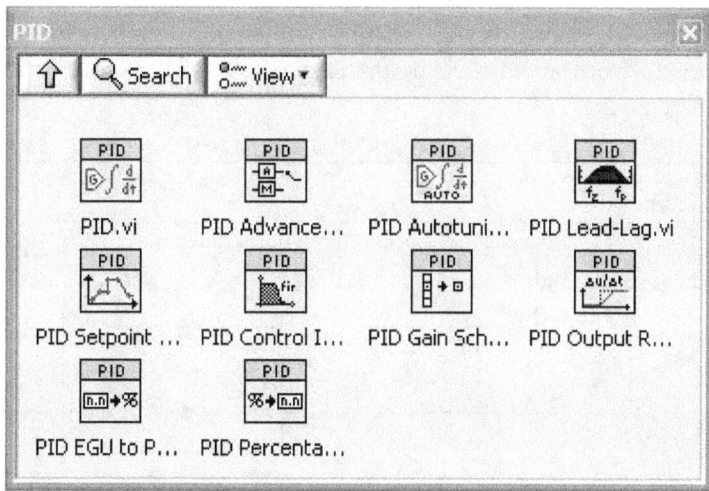

Example 5: PID Control : Below the figure has seen how we can use the PID Advanvanced.vi in order to control a simulated Model.

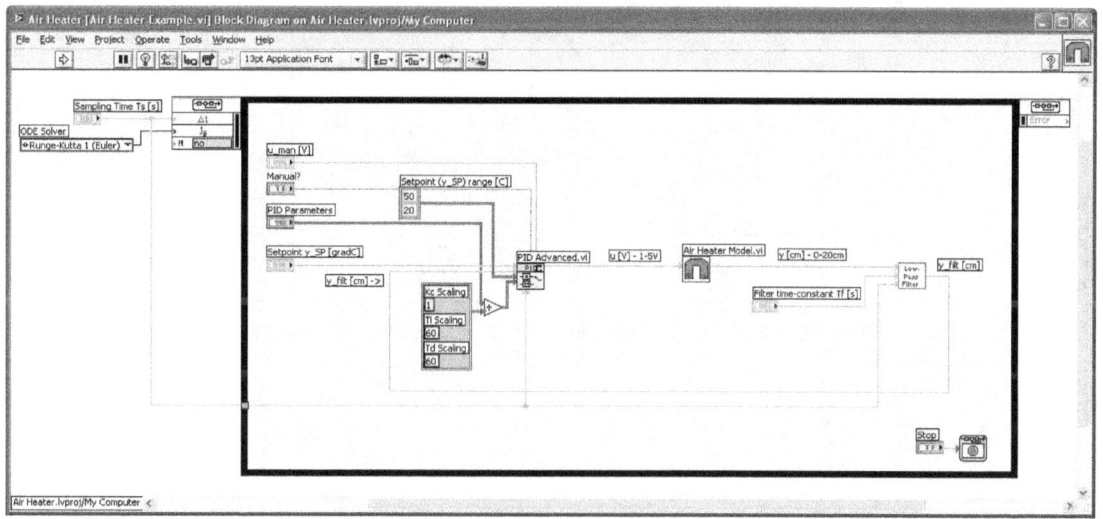

The LabVIEW PID and Fuzzy Logic Toolkit include a VI for auto-tuning.

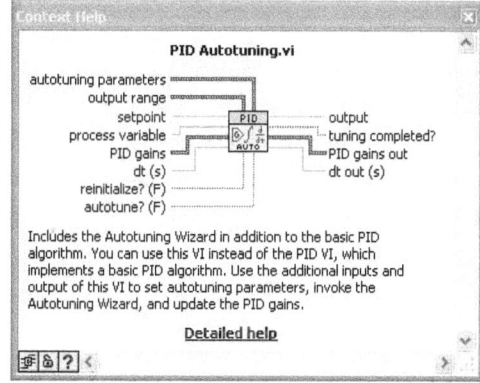

10.9 Control Design

Control design is a process that involves developing mathematical models that describe a physical system, analyzing the models to learn about their dynamic characteristics, and creating a controller to achieve certain dynamic characteristics.

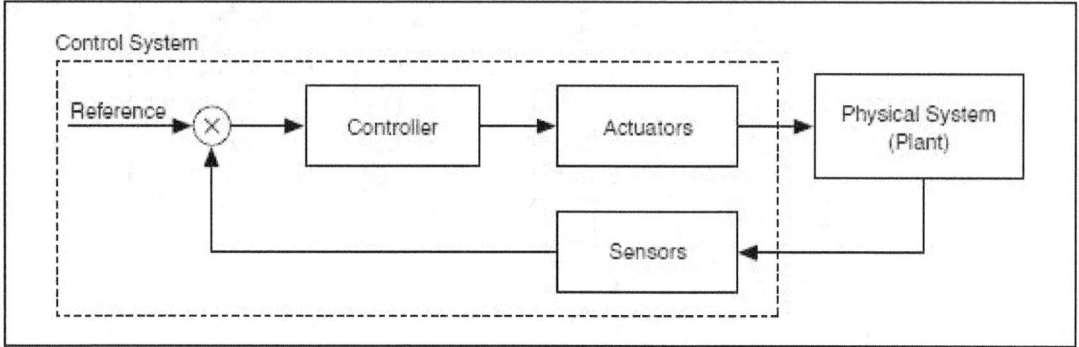

Control Design palette:

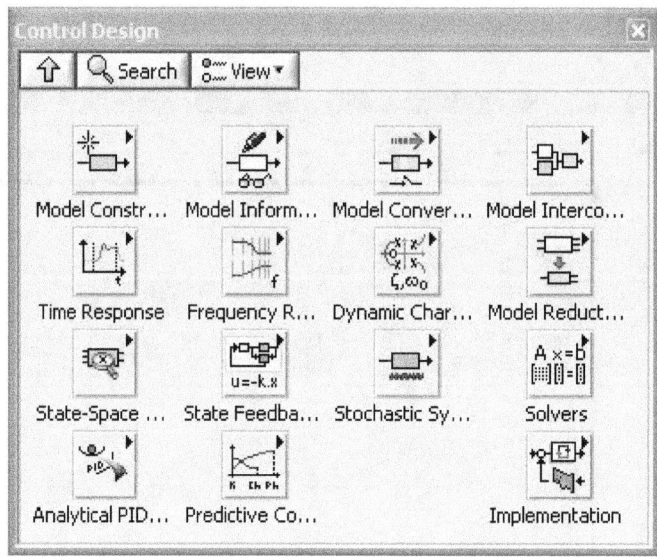

10.10 System Identification

The "System Identification Toolkit" combines data acquisition tools with system identification algorithms for accurate plant modeling. It may take advantage of LabVIEW intuitive data acquisition tools such as the DAQ Assistant to stimulate and acquire data from the plant and then automatically identify a dynamic system model. Also, it converts system identification models to state-space, transfer function, or pole-zero-gain form for control system analysis and design.

The toolkit includes built-in functions for common tasks such as data preprocessing, model creation, and system analysis. Using other built-in utilities, you can plot the model with intuitive graphical representation as well as store the model. System Identification palette:

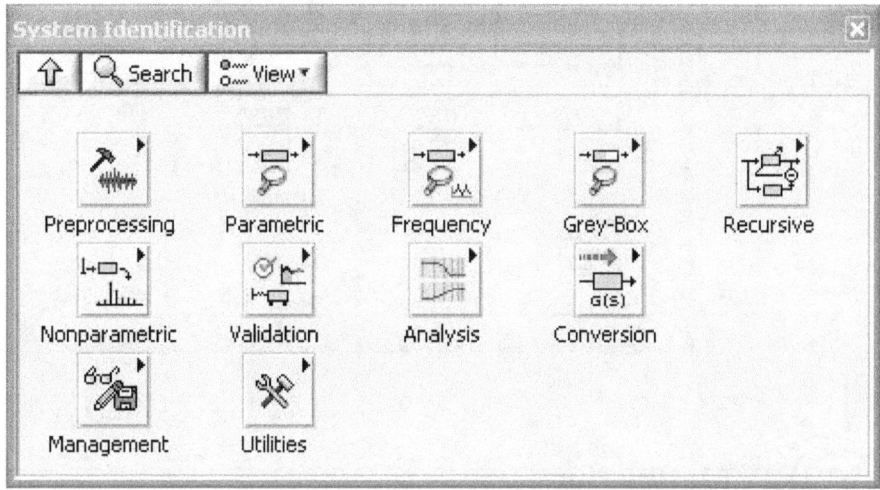

10.11 Fuzzy Logic

Fuzzy logic is a method of rule-based decision making used for expert systems and process control. Fuzzy logic differs from traditional Boolean logic in that fuzzy logic allows for partial membership in a set. It can be used fuzzy logic to control processes represented by subjective, linguistic descriptions. A fuzzy system is a system of variables that are associated using fuzzy logic. A fuzzy controller uses defined rules to control a fuzzy system based on the current values of input variables, the sketch is shown in figure 10-12 describes the fuzzy logic process.

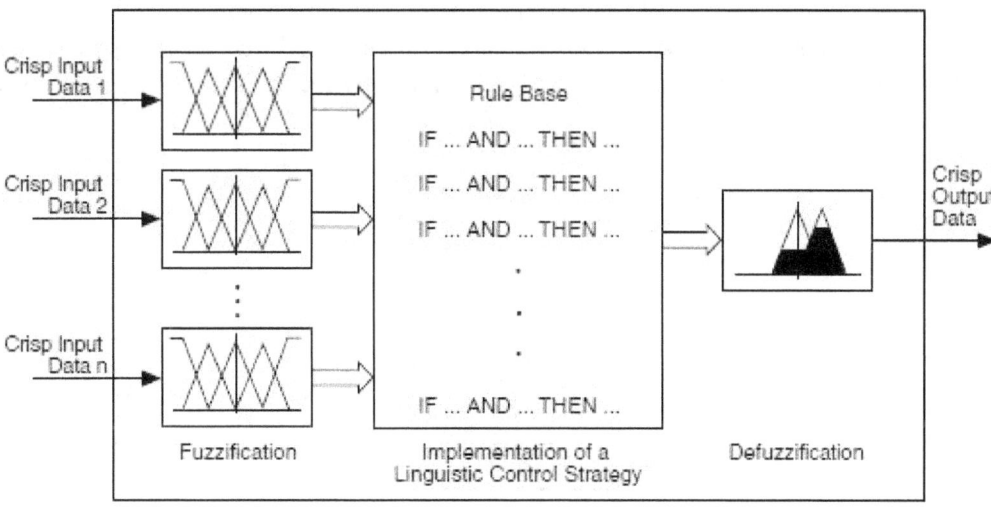

Figure 10-12

261

The Fuzzy Logic palette in LabVIEW:

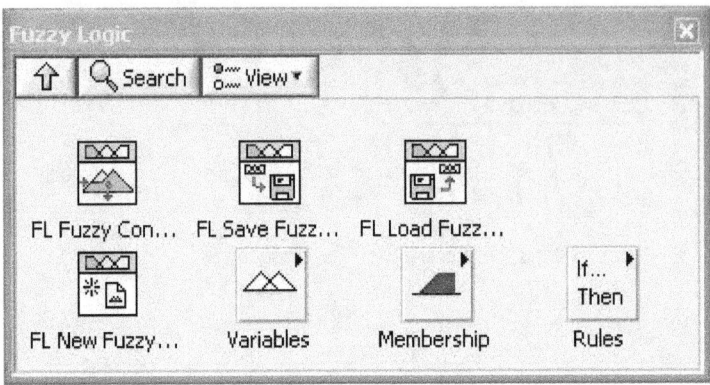

10.12 LabVIEW MathScript

The "LabVIEW MathScript Window" is an interactive interface in which you can enter .m file script commands and see immediate results, variables and commands history. The window includes a command-line interface where you can enter commands one-by-one for quick calculations, script debugging or learning. Alternatively, you can enter and execute groups of commands through a script editor window.

A variable display updates to show the graphical / textual results and a history window tracks your commands. The history view facilitates algorithm development by allowing you to use the clipboard to reuse your previously executed commands.

Therefore, it can be use the "LabVIEW MathScript Window" to enter commands one at time. Also, can enter batch scripts in a simple text editor window, loaded from a text file, or imported from a separate text editor. The "LabVIEW MathScript Window" provides immediate feedback in a variety of forms, such as graphs and text.

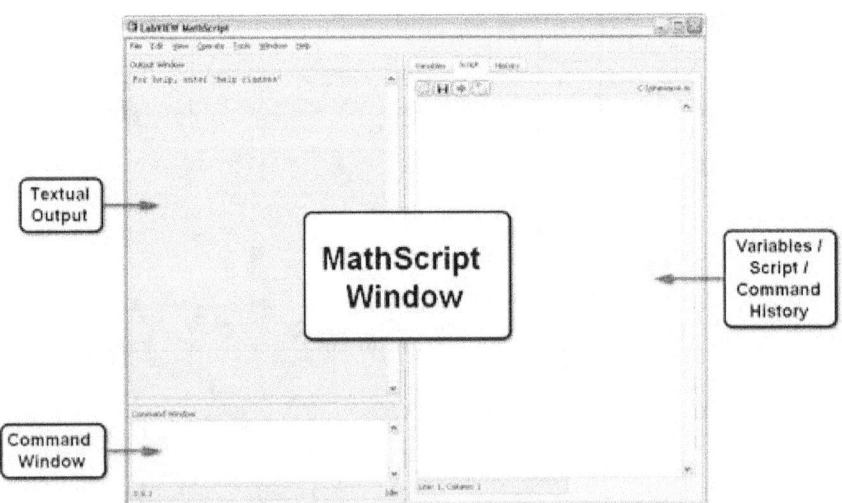

Example 6: Matrices

Defining the following matrix

The syntax is as follows:

```
>> A = [1 2;0 3]
```

Or

```
>> A = [1,2;0,3]
```

for an example, want to find the answer to

```
>>a=4
```

```
>>b=3
```

```
>>a+b
```

MathScript then responds:

ans = 7

MathScript provides a simple way to define simple arrays using the syntax:

"init:increment:terminator". For instance:

```
>> array = 1:2:9

array =

   1 3 5 7 9
```

Example 7:

Defines a variable named array (or assigns a new value to an existing variable with the name array) which is an array consisting of the values 1, 3, 5, 7, and 9. That is, the array starts at 1 (the init value), increments with each step from the previous value by 2 (the increment value), and stops once it reaches (or to avoid exceeding) 9 (the terminator value).

The increment value can actually be left out of this syntax (along with one of the colons), to use a default value of 1.

```
>> ari = 1:5

ari =

1 2 3 4 5
```

Assigns to the variable named ari an array with the values 1, 2, 3, 4, and 5, since the default value of 1 is used as the incremented values.

Note: that the indexing is one-based, which is the usual convention for matrices in mathematics. This is typical for programming languages, whose arrays more often start with zero.

Matrices can be defined by separating the elements of a row with blank space or comma and using a semicolon to terminate each row. The list of elements should be surrounded by square brackets: []. Parentheses: () are used to access elements and sub arrays (they are also used to denote a function argument list).

```
>> A = [16    3          2 13; 5 10 11 8; 9 6 7 12; 4 15 14 1]

A
=

16   3   2   1
             3
 5   1   1   8
     0   1
 9   6   7   1
             2
 4   1   1   1
     5   4

>> A(2,3) ans =

11
```

Sets of indices can be specified by expressions such as "2:4", which evaluates to [2, 3, 4]. For the example, a sub matrix taken from rows 2 through 4 and columns 3 through 4 can be written as:

>> A(2:4,3:4)

ans =

11 8

7 12

14 1

A square identity matrix of size n can be generated using the function eye, and matrices of any size with zeros or ones can be generated with the functions zeros and ones, respectively.

```
>> eye(3)
ans =

   1  0  0

   0  1  0

   0  0  1


>> zeros(2,3)
ans =

   0  0  0

   0  0  0
>> ones(2,3)
ans =

   1  1  1

   1  1  1
```

Here are some useful commands:

Command	Description
eye(x), eye(x,y)	Identity matrix of order x
ones(x), ones(x,y)	A matrix with only ones
zeros(x), zeros(x,y)	A matrix with only zeros
diag([x y z])	Diagonal matrix
size(A)	Dimension of matrix A
A'	Inverse of matrix A

This chapter has only explain the basic concepts of creating plots in MathScript.

Example 8: Plotting

Function plot can be used to produce a graph from two vectors x and y. The code:

x = 0:pi/100:2*pi;

y = sin(x);

plot(x,y)

There are so many examples shown the plotting of a graph in LabVIEW , the national instrumentation website has plenty of it. The following examples have taken some of them. **Quiver and 3D Contour Plots in the MathScript Window**: demonstrates the advanced plotting capabilities of MathScript. Using the MathScript node, users can display a 3D Contour plot and a Quiver plot in pop-up windows as seen figure 10-13.

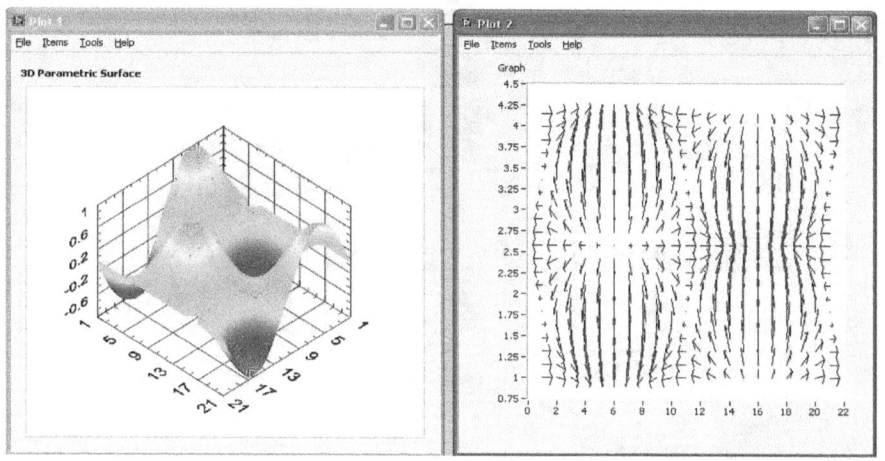

Figure 10-13

> **Interacting with LabVIEW and the MathScript Window Using Global Variables**:

Demonstrates run time interaction between LabVIEW and the MathScript interactive window using global variables, as shown in figure 10-14.

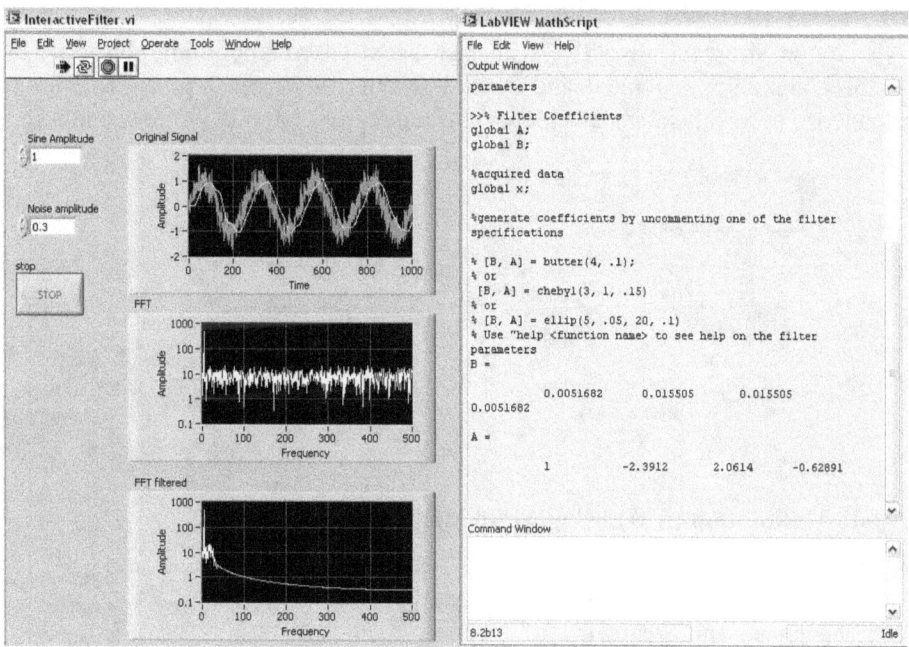

Figure 10-14

> **Load and Save files with LabVIEW and the MathScript Node**: demonstrates LabVIEW's easy to use File I/O with MathScript's built in DSP functions as seen in figure 10-15.

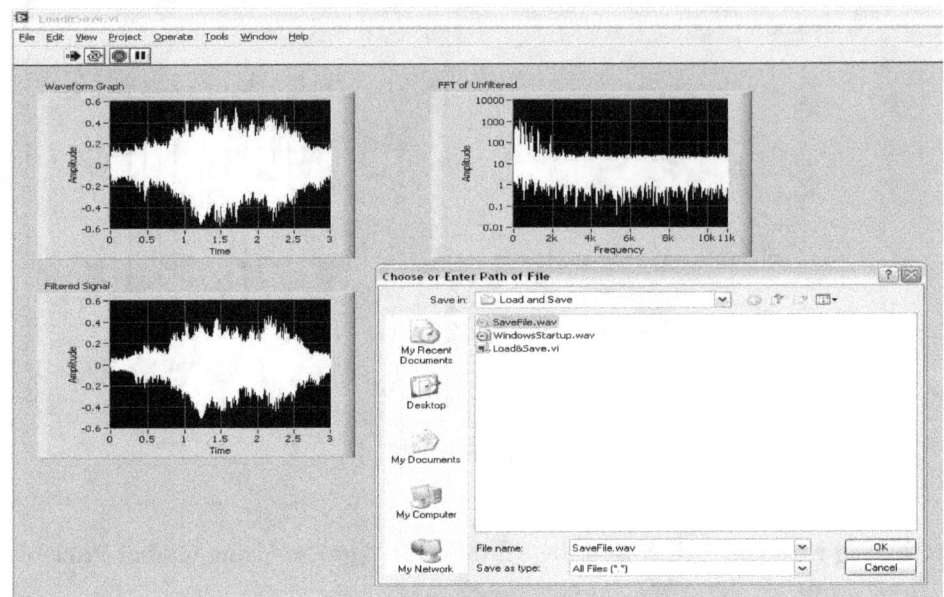

Figure 10-15

> ## Smoothing Algorithm using MathScript:

Using this example, you can interactively change a text-based smoothing algorithm and observe its effects. A noisy signal is generated and passed through a MathScript Node with the smoothing algorithm loaded. The results are viewable on the interactive LabVIEW front panel in 3D graphs, as seen below.

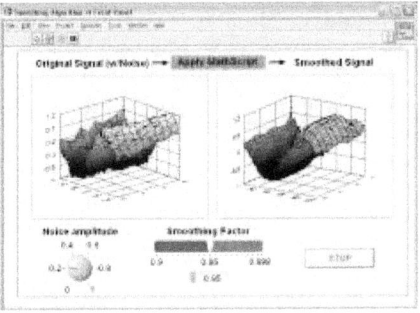

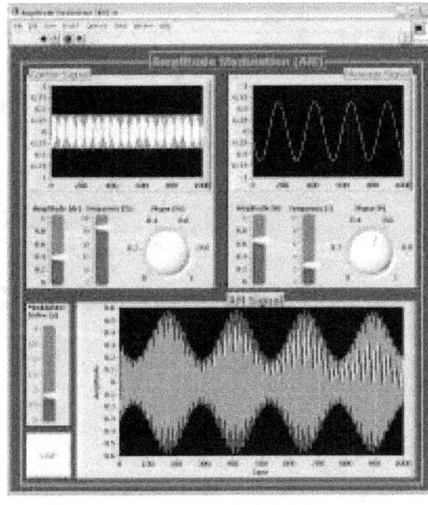

> ## Amplitude Modulation (AM) using MathScript:

In this example, a carrier signal and message signal are amplitude modulated to create an AM signal, using MathScript. The phase and frequency of the carrier and message signals can be modified on the fly, while the resulting AM signal is updated immediately as seen in the figure.

268

Chapter 11

Vision Systems

Machine vision (MV) is the process of applying a range of technologies to provide imaging-based automatic inspection, process control and robot guidance in industrial applications. The generally accepted definition of machine vision is "**the analysis of images to extract data for controlling a process or activity**".

The primary uses for machine vision are automatic inspection and robot guidance. The main categories into which MV applications fall are quality assurance, sorting, material handling, robot guidance, and calibration.

11.1 Image Processing and Analysis

After the image is taken we need to do some image processing and analysis. Below there is a list of the most used functions:

11.1.1 Thresholding

Converting a grayscale image to black and white.

- Pattern recognition and matching
- Texture recognition
- Barcode reading

There is different kind of barcodes , and the barcodes are seen everywhere today. We have 1D barcodes, 2D barcodes, QR codes, etc. Also, today barcodes have a lot of applications in the markets and services overall the world. A barcode is an optical machine-readable representation of data, which shows data about the object by varying the widths and spacing's of parallel lines, and may be referred to as linear or 1 dimensional (1D) barcode. Below we've seen a standard barcode:

Later they evolved into rectangles, dots, hexagons and other geometric patterns in 2 dimensions (2D). Although 2D systems use a variety of symbols, they are generally referred to as barcodes as well. Below we see a so-called QR code:

The QR code is one of the most popular types of two-dimensional barcodes. Barcodes originally were scanned by special optical scanners called barcode readers; later, scanners and interpretive software became available on devices including desktop printers and smart phones.

11.1.2 OCR

OCR or Optical Character Recognition is the mechanical or electronic translation of scanned images of handwritten, typewritten or printed text into machine-encoded text. It is widely used to convert books and documents into electronic files. Typically multifunction printers with scanner functionality include some software for OCR.

11.1.3 Gauging

Measuring object dimensions.

- Position
- Edge Detection
- Color analysis
- Filtering

11.1.4 Counting and Classification

Classification is a tool for identifying an unknown object by comparing its significant features to a set of features that represent known samples. Typically , we could have some bolts and screws we want to classify or count on an assembly line, as seen in figure 11-1.

Figure 11-1

In these situations we typically have a template image of each of the objects we want to classify or count that is used for comparison.

11.2 Vision Cameras

There exist different cameras used in machine vision. We can divide into 3 categories based on their connections to the PC:

- USB cameras

- IEEE 1394 (FireWire) cameras

- GigE (Ethernet) cameras

In this tutorial we will use a **Basler scA640-70gc** GigE camera.

11.2.1 GigE (Ethernet) Cameras

The camera is connected to the computer using a standard Ethernet cable. Below we see a standard GigE Ethernet camera like the figure shown below:

271

11.3 Vision Systems in LabVIEW

National Instruments offers different kind Vision software depending on your application and your needs:

- **NI Vision Acquisition Software**

- **NI Vision Development Module**

- **NI Vision Builder for Automated Inspections**

This section will discuss the different packages more in details. If someone has installed all these 3 packages, then it ends up with the following palette in LabVIEW as in figure 11-2:

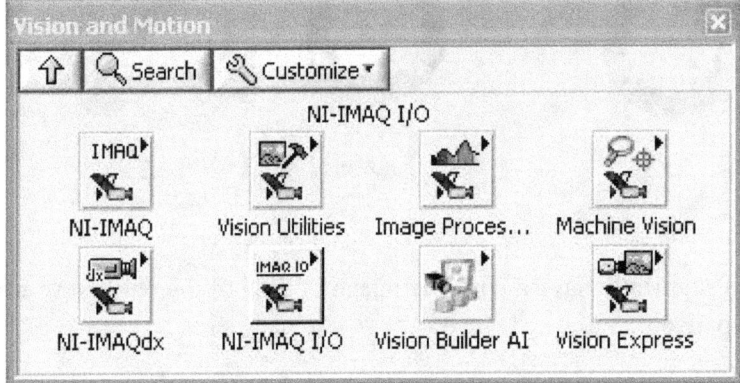

Figure 11-2

11.3.1 NI Vision Acquisition Software

The **NI Vision Acquisition software** is the basic software you need if you want to create Vision applications for LabVIEW or the .NET platform. The NI Vision Acquisition software includes the necessary drivers, such as NI-IMAQ and NI-IMAQdx. The **NI-IMAQdx** driver software gives you the ability to acquire images with IEEE 1394 (FireWire), GigE Vision (Ethernet), and USB cameras.

11.3.2 Vision Development Module

For more advanced machine vision and image processing you will need the **Vision Development Module**. The Vision Development Module contains hundreds of image processing and machine vision functions, both for LabVIEW and the .NET platform. This package includes built-in functions for:

- Pattern matching
- Texture recognition
- Counting and Classification
- OCR (Optical Character Recognition)
- Bar Code readers
- Image Filters

The figure 11-3 is shown the "**Image Processing**" palette in LabVIEW:

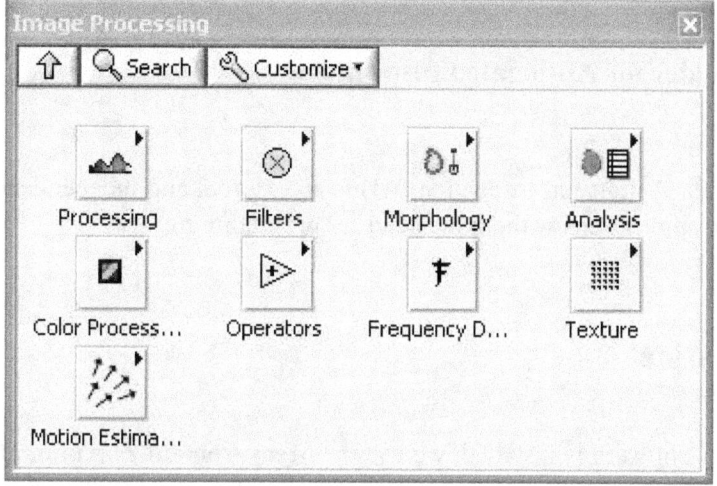

Figure 11-3

Also, the figure 11-4 shows the "**Machine Vision**" palette in LabVIEW:

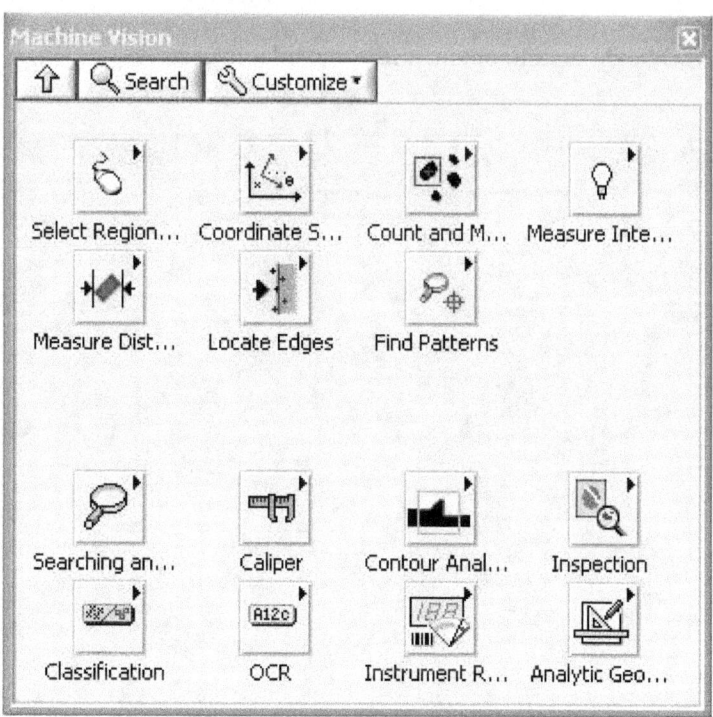

Figure 11-4

11.3.3 Vision Builder for Automated Inspections

NI Vision Builder for Automated Inspection (AI) is an external and independent application for building and machine vision applications without the need for programming.

11.4 Configuration

When the necessary software is installed, we use the **Measurement & Automation Explorer (MAX)** to get started. When we plug in the camera using an Ethernet cable into the computer, the camera should appear in the list. It can be connect a camera to a local Windows machine or a LabVIEW Real-Time target machine. We will focus on connection the camera to a local machine. Complete the following steps to connect a GigE camera or an IEEE 1394 camera to a local Windows machine:

- Connect the camera to the Ethernet port on the local machine.

- In the MAX configuration tree, expand Devices and Interfaces to obtain a list of installed devices.

- Expand NI-IMAQdx Devices to obtain a list of available cameras.

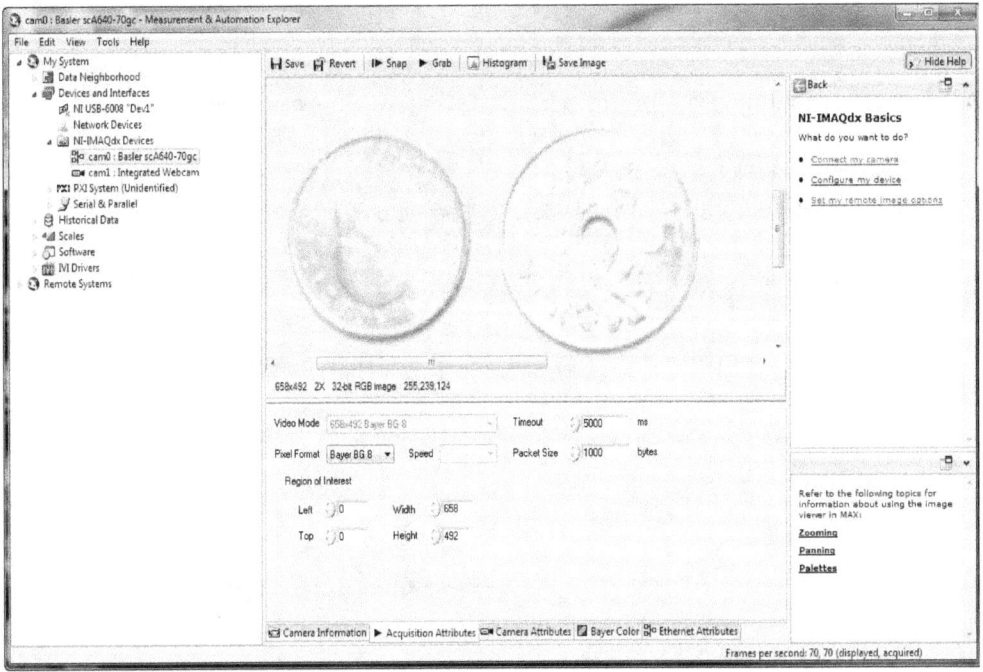

When the camera is successfully connected, you can configure and test the camera in MAX. Click the camera name to select the appropriate camera (in this tutorial we use a "Basler scA640-70gc" camera). Click **Snap** to acquire a single image or click **Grab** to continuously acquire an image. If everything works, you should be able to see an image inside camera's window, the following Figure shows this action.

Though you might get an error like the figure below:

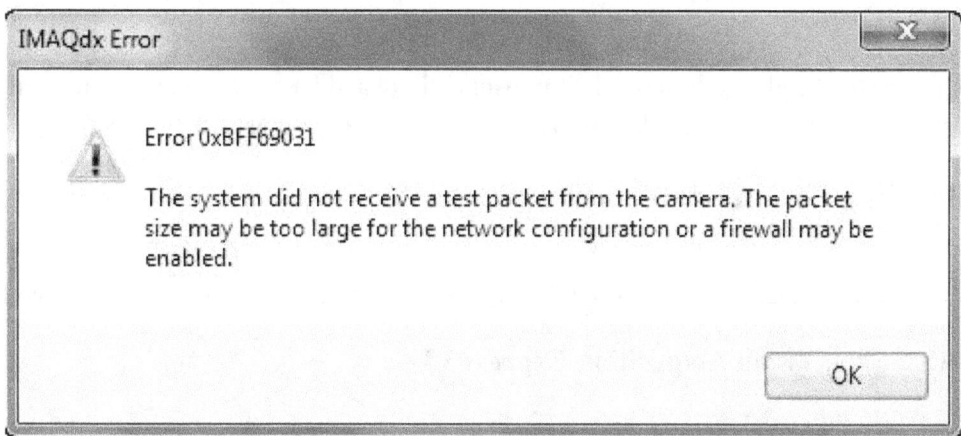

Which means you should reduce the **Package Size** in the configuration and/or configure the **Firewall** on your computer. When it comes to the Firewall, the easiest thing to do is to turn the whole Firewall off in order to make sure the camera works, see the following figure 11-5.

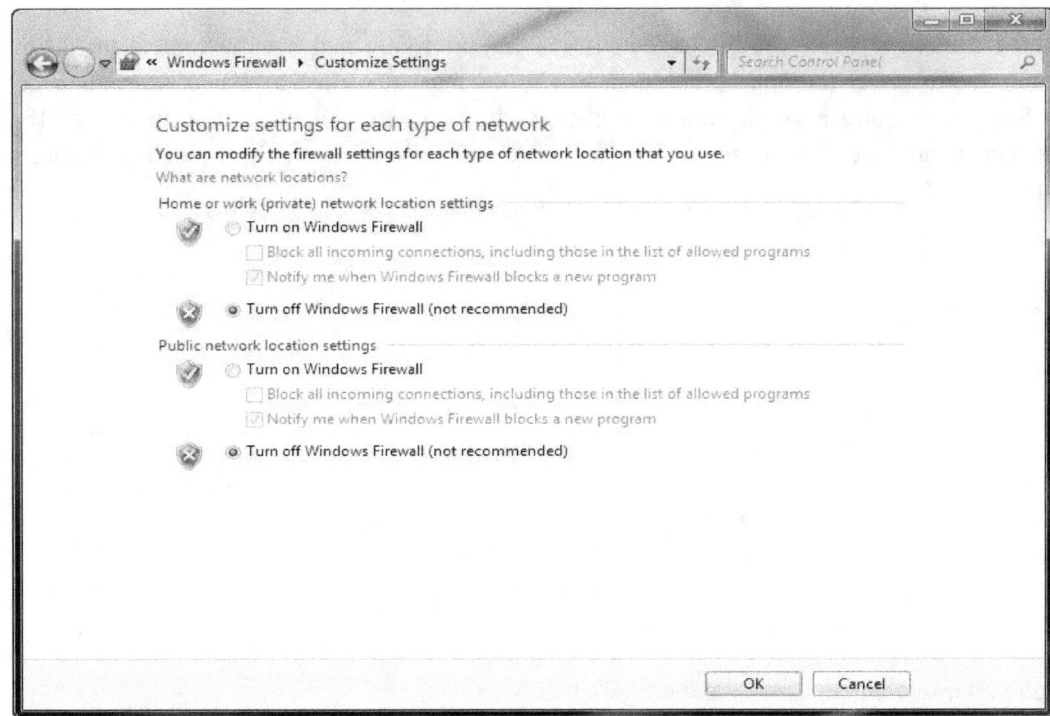

Figure 11-5

But it is not recommended to turn off the Firewall entirely and let the computer be unprotected for a long time. If still not working, you should also try to turn of the **Anti Virus** software temporary.

11.5 Building Vision Systems in LabVIEW

This section will demonstrate how to acquire images from the camera using LabVIEW code.

11.5.1 Using the Vision Acquisition Express VI

The simplest way to acquire images from LabVIEW is to use the Vision Acquisition Express VI. We find the "Vision Acquisition Express VI" in the Vision Express palette in LabVIEW as seen in figure 11-6:

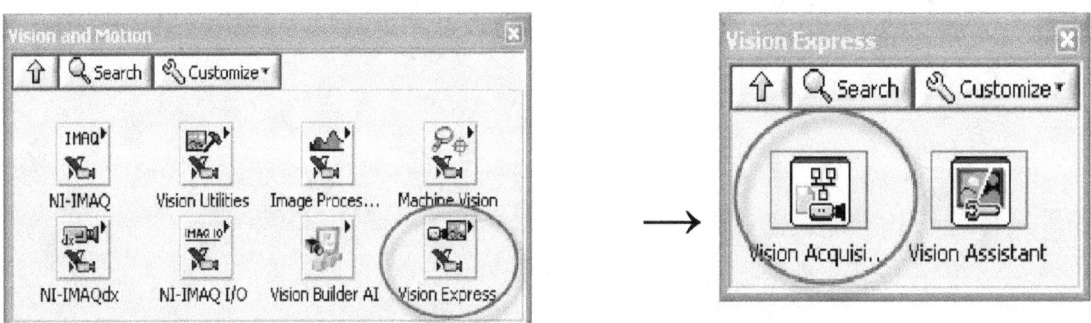

Figure 11-6

When we drag the "Vision Acquisition Express VI" to the lock diagram, a wizard will appear, figure 11-7 shows this VI :

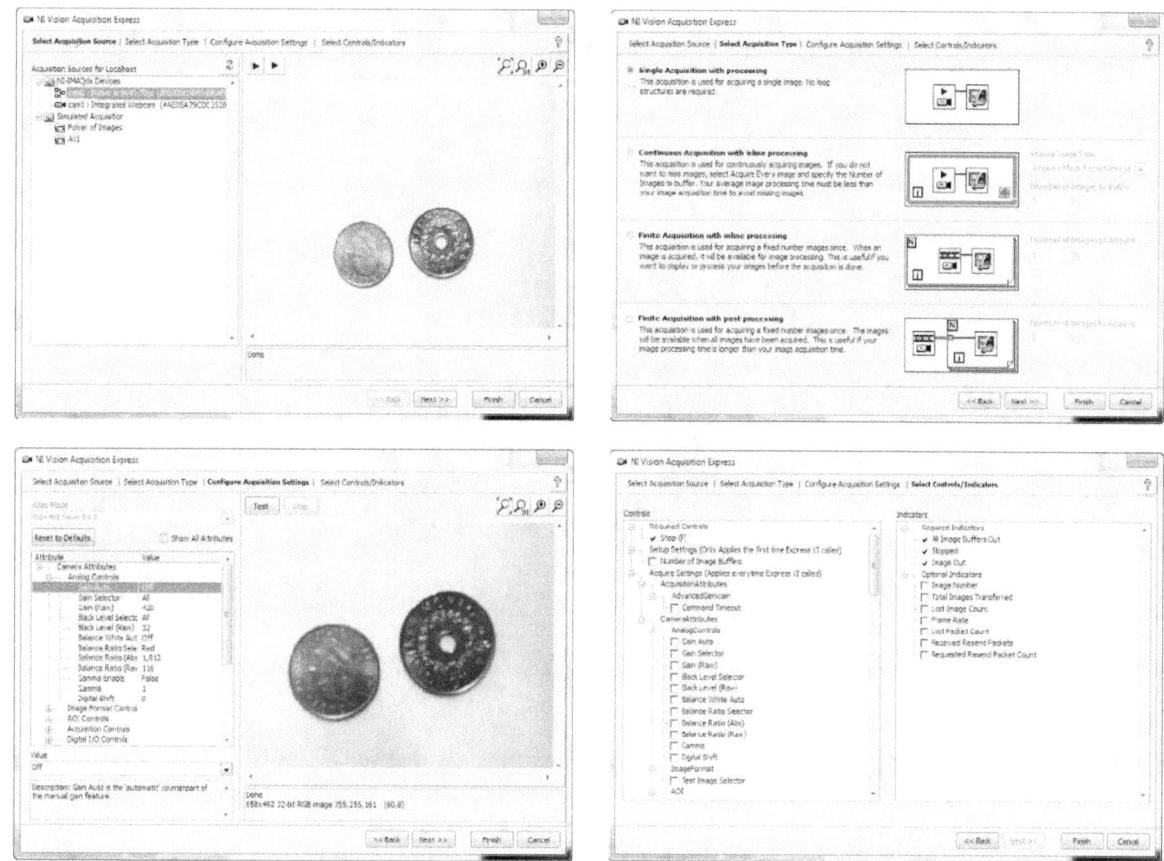

Figure 11-7

The finished LabVIEW program will simply look like the Figure 11-8:

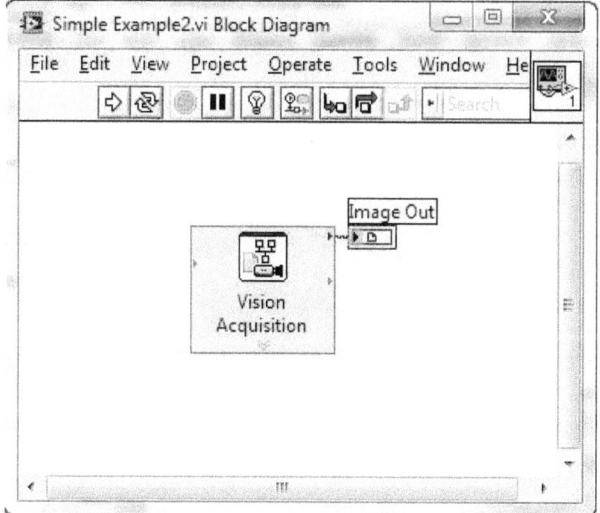

Figure 11-8

The image will be acquired on the Front Panel in the "Image Out":

11.5.2 Using the IMAQdx VIs

Below we've seen the NI-IMAQdx palette in LabVIEW:

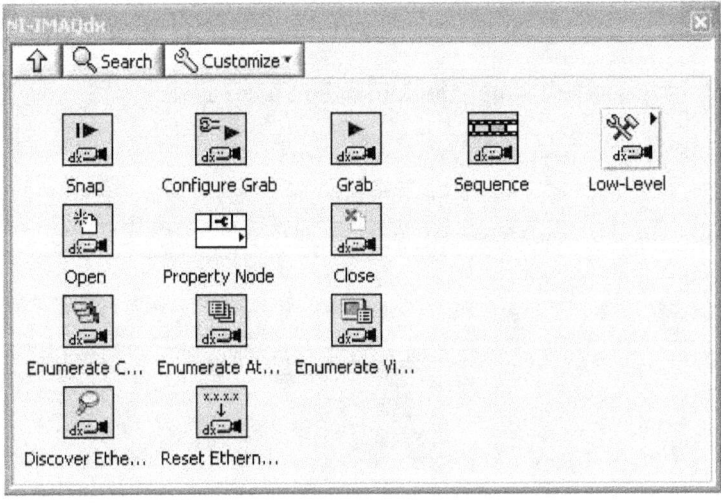

Also, we've seen a simple example where we use the IMAQdx VIs to create an application where we acquire a single image from the camera as shown in figure 11-9 and 11-10.

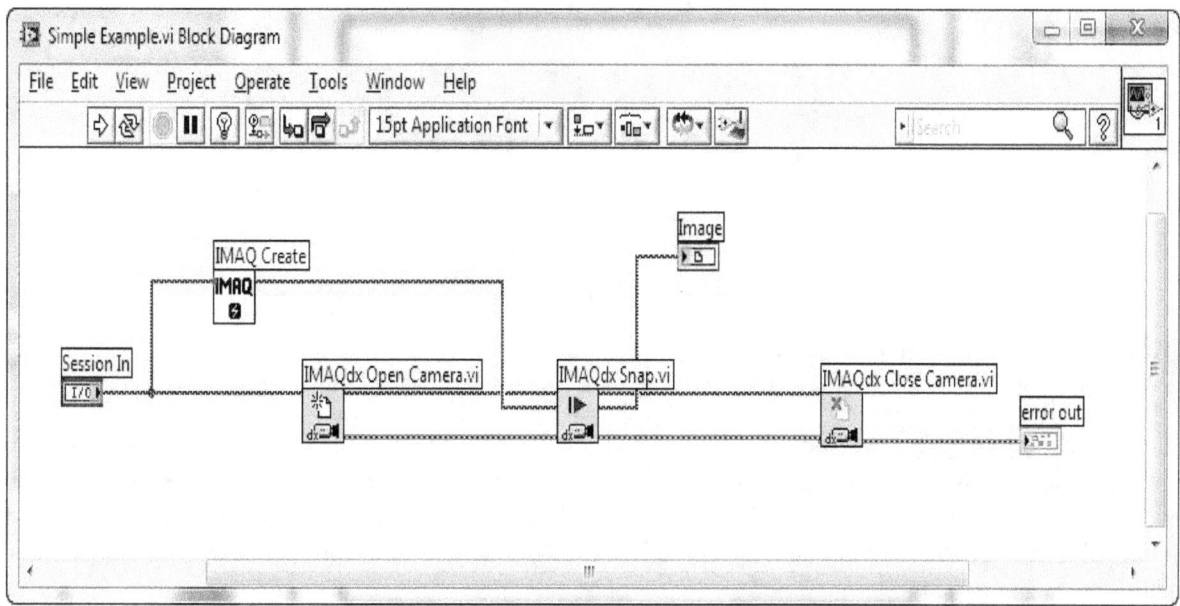

Figure 11-9

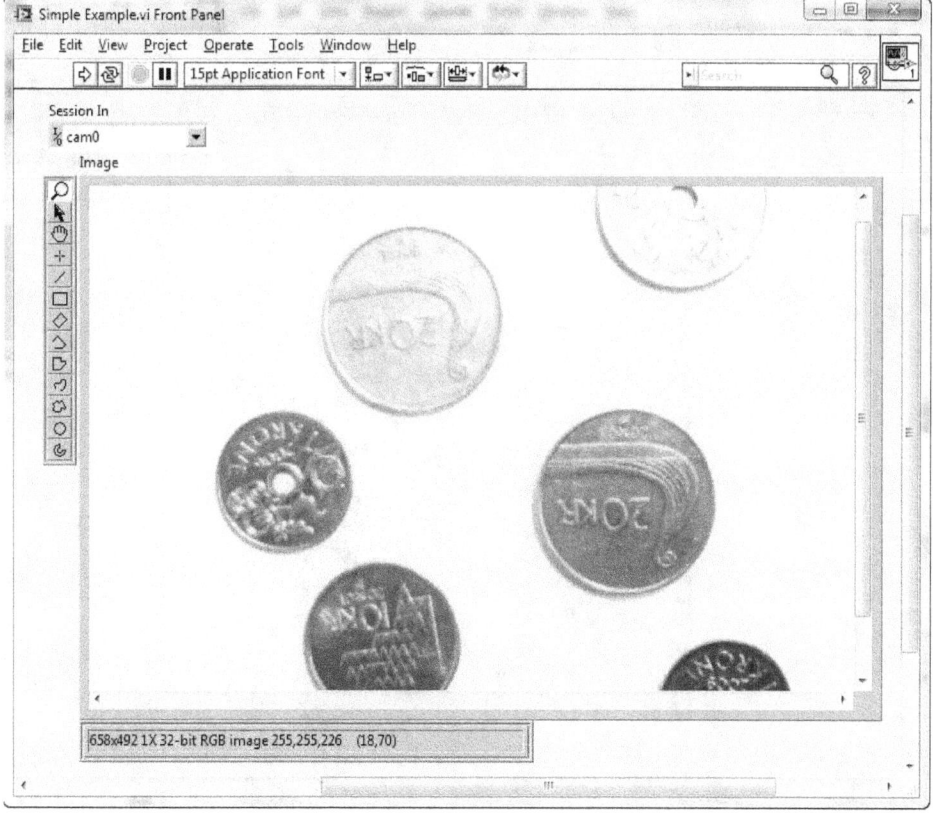

Figure 11-10

279

On the Front Panel we can use different containers for showing images on the screen as shown below:

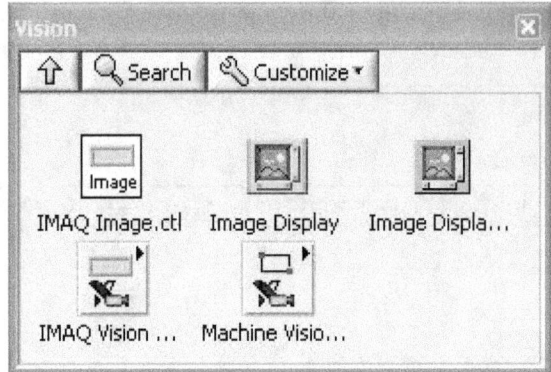

11.5.3 Open Images from a File

When working with Vision systems it is important to be able to save the images to a file or open an existing image from a file. In figure 11-11 and 11-12, we've seen an example of how we can open and load an image from a file into LabVIEW. In the example above it will pop-up an open File dialog box, but we can also specify the file path directly in the LabVIEW code. On the Front panel we will see the image inside an image container.

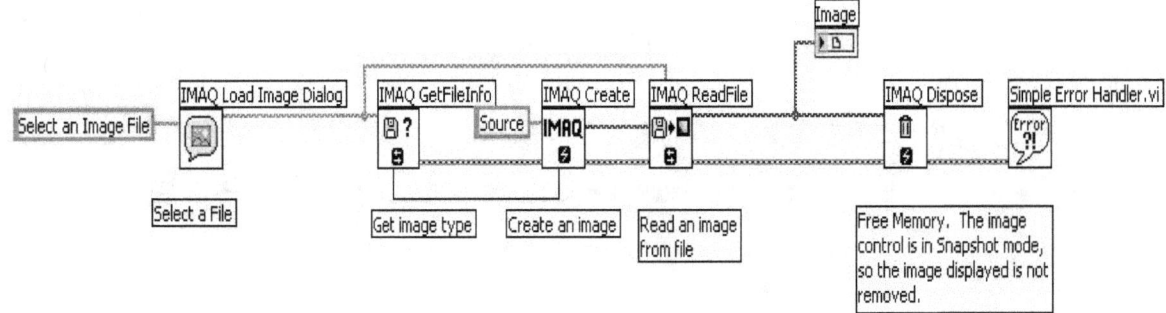

Figure 11-11

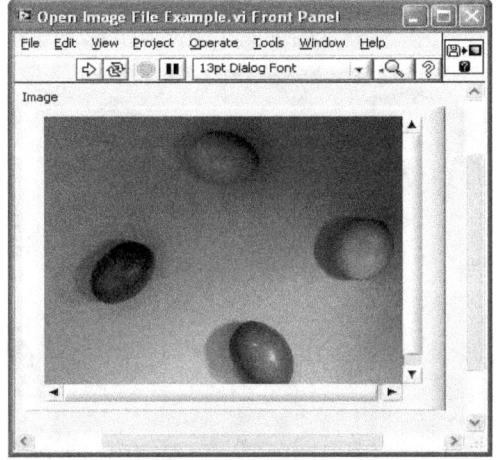

Figure 11-12

280

11.6 Vision Functionality

NI Vision for LabVIEW is organized into the following main function palettes: **Vision Utilities**, **Image Processing**, **Machine Vision**, and **Vision Express**. This section describes these palettes and their uses. As shown in figure 11-13, we've seen the Vision and Motion palette in LabVIEW:

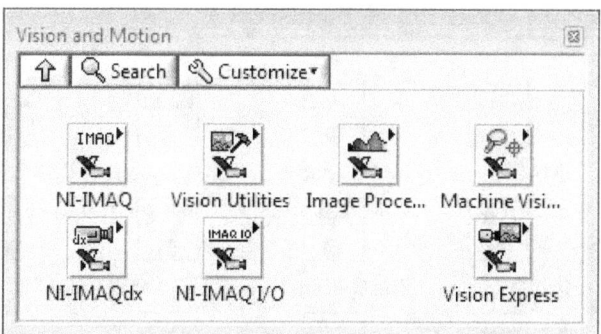

Figure 11-13

Adding that we've seen the "Image Processing" (Vision and Motion → Image Processing) palette in LabVIEW:

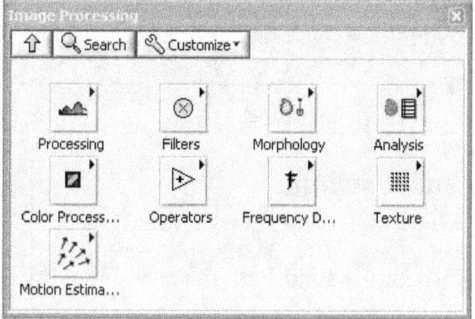

The figure 11-14 on the right has shown the "Machine Vision" (Vision and Motion → Machine Vision) palette in LabVIEW:

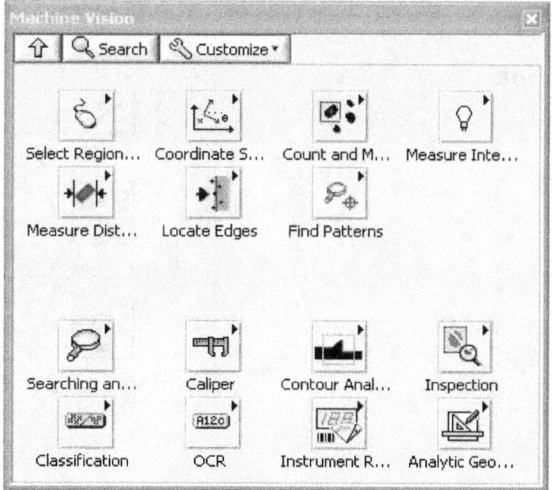

Figure 11-14

11.6.1 Thresholding

Vision and Motion → Image Processing → Processing

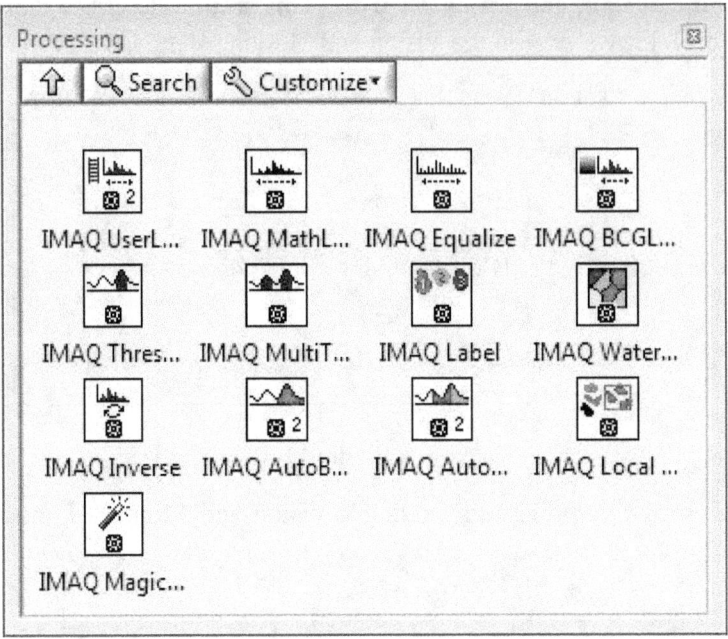

11.6.2 Pattern recognition and matching

Vision and Motion → Machine Vision → Find Patterns

11.6.3 Texture recognition

Vision and Motion → Image Processing → Texture

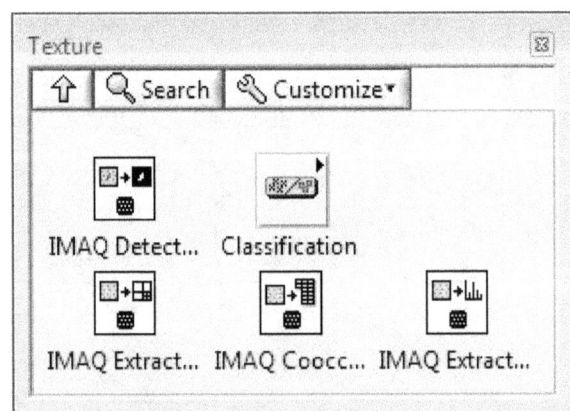

11.6.4 Barcode reading

Vision and Motion → Machine Vision → Instrument Readers

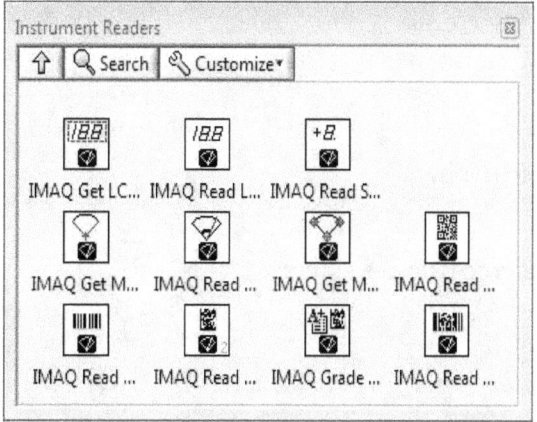

11.6.5 OCR

Vision and Motion → Machine Vision → OCR

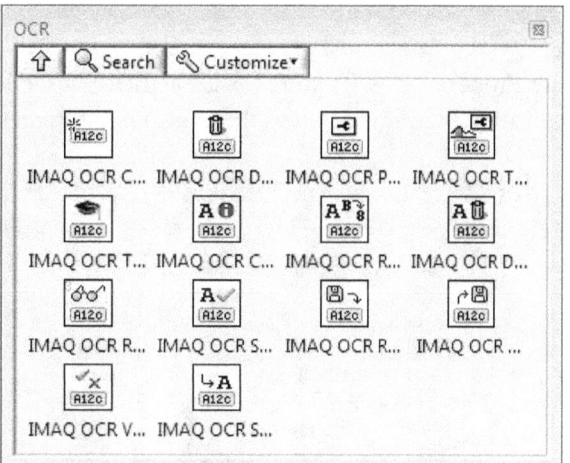

11.6.6 Gauging

Vision and Motion → Machine Vision → Measure Distances

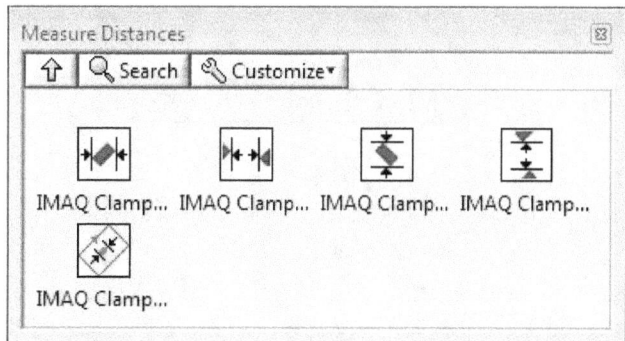

11.6.7 Position & Edge Detection
Vision and Motion → Machine Vision → Locate Edges:

11.6.8 Color analysis & Filtering

Vision and Motion → Image Processing → Filters:

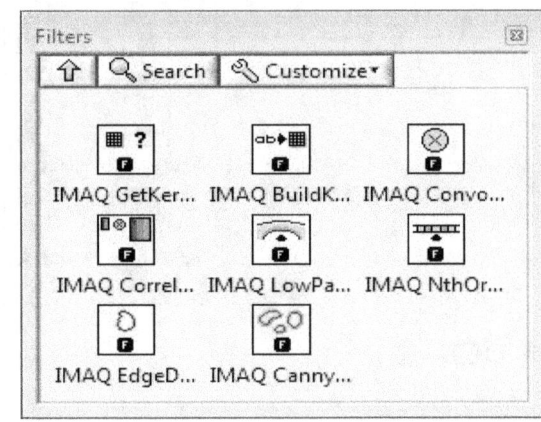

11.6.9 Counting and Classification

Vision and Motion → Image Processing → Texture → Classification Or Vision and Motion → Machine Vision → Classification, the following Figure shows the classification palette.

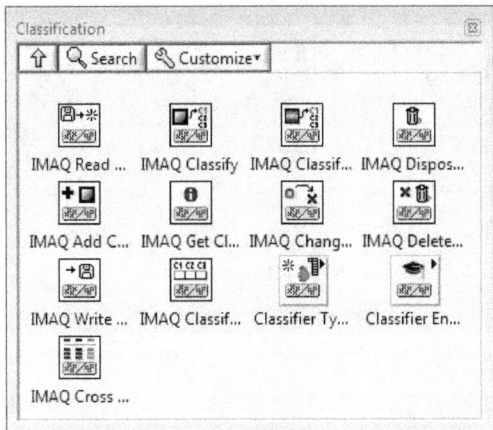

And Vision and Motion → Machine Vision → Count and Measure Objects:

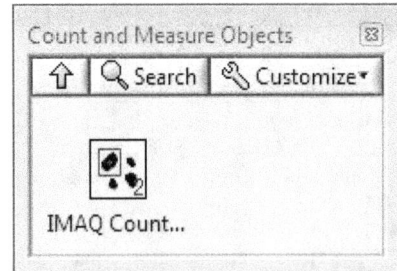

Chapter 12

Tips and Tricks in LabVIEW

LabVIEW is a quite interesting programming language, and despite its odd first impression is very powerful. All LabVIEW field engineers nightmare is to know that the users think that LabVIEW is a simulation environment. Of course it has a very strong simulation part, but the main target is the real hardware controlling.

This chapter will discuss some motivations for programming faster. Then cover several specific tips for the LabVIEW programming language faster. At the end, discuss miscellaneous examples of both new LabVIEW 2013 features and Remote LabVIEW applications that users may not have heard about yet.

12.1 Motivation

One of the best things about LabVIEW being a graphical programming language is the ability to visualize the block diagram that will accomplish a particular task. Unfortunately, since you can't magically convert your mental image to a .vi file, you must construct the VI you're visualizing. Any tips related to programming faster in LabVIEW are ultimately going to come to identifying, and avoiding, bottlenecks in the process of constructing that VI. There are several methods to use into our applications in LabVIEW as the following:

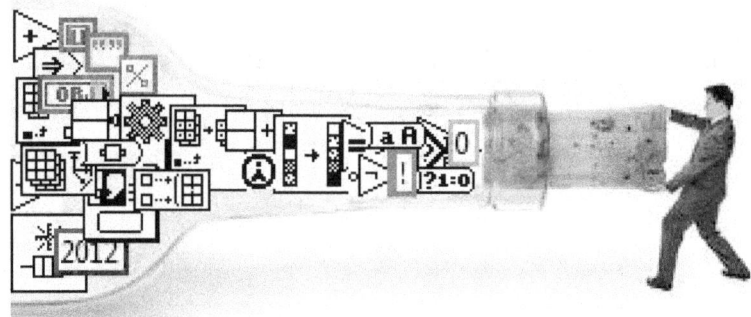

A) Quick Drop is the fastest way to create VIs. For LabVIEW developers who know the names of the objects they need (which I contend is the vast majority of LabVIEW developers), navigating the palettes to find those objects is a huge bottleneck. These are some of Quick Drops might used in many versions:

- **LabVIEW 8.6**

Feature introduced – Press *Ctrl-Space* to invoke. Drop items from Controls and Functions palettes by name Create shortcuts for commonly-dropped items

'cs' instead of 'Case Structure'.

- **LabVIEW 2009**

Drop items from any open project Quick Drop Keyboard Shortcuts (QDKS)

- **LabVIEW 2010**

More QDKS

- **LabVIEW 2011**

Instantly usable on first launch (sort of)

- **LabVIEW 2012 and 2013**

Miscellaneous minor features/improvements

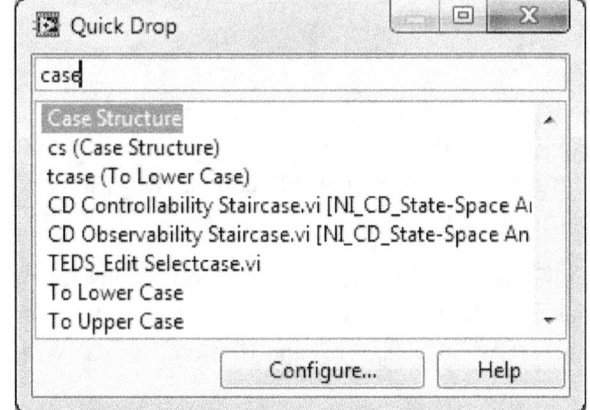

There are additional tips to benefit Quick Drops users;

- **Super Quick Drop** – Click in the VI to dismiss Quick Drop, and the object is dropped where you clicked
- **Quick Drop Fast Search** – Add 'Quick Drop Fast Search=True' to your INI file for more responsive typing in the Quick Drop window. If you find Quick Drop is taking too long to load or isn't responsive enough as you type, try this INI token. This INI token is exposed in the QD configuration dialog in LabVIEW 2013.
- **Common shortcuts** – Google '*quick drop palette object shortcuts*' to download and use some common object shortcuts

New addition of "Wire Multiple Objects Together" shortcut to LabVIEW 2014.

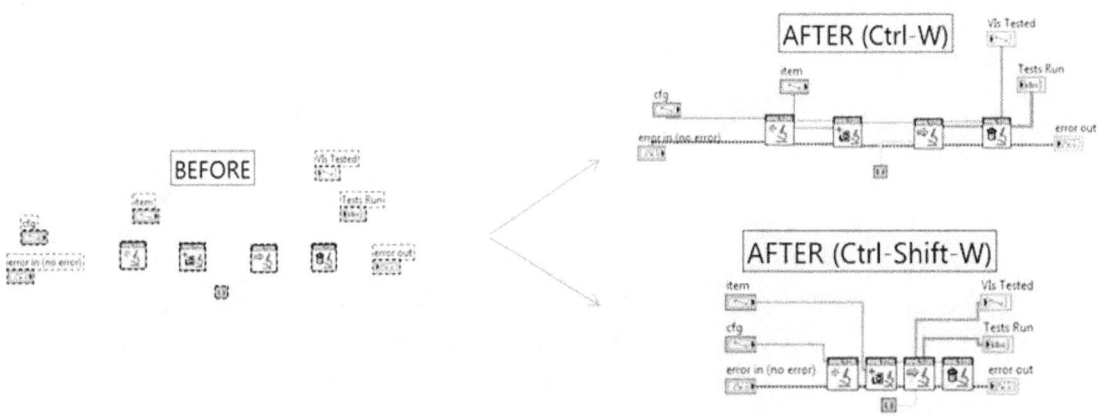

B) Avoid right-click menus if possible.

Adding items to enums and rings

1. Hold the Ctrl key and click in the enum/ring to start editing text.
2. After typing the first enum/ring string, press **Shift-Enter** to create the second string.
3. Repeat until all items are added.

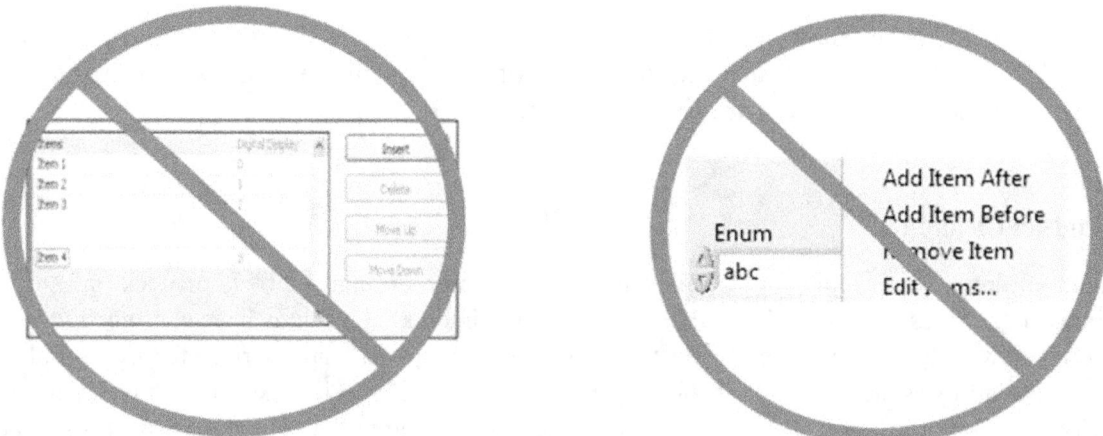

Right-click menus can be bottlenecks as you search for the correct item in the list.

Adding cases to case structures

1. Click in the selector ring of the case structure.
2. Press **Shift-Enter** to add a new case.
3. Press **Ctrl-Shift-Enter** to duplicate the current case.

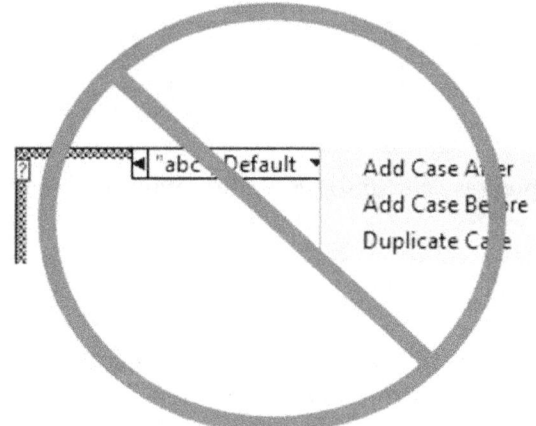

C) Avoid repetitive tasks

- Use the unofficial Quick Drop Ctrl-Shift-G shortcut to add code snippets as droppable QD items, this allows you to drop frequently-used segments of code in a single operation
- Add your own QDKS, to see what the QD community has come up with.
- Customize 'Create SubVI' behavior (LV 2011 and later)
- Add your own project templates (LV 2012 and later)

The Ctrl-Shift-G shortcut allows you to specify a section of code that you want to create a "Place VI Contents" VI out of. Once you do this, then the section of code is available in the user palettes, and (more importantly), as something you can drop in Quick Drop.

D) Diagram Cleanup

Diagram cleanup is a great way to alleviate the "diagram arrangement" bottleneck in LabVIEW programming. It was introduced in LabVIEW 8.6, but the internal algorithms have continued to improve with each new LabVIEW release. Note that diagram cleanup isn't appropriate for every LabVIEW diagram...its utility is maximized with single-screen diagrams with limited nesting, which should be the majority of the VIs you write. For heavily-nested diagrams, and large top-level diagrams, it is still best to arrange those yourself.

> **Use for:**
- Small diagrams (less than 1024x768)
- Diagrams with minimal nesting
- Non-user-visible diagrams

Most of your diagrams

> **Do not use for:**
- Top-level architecture diagrams
- Heavily-nested diagrams
- Diagrams users will see

Diagrams where the arrangement is critical to understanding the operation of the VI

E) Develop with LabVIEW Projects

When you use the LabVIEW Project, you will typically spend less time in file browsing dialogs. It has found file dialogs to be a huge bottleneck in LabVIEW programming, as I typically know exactly where the files are that I need, but I still need to navigate the (often tedious) file system to find them. Having all of the files related to a given LabVIEW application in a single window helps eliminate that bottleneck. Additionally, the Quick Drop window provides easy access to drop controls, VIs, and other objects from any currently-open project into your VIs. This functionality is available in Quick Drop in LabVIEW 2009 and later.

Quicker access to your VIs
(less file browsing)

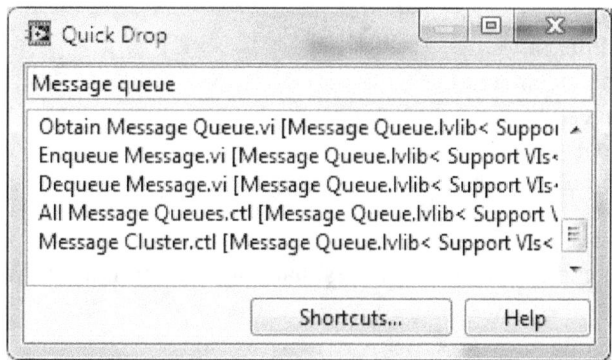

Droppable project items in
Quick Drop (LV 2009+)

F) Connector Panes

In LabVIEW 2010, we received the ability to quickly swap two terminal positions on a VI's connector pane. To use this feature, simply click on a terminal in the connector pane, then control-click in another terminal. The terminals will switch places. If either of the terminals is an empty terminal, then the wired terminal will move to that empty terminal. In LabVIEW 8.5, the 'Connector pane terminals default to Required' setting was added to Tools > Options > Front Panel. This setting changes the default terminal connection type on subVI terminals to 'Required' instead of 'Recommended'. This feature can save hours of debugging time for those situations where you added a new terminal to a subVI, but forgot to wire it in the calling VI. Note that this feature ignores 'error in' terminals.

➢ Fast Connector Pane Switching (LV 2010)

➢ Required terminals by default (LV 8.5)

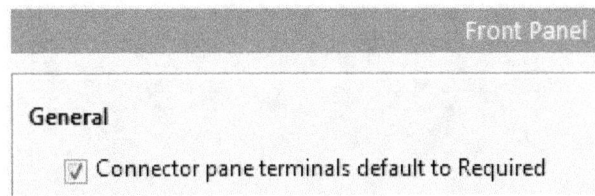

12.2 New LabVIEW 2013 features

There are some of the features you've probably heard about:

- ○ Event Structure and API Improvements
- ○ Improved Excel Integration
- ○ Mouse Wheel Enhancements
- ○ Bookmark Manager
- ○ Attached Comments
- ○ Overhaul of Shipping Examples

The following steps are some of the main features that may used in LabVIEW 2013:

1. Quick Drop Enhancements

- • Typo Help (No more Auto complete)

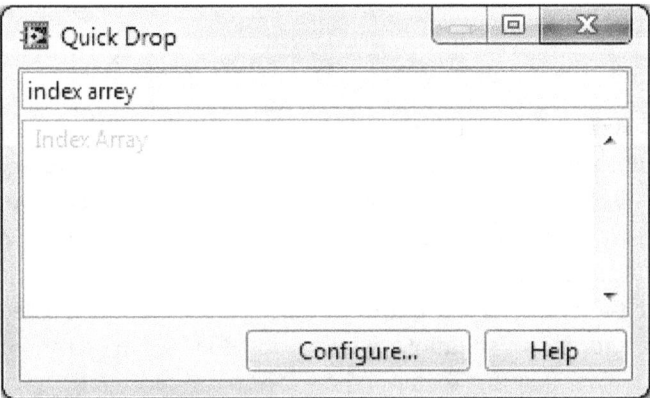

- Insert respects selected segments

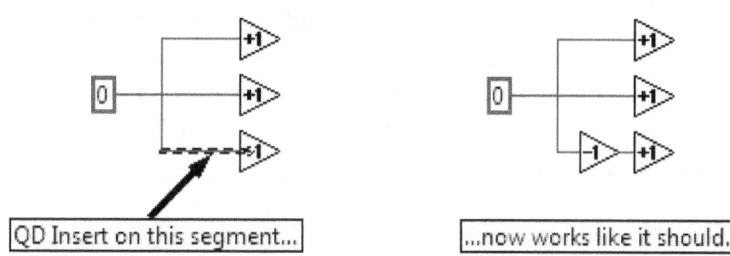

No More Auto complete: When you type in the Quick Drop text box, your text will no longer be auto-completed. Instead, the text you type now acts as a filter for the content in the results list. The first item in the list is always selected by default (but you can still navigate the list with the arrows), and the text you type is never changed in any way by the Quick Drop window.

Typo Help: If you mis-type the name of an item, Quick Drop will display the last item that did match what you typed, and this item will be dropped, regardless of the typo.

Insert Respects Selected Segment: When you perform a Quick Drop Insert (Ctrl-Space-Ctrl-I), the new object will be inserted on the selected segment of the wire. Previously, the insert always took place on the first segment of the wire after its source, regardless of segment selection. Note that this improvement does *not* currently apply to a smart insert on multiple wires (Ctrl-Space-Ctrl-Shift-I).

2. VI Server/VI Scripting

- Get GObject Label.vi – This new VI allows you to retrieve the label text of any GObject. Prior to having this VI, you had to cast the GObject into the proper class before reading its Label. Text property.

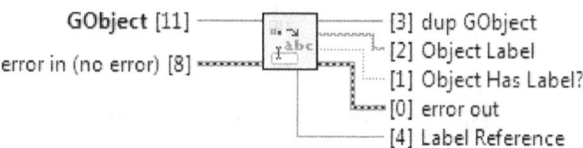

- Event Structure scripting

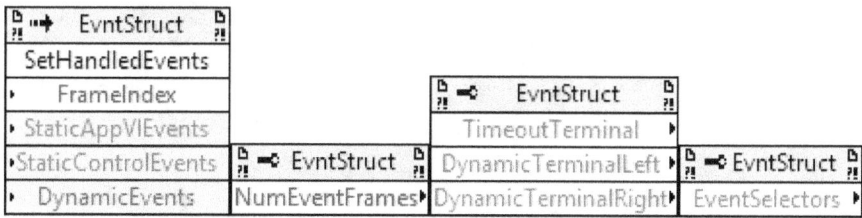

New properties/methods – This is just a sampling…the full list is in the LabVIEW Upgrade Notes. Some of these properties/methods aren't actually "new", but rather were formerly private, internal properties/methods that have been updated for public use. Event Structure scripting – Prior to LabVIEW 2013, there was no Event Structure-specific interface in LabVIEW Scripting, making programmatic manipulation of Event Structures impossible.

3. **Dialog Improvements**

Find VIs on Disk   Project Find

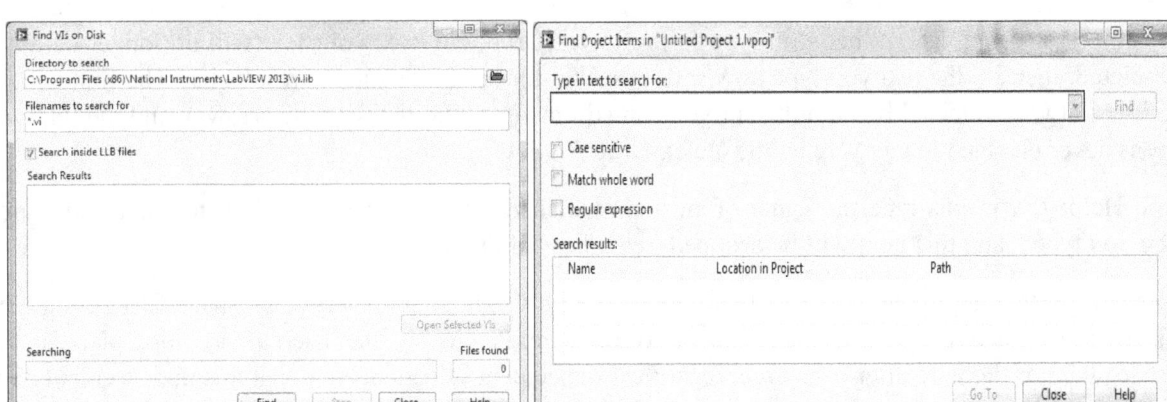

- Find VIs on Disk – This dialog has been updated to persist its window size and position between uses (and between LabVIEW sessions). It also remembers the last searched directory, and has more intuitive key navigation for quick browsing of found files.

- Project Find – You can sort results by clicking column headers in the results table. This makes it much easier to parse large numbers of results.

12.3 Very long HEX String to ASCII Conversion

It is shown a method in LabVIEW how to convert a very long HEX string to ASCII. This could be useful if someone wants to analyze memory dumps or binary files, which are mostly in HEX, but to be interpretable, we need to convert it in some form, which we can understand, and mostly we convert it in ASCII. As we can see in figure 12-1 the program is working.

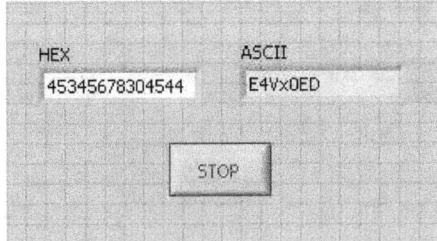

Figure 12-1 HEX to ASCII Front Panel

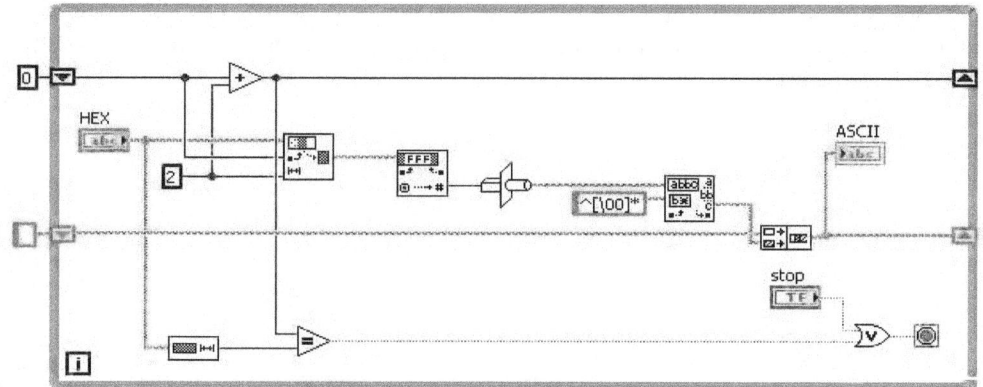

Figure 12-2 HEX to ASCII Block Diagram

In figure 12-2, we have the solution for the problem. First of all we have to enter a very long HEX string. With the String Subset VI we have to fragment the string into groups with 2 HEX numbers (8 bits). The first Shift Register with 0 initialization will always increase the fragmentation index with 2, this way we will reach at the end minus 2 (n-2) of the HEX word. There is a condition too, if we reach at the end of the HEX string to exist the While loop.

From String Subset we convert the HEX string to number, the number is converted to ASCII with the Type Cast VI. For some strange reason the Type Cast VI puts an ASCII space before the converted ASCII character, this space is a \00 type space, so this will have to be deleted, it's deleted with the Match Pattern VI using the ^[\00]* command and wired to the after substring output (after the 00), and we will obtain the good converted ASCII value. This value will be concatenated with the following strings using the second Shift Register with the empty string initialization.

12.4 Very long ASCII to HEX string conversion

If we made the HEX to ASCII conversion, than we have to do the reverse operation too. In LabVIEW ASCII to HEX the conversion is not that simple. Unfortunately there is no VI which will do the job, so the user has to make his own VI. On possible method will be presented next. In figure 12-3 we've seen the working program.

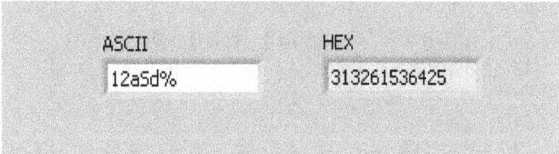

Figure 12-3 ASCII to HEX Front Panel

In figure 12-4, We have the ASCII input, which will be converted in a number, with byte data type. Byte is more than enough, because we have a total of 128 ASCII characters, in the other hand in LabVIEW there is no ASCII string to number conversion in with other data type, so this is our only solution.

The numbers are converted into HEX with the length of 2 for each HEX number. The string to number conversion makes an array, this way the HEX values will be an array of HEX string with the length of 2. We want to make a long HEX string, not an array of strings. We made a For loop with the number of iteration equal to the length of the HEX array. We indexed the array and with the use of Shift Registers we concatenated it into a long HEX string.

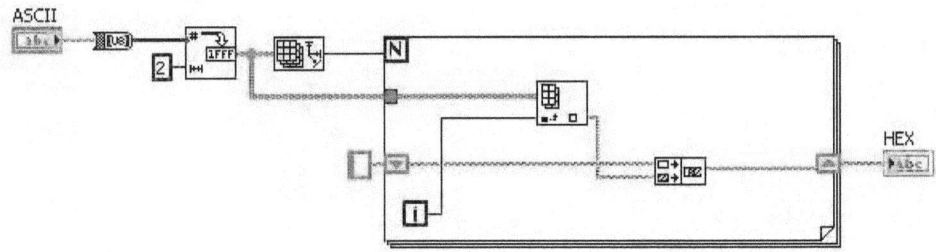

Figure 12-4 ASCII to HEX Block Diagram.

12.5 Creating a toggle button from a push button

This application would be useful when we want to have an LED light to be ON, for very short impulses, like the sound of claps of falling coins, otherwise we would see only a LED blink and we would have to be very attentive. The idea would be to keep the signal on "1" logic for a longer period with shift registers. There could be many variants, but the hardest variant will be presented, the one where after the first "1" logic the light is ON and after the second "1" logic the light is OFF. For the Button the "Latch When Released" Mechanical Action was used, so it's a push-button. As we can see the program's name is Button Latch (figure 12-5.), we need a latch to keep the LED ON after only one impulse and to turn the LED OFF after the second.

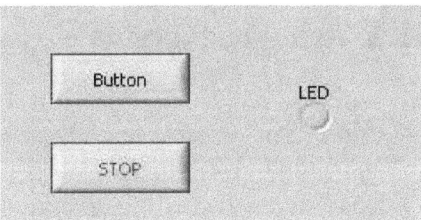

Figure 12-5 Button Latch Front Panel

In figure 12-6 we can see that Shift Registers were used. As we can see the initialization value for the Shift Registers is FALSE. The Button is connected to the Implies VI, which computes an OR logic between the negated x (first) input and y (second) input. In case of Button press, we will have FALSE in x input and FALSE in y input, so FALSE output, this means that from the Select VI the FALSE output is activated, which has an input of inverted FALSE (which is TRUE), so the LED will be ON.

If the Button is pressed again, then we will have FALSE in x input and FALSE at y input of the Implies VI, so at the Implies function will have FALSE output. The input of the second Shift Register will be TRUE, because before this the LED in ON. With the FALSE Implies VI we activate the inverted TRUE (which is FALSE), so the LED will be OFF.

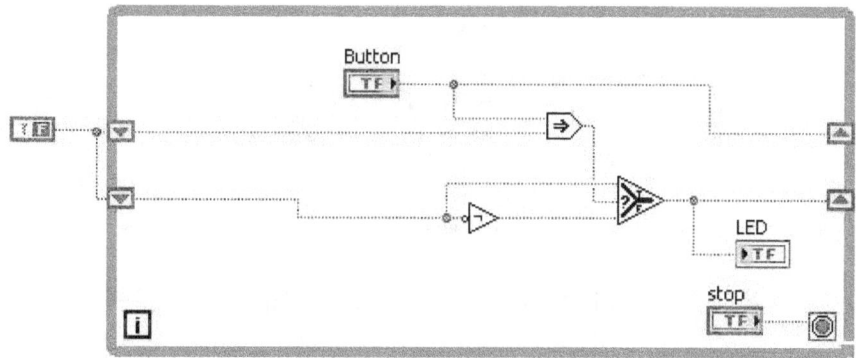

Figure 12-6 Button Latch Block Diagram

12.6 Append file (Write to file line by line)

Surprisingly in LabVIEW we can't find a simple method to append a file. If a program writes to a file for the second time, it will overwrite the previous data. This is what we want to avoid, when we want to make complex logs or even when we want to save the parameters of a program during execution.

This program is useful when we make some moves, like mouse, joystick moves, and we want to save the coordinates into a file and maybe the file will be closed an opened in the program more times. We can create programs to move robotic arms with mouse, or to save the specific parameters of measuring equipments. Figure 12-7 shows that after multiple runs of the program, the previous text is not erased.

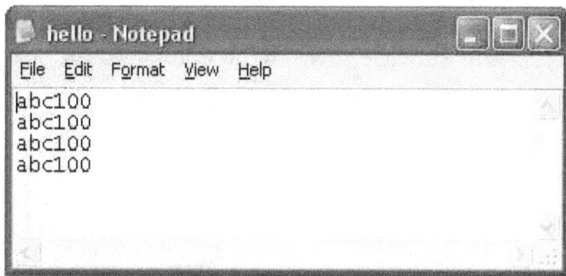

Figure 12-7 Text File with Appended Text

In figure 12-8 we have the standard dataflow of the programming with the settings and the writing to the file, between the opening/creating and closing the file. At the end it's good to put a Simple Error Handler. The file opening/creating VI is set to open or create and a certain path is given to it.

The file setting VI is set to have an offset with the length of 0, and is set to write at the end of the file. The writing VI has an input with a text which is concatenated with a new line constant, this way after any running of the program we will have the appended text in a new line, keeping the old information.

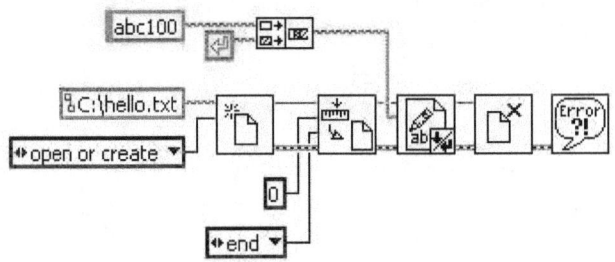

Figure 12-8 Text Append Block Diagram

12.7 Create an arbitrary signal

As we know LabVIEW has a lot of built in signals, but it's impossible to have any arbitrary signal. It has the possibility to create arbitrary signals, but we have found it not really flexible in some situations. It has an arbitrary signal creator with Express VIs, but we always avoid using them, because you can have more control over traditional VIs. The chosen signal was the trapezoidal signal, which is mostly used when controlling motors. Nobody wants a square pattern for the motors acceleration. In cars, in lifts the motor usually accelerates in a trapezoidal pattern.

As we can see in figure 12-9, the program is working. The hardest part is configuring the parameters. The parameters are configured using equation (1).

$$N_p.U_{minr} = A-U_o, \ N_p.U_{minf} = A-U_o \ \dots\dots\dots\dots\dots\dots\dots\dots\dots(1)$$

Where N_p is the number of points; U_{minr}, U_{minf} are the minimal voltages for rise and fall; A is the amplitude of the signal; U_0 is the start voltage. In our case it will be: 0,0003 * 10000 = 5 – 2, this way we created the trapezoidal signal.

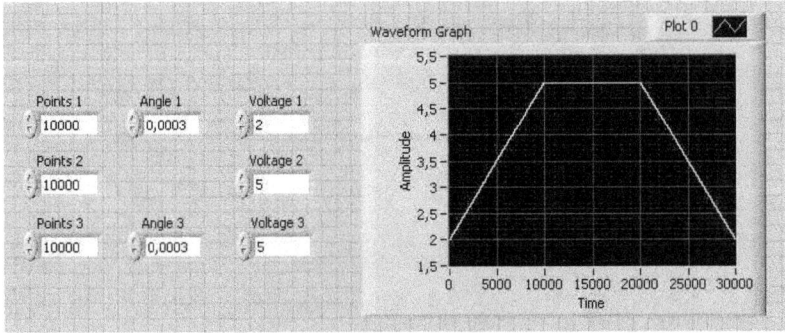

Figure 12-9 Trapezoidal Signal Front Panel

296

The Block Diagram (figure 12-10) it is not so complicated, the main thing is the idea to build arrays of rising edge, continuous part and falling edge. The building of the arrays is made with For loops, shift registers and the exit from the loops is made with enable indexing, which is by default at For loops. Maybe the hardest part is the building of the array, which builds three 1D arrays intro one 1D array. Normally LabVIEW makes the built array in multidimensional array, but we need a 1D array, because we want one single signal. This can be done by changing the enable indexing to disable indexing and then change it back to enable indexing again at the exit from the For loop of the horizontal signal.

12.8 Virtual instruments for the PXI chassis

In the remote LabVIEW a six PXI experiments working and controllable trough a web browser.

12.8.1 Transfer characteristic of a NAND gate

The first experiment makes the transfer characteristic of a gate; we made it for a NAND gate. In figure 12-11. we can see the block schematics of the transfer characteristics of the NAND gate. We generate a rising ramp signal at the NAND gates one input and supply a constant voltage to the other input and measure the voltage at the output, this way we have the transfer characteristics of the gate.

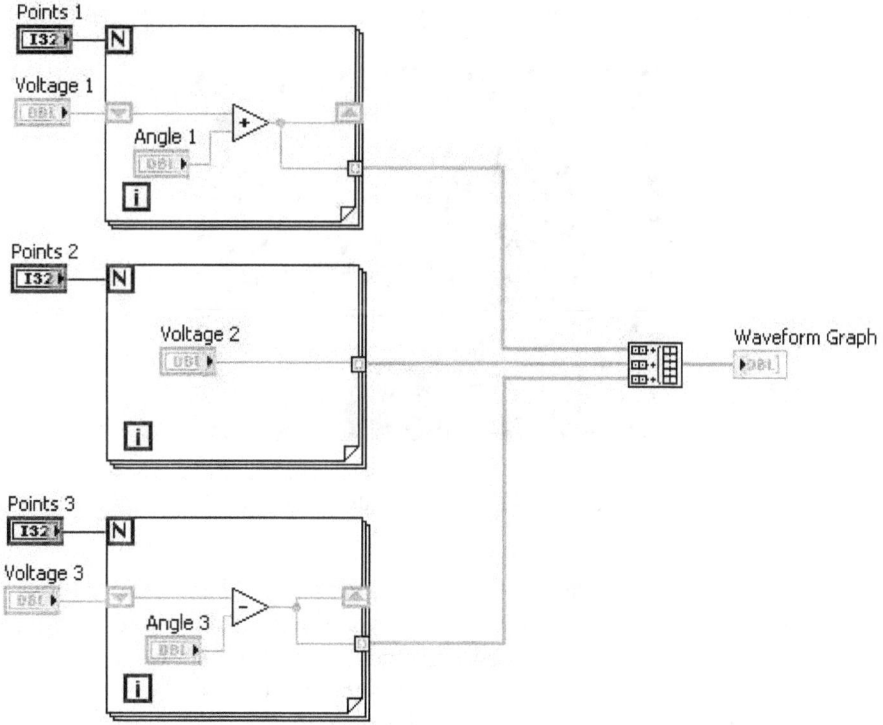

Figure 12-10 Trapezoidal Signal Block Diagram

297

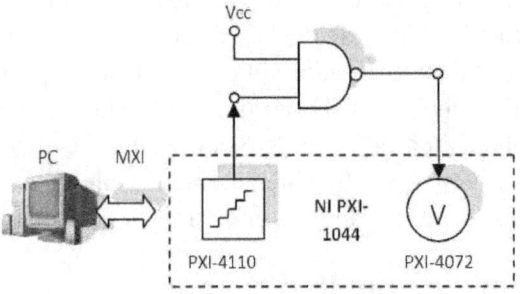

Figure 12-11 Block Schematics of the Experimental Setup

The transfer characteristic is the input voltage as a function of the output voltage like shown in equation (2).

$$U_{out} f(U_{in}) \dots\dots\dots\dots\dots\dots\dots(2)$$

In figure 12-12, we can see the experimental setup of the NAND gate on a PCB with protection and connectors for accessing. The used NAND gate is integrated in a 74ACT00 IC. The used NI equipment for this experiment was the NI PXI-4110 power supply and the NI PXI-4072 digital multi meter.

Figure 12-12 NAND Gate with PCB

In figure 12-13, we can see the Front Panel of the experiment. As we can see we have the digital multi meter (DMM) configuration with reference ID, a range at 5 V and resolution at 6½ digits. We have two channels (0 and 1) of the power supply activated with the current limit set to 100 mA and a slider to set the voltage level from 0 – 4 V. The most important part is the XY Graph where we see the actual transfer characteristic of the gate, we have also indicators of the temporary value on both axes of the graph. We have also cursors to measure certain values. As we can see is not so complicated to program, the hardest part is maybe that we use traditional NI acquisition cards, not the newer DAQ cards.

In figure 12-14, we have the Block Diagram of the program. All the instruments are programmed in this pattern, first we have to create the channel and after it close it. Between these two VIs we have the configuration or acquisition and generation. Mostly we have the creation, configuration and closing outside the loop and the acquisition or generation part in the loop.

In out experiment we first create the channel for the DMM and than configure its range ad digits. After it we will create the power supply's channel we configure the voltage and current for the first channel and enable the output, after we will do the same or the other channel. For channel 1 we will measure the output voltage. We will measure the voltage with the multi meter at the output of the NAND gate. Finally we will close an reset the two equipments. We can remark that LabVIEW uses the read expression for acquisitioning and the write expression for generation, this way the used icons for the VI are mostly a pair of glasses for reading and a pencil for writing.

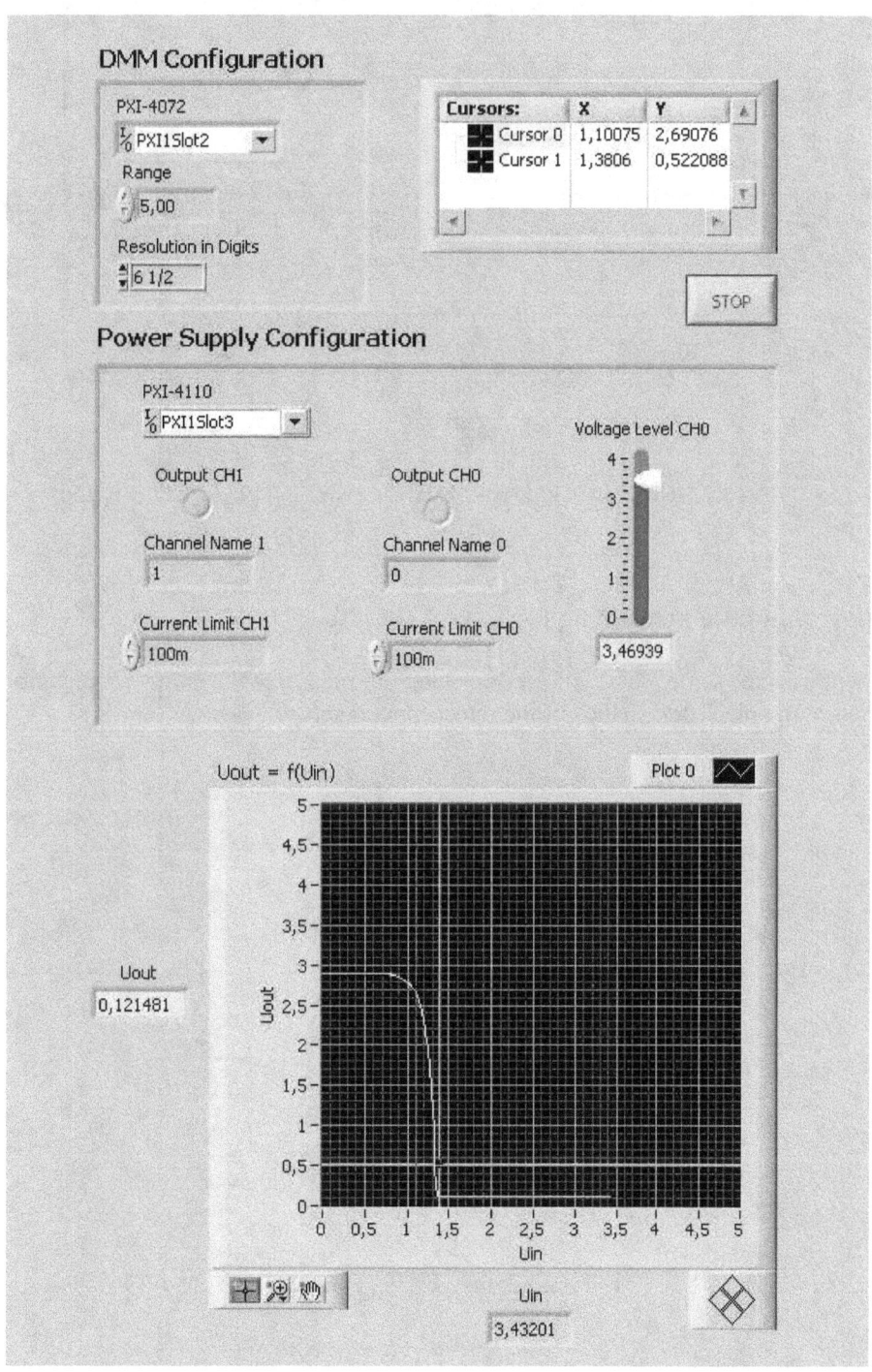

Figure 12-13 Front Panel of the Transfer Characteristic Program

299

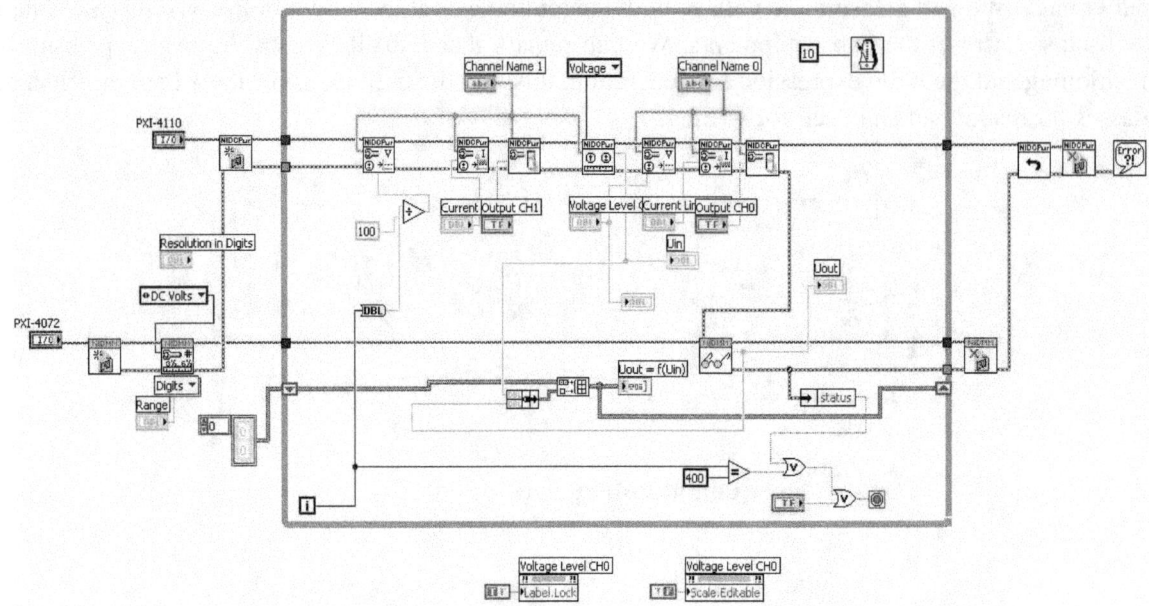

Figure 12-14 Block Diagram of the Transfer Characteristic Program

12.8.2 Propagation time measuring for a NAND Gate

This experiment uses the same 74ACT00 NAND gate, but measures the gate's propagation time. In figure 12-15. we can see the block schematics of the experimental setup.

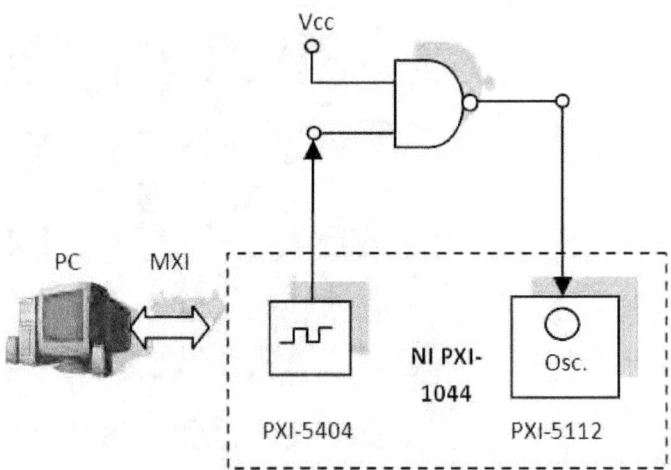

Figure 12-15 Block schematics for the propagation time measuring.

We generate a square waveform to the gate's one input with a signal generator and measure its output with an oscilloscope. The propagation time is given by the formula from equation (3).

$$t_p = (t_{pHL} + t_{pLH}) / 2 \quad \ldots\ldots\ldots\ldots\ldots\ldots\ldots\ldots\ldots\ldots\ldots\ldots\ldots\ldots(3)$$

In the standard TTL gate's case t_{pLH} = 12 ns and t_{pHL} = 8 ns, so tp = 10 ns. The experimental setup can be seen on figure 12-12. The setup is on the same PCB as for the transfer characteristic. The used NI equipments are NI PXI-4110 power supply, the NI PXI-5112 oscilloscope and the NI PXI-5412 signal generator. In figure 12-16, you can see the Front Panel of the experiment.

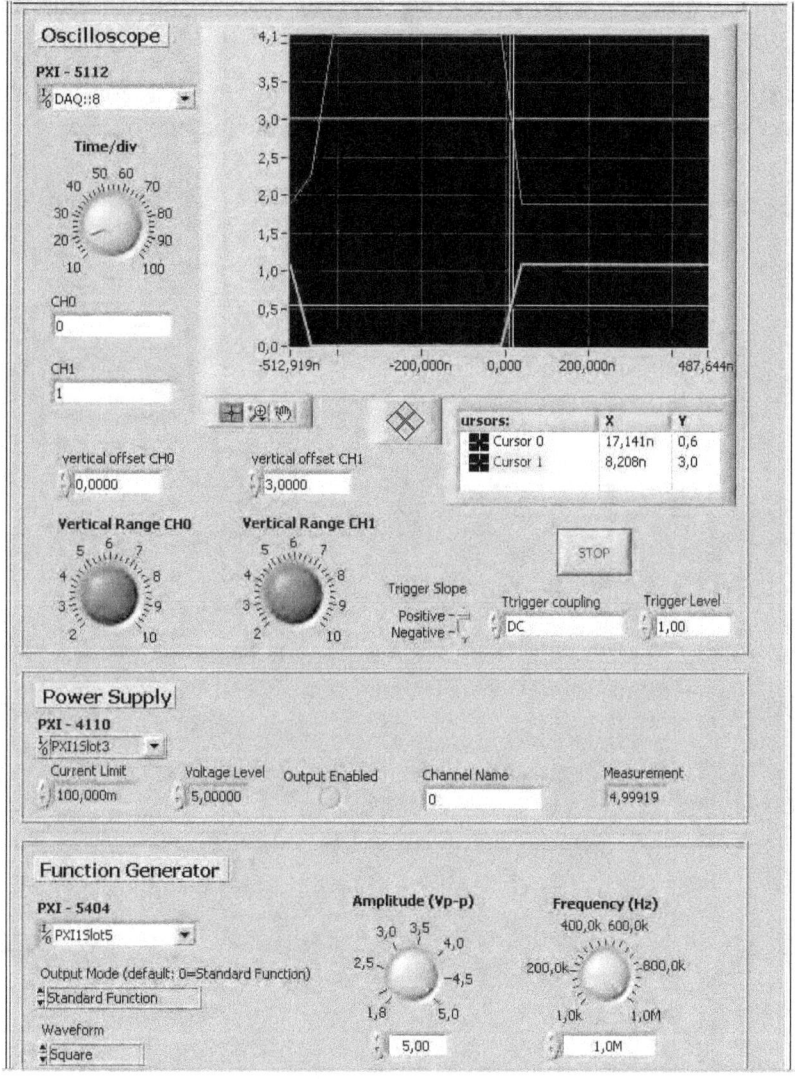

Figure 12-16 Front Panel of the Transfer Characteristic Program

As we can see we have the configuration of the oscilloscope, the power supply and the function generator. The oscilloscope has a configured resource ID and the Time/div setting. It has also both channels activated with 0 V offset for channel 0 and 3 V offset for channel 1 and Volts/div (Vertical Range) dial.

301

It also has some triggering configurations like slope on positive, DC coupling and 1 V level. We have the both graphs (input and output of the NAND gate) on the waveform graph and cursors to measure the propagation time.

The power supply has a basic configuration just for supplying current to the IC. It has resource ID, current limit to 100 mA, 5 V voltage level, output enabled indicator, an indicator showing 0 as the selected channel, and a measurement of the output of the activated channel. The function generator has resource indicator, the settings for output mode, which is on standard, square waveform type, and the setting for amplitude, which is at 5 V and the setting for frequency, which is at 1 MHz.

In figure 12-17,we can see the Block Diagram of the program. As we can see we have the initialization for the function generator, the setting of the output mode, the amplitude and frequency setting, the enabling of the output and the starting of the generation, the oscilloscope is initialized and has standard initialization.

Finally the power supply is initialized. We've enter in the While loop , it configures the voltage and the current of the power supply; we enable its output and measure the output voltage.

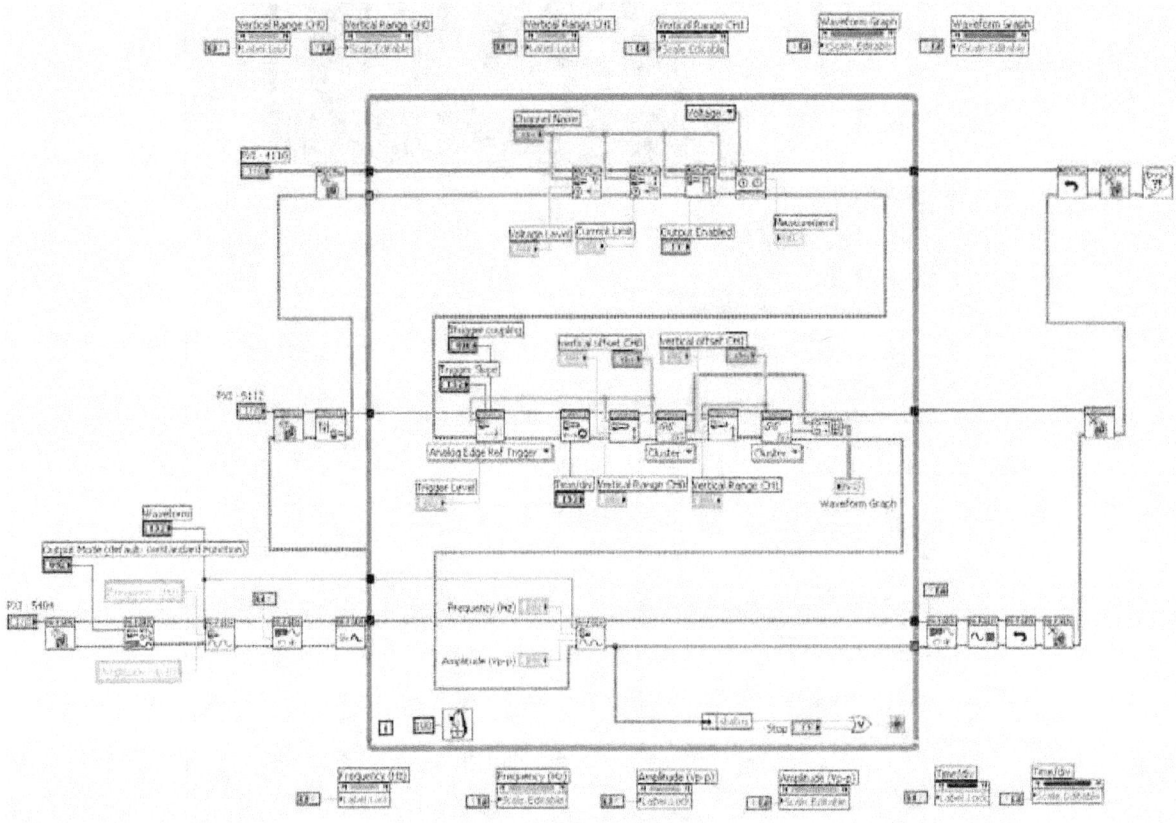

Figure 12-17 Block Diagram of the Transfer Characteristic Program

We continue to the oscilloscope, we configure the trigger settings, the timing settings. After we make the settings for the vertical range and start the readings of the values for representing on the graph both of the channels (0 and 1). We unite the two channels with a Build Array and output it on a Waveform Graph. For the function generator we can control the frequency and amplitude during the program execution, this way we have it inside the loop too.

302

When we exit the loop we disabled the output, we stopped the generation we resettled the device. Finally we close the function generator, the oscilloscope, we reset and close the power supply and end the whole execution with a simple error handler to have the error messages in case if something goes wrong.

12.8.3 Duty cycle analyzer

The duty cycle analyzer experiment has a more complicated setup. We have the 33250A signal generator from Agilent and the NI PXI-6541 logic analyzer from National Instruments. In figure 12-18, we have the block schematics. We have the Agilent signal generator, which generates a square signal to the Duty Cycle Analyzer.

The signal is generated in the logic control and in the PLL. The signal that exists the PLL will be entering in a counter and after I in a display module. The signal is gathered with a logic analyzer from National Instruments and shown on a graph.

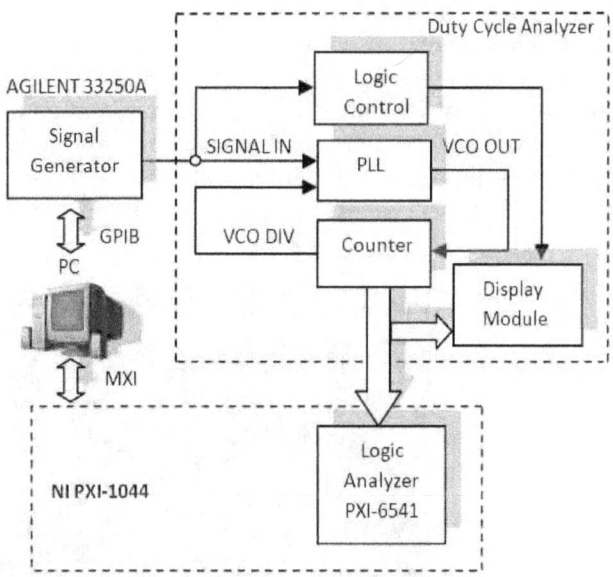

Figure 12-18 Block Schematics of the Logic Analyzer Experiment

We have also an external circuit which can be seen on figure 12-19. This external circuit is a duty cycle analyzer on which 14 different signals are analyzed with the experiment. The circuit has more modules like a logic control, a PLL and a counter and the duty cycle of the square signal is shown on the dual digit seven segment displays.

Figure 12-19 Experimental Setup of the Logic Analyzer Experiment

In figure 12-20, we can see the front panel of the Duty Cycle Analyzer program. For the logic analyzer we have a resource ID, 14 activated channels, and the clock rate at 300 kHz and 10000 acquired samples. For the Agilent signal generator we have resource ID, square waveform type, 100 Hz for frequency, 5 V amplitude, 25 % duty cycle, 0 V offset. The first signal (Clk) represents the duty cycle. The second signal (VCO_Out) is the representation of the PLL. The third signal (VCO_Div) is one input for the PLL. Q0 – Q7 is the counterpart. Mst_QA, LD_Dcd, Mst_nQ represent logic levels for testing purposes.

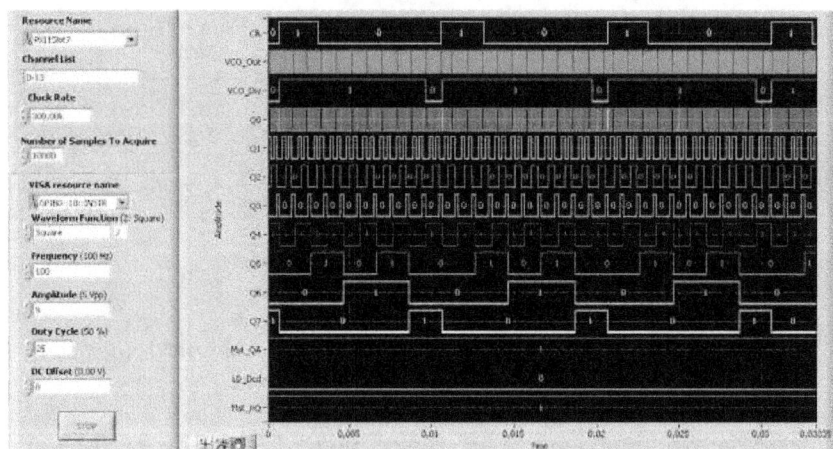

Figure 12-20 Front Panel of the Logic Analyzer Experiment

In figure 12-21, we have the block diagram of the experiment and the initialization of the logic analyzer. In a while loop we have the configuring of the channels, settings of the clock, setting of the buffer (Number of Samples To Acquire) and finally we represent the data. The Agilent signal generator is initialized, the amplitude, frequency, offset and waveform type is configured. We continue with the duty cycle is configuration, the output is enabling and at the end we close the signal generator. Finally the logic analyzer is closed too, and the program is ended with a simple error handler.

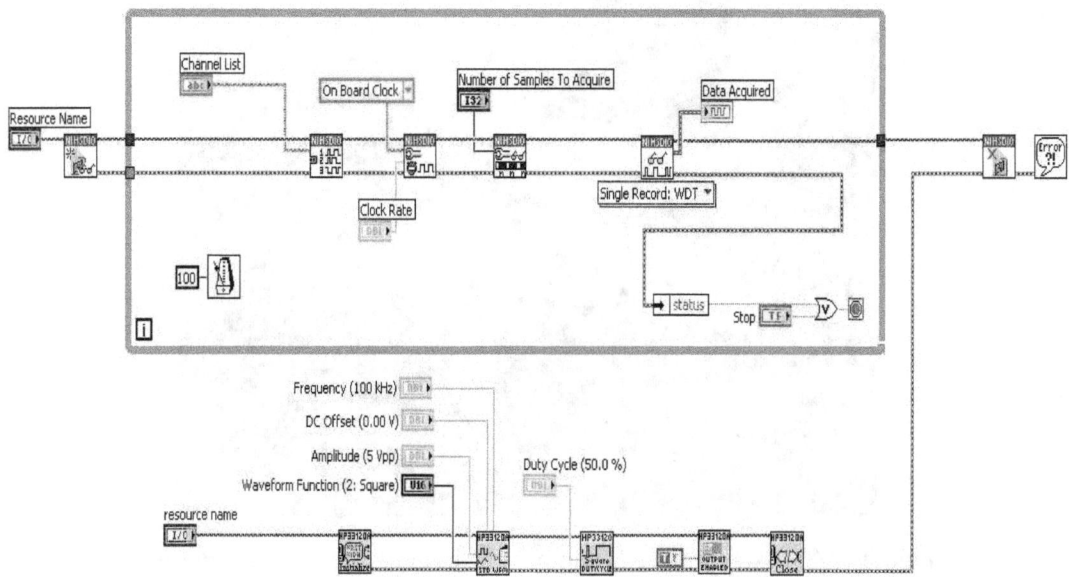

Figure 12-21 Block Diagram of the Logic Analyzer Experiment

12.9 Simple motor control

In figure 12-22, we can see the Block Diagram of the experiment. This experiment uses only PXI instruments and thee PICDEM Mechatronics board from Microchip for amplifying and PID control. The signal is amplified for the motors with the PICDEM Mechatroncis board; this board has also an implemented PID algorithm.

The motor has a disk with slots, which rotates between an Opto-coupler pair, similar to old mouse with ball. The signal from the Opto-coupler is sent in a signal; amplifying and conditioning circuit. The output signal is sent to the NI PXI-6608 counter and with a simple formula the RPM is calculated. The signal that the Opto-coupler reads should be similar to the generated signal, in our case both trapezoids should look the same.

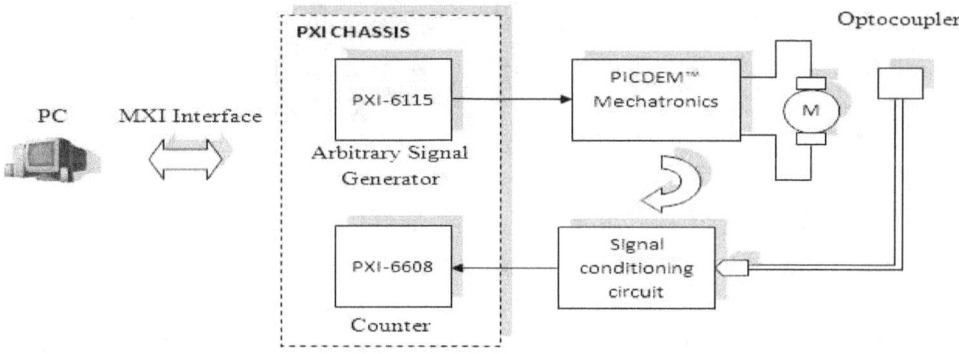

Figure 12-22 Block Schematics of the Simple Motor Control Program

305

In figure 12-23, we can see the PICDEM Mechatronics board and the motor with the disk with two slots.

Figure 12-23 Experimental Setup of the Simple Motor Control Program

In figure 12-24, we have the Front Panel of the program in LabVIEW with the trapezoidal signal generation on Waveform Graph and the RPM readings represented on Waveform Graph.

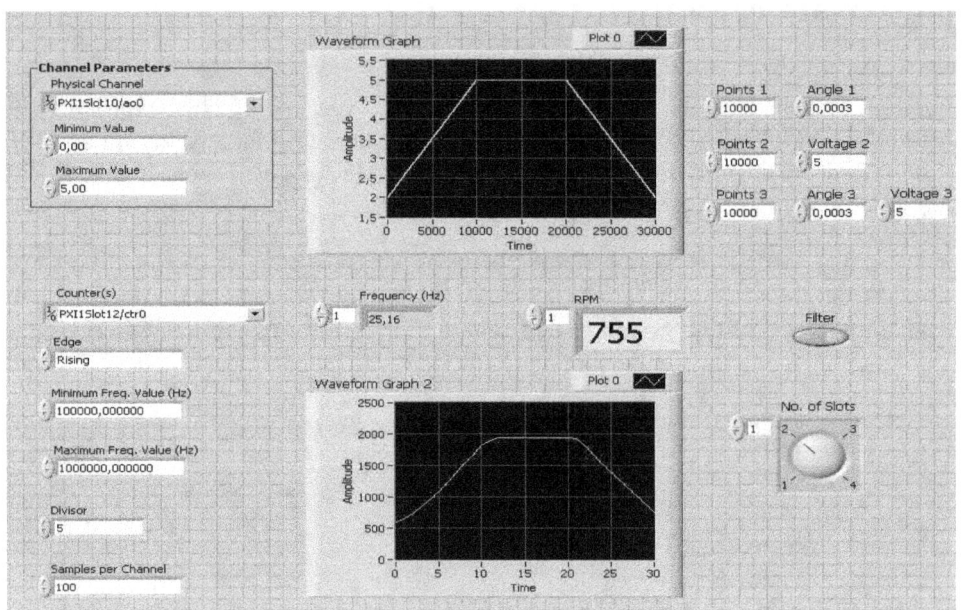

Figure 12-24 Front Panel of the Simple Motor Control Program

In the upper part of the Front Panel we have the signal generation configuration, where we have the resource ID of the signal generation DAQ and the minimum value a t 0 V and the maximum value at 5 V.

The second part of the Front Panel is the counter configuration with resource ID, rising edge, minim frequency at 100 KHz and maximum frequency at 1 MHz, 100 samples per channel and the sampling divisor is 5.

The speed of the AC motor is determined primarily by the frequency of the AC supply and the number of poles in the stator winding, according to the relation: RPM = 2 * F * 60/p where RPM = (Synchronous) Revolutions per minute F = AC power frequency p = Number of poles, usually an even number but always a multiple of the number of phases. Therefore, there are indicators for frequency and RPM , also a median filter can be deactivated with a button and have a slot setting dial and must be set to be equal with the number of slots of the disk present on the motor.

In figure 12-25, we can see the Block Diagram of the program. We have two parallel while loops. The first loop is for the trapezoidal signal generation. We have first the creation of the channel, after we have the start of the generation process. Next we enter the While loop, this is the method how a signal is generated by a DAQ board; we provide the numbers (samples) of the signal and the use the write signal VI, which has the icon with the pencil. After we exit the loop with closing the instrument (delete task) and we finish the program with the simple error handler.

The second loop is for the counter configuration. We have channel creation for the counter, the timing settings (sample configuration), the start of the execution continues, followed by the signal reading, which is filtered and with a specific formula we calculate the RPM and represent it on a Waveform Graph. Finally we close the instrument and we handle errors. We put everything in the loop to be controllable during execution.

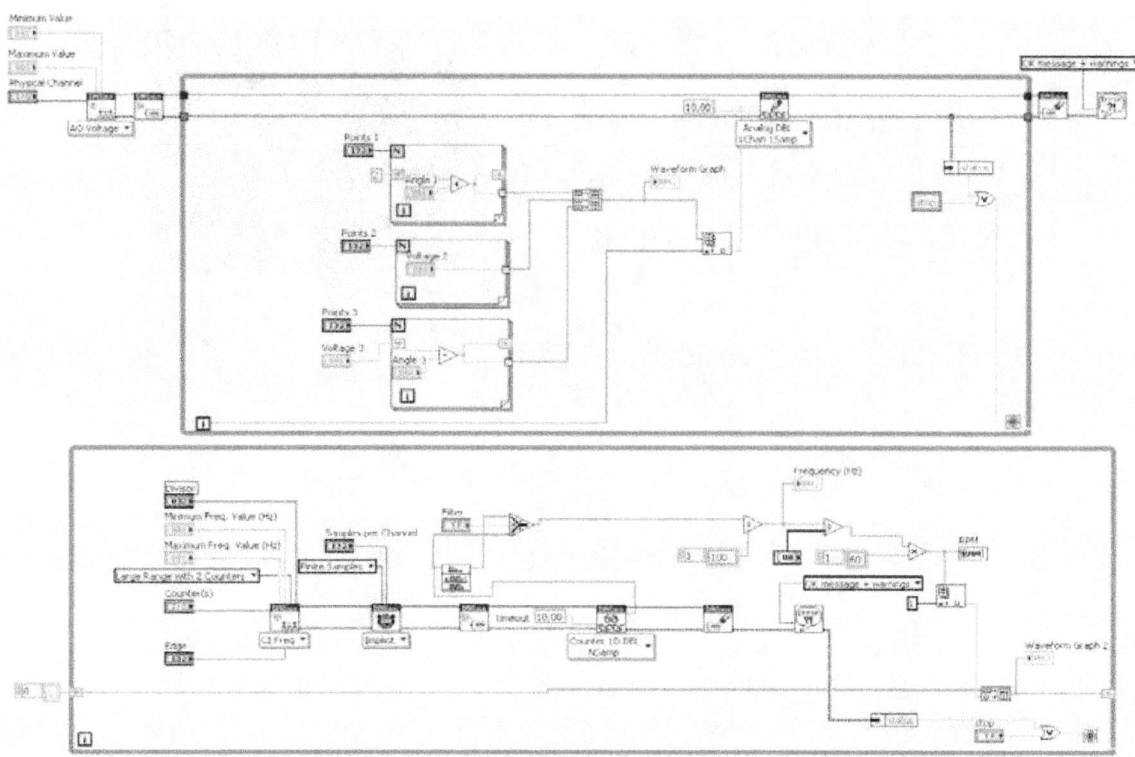

Figure 12-25 Block Diagram of the Simple Motor Control Program

307

12.10 Simple temperature measuring

In figure 12-26, we can see the setup for the temperature measurement. We have an LM35 centigrade sensor and an NI USB-6251 acquisition board. We also made the experiment with the NI USB-6009 smaller acquisition board and worked very well. The idea of this experiment is to measure voltage given from the temperature sensor and with the formula from the datasheet we convert the voltage to temperature.

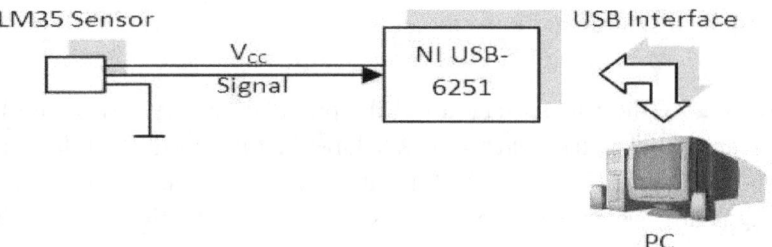

Figure 12-26 Block Schematics of the Simple Temperature Measuring

In figure 12-27, we have the Front Panel of the program. We have resource ID, timing rate at 1 and the actual temperature shown on an indicator and on a Waveform Graph.

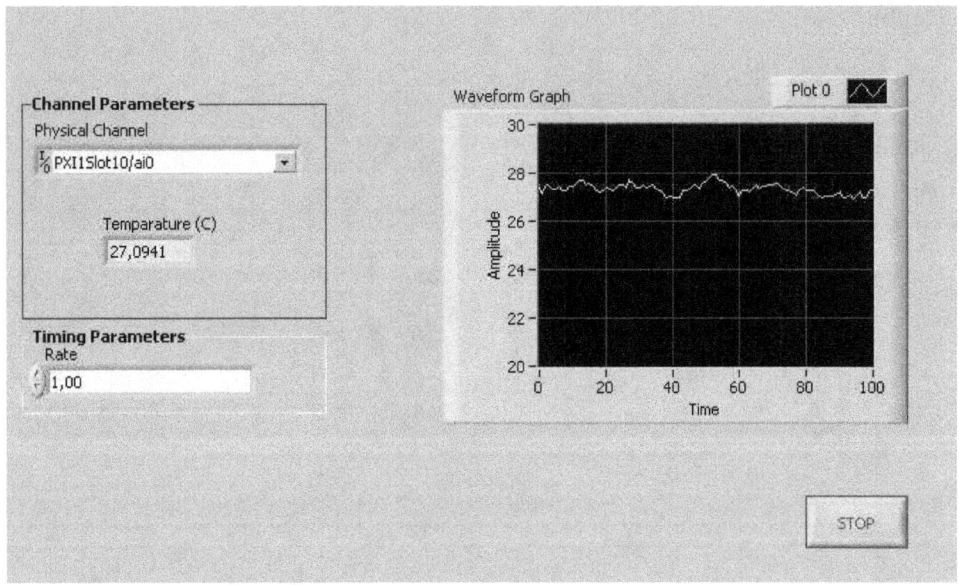

Figure 12-27 Front Panel of the Simple Temperature Measuring

In figure 12-28, we have Block Diagram of the program. We start with the channel creation, the setting of the timing (samples) and the start of the process. We enter in the While loop, here we have the reading of the samples and some mean value calculation over 10 samples.

The temperature at the output of the LM35 centigrade sensor is its output voltage multiplied by 100. Finally we close the instrument and handle errors.

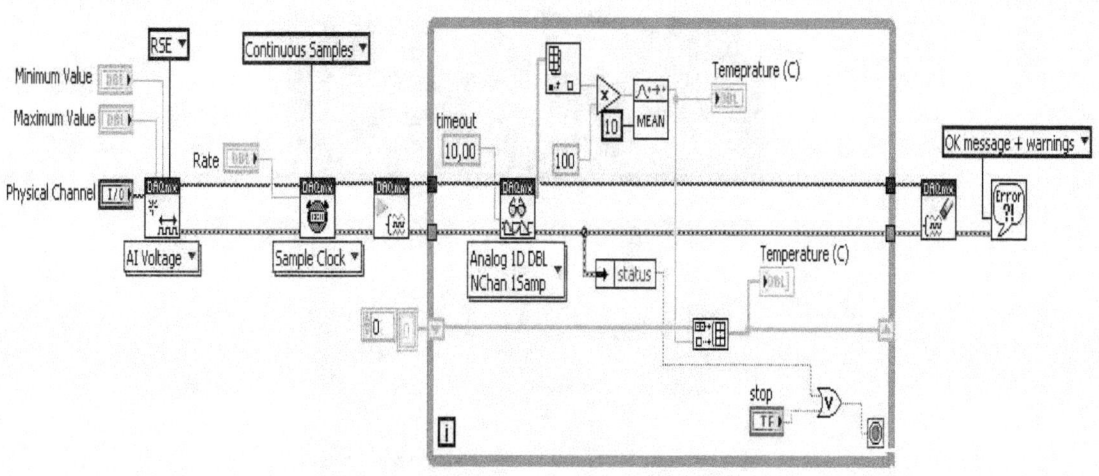

Figure 12-28 Bock Diagram of the Simple Temperature Measuring

12.11 Power supply testing

This experiment represents a more complex functional test. If we have many samples it's not the best solution. For many samples graphs are not indicated, but for only one sample this is the best test. With a graphs we can see really if the power supply is working. The signal is generated with an Agilent 33250A function generator and the input and output signals are viewed using an NI PXI-5112 oscilloscope (figure 12-29).

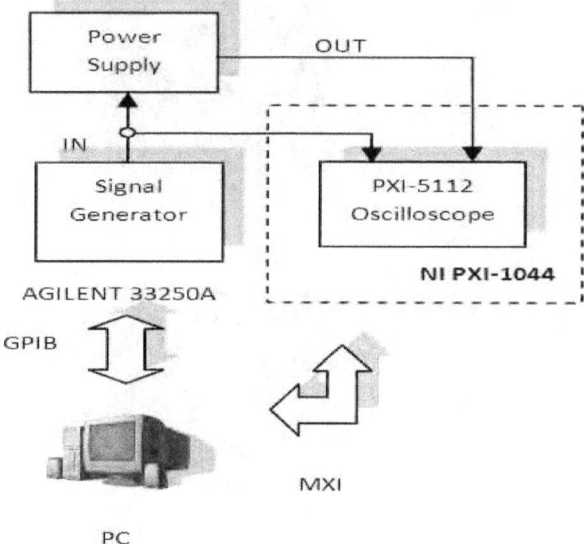

Figure 12-29 Block Schematics of the AC-DC Power supply

In figure 12-30, we can see the AC-DC power supply. In figure 12-31, we have the Front Panel of the Agilent 33250A signal generator. As we can see we have resource ID, the waveform type is sine wave, the frequency 100 kHz, the offset is 0 V and the amplitude is 5 V.

Figure 12-30 Power Supply

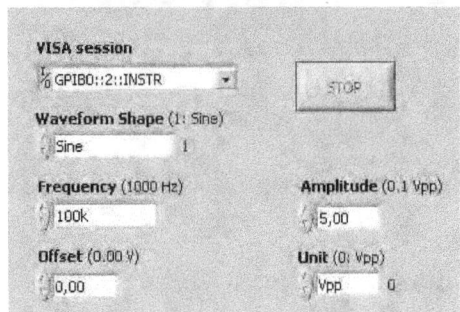

Figure 12-31 Front Panel for the Agilent 33250A Function / Arbitrary Waveform Generator

In figure 12-32, we can see the Front Panel of the oscilloscope application. We can see the resource ID of the oscilloscope and the horizontal adjust dial. We have two channels activated with two graphs and two vertical adjust dials. The first graph represents the input AC signal and the second graph represents the output DC signal. From this graphs we've seen that the power supply works correctly and we have an AC – DC power supply.

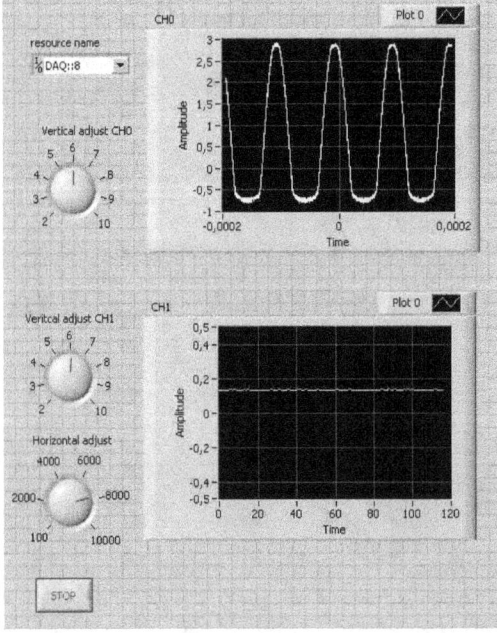

Figure 12-32 Font Panel for the Power Supply Testing with Oscilloscope

In figure 12-33, we have the 33250A Agilent signal generator programming. We have between the initialization and closing a While loop. In the loop we have the waveform configuration VI and the output enable VI.

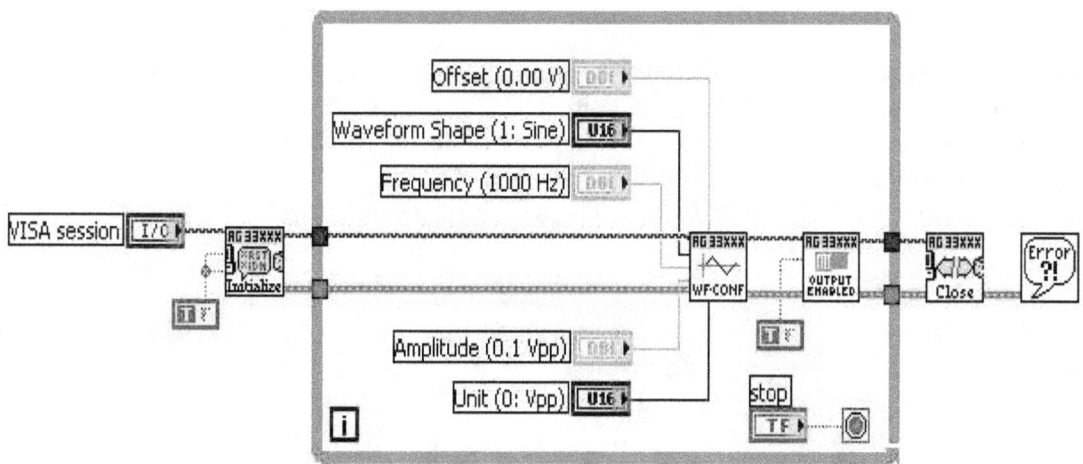

Figure 12-33 Block Diagram for the Agilent 33250A Function / Arbitrary Waveform Generator

In figure 12-34, we have the Block Diagram of the oscilloscope application. We start with the channel creation. We enter in the while loop. We have some timing settings for the horizontal adjust and the vertical adjust and readings for the both activated channels (channel 0 and channel 1). We put a Bessel filter to the output signal of the voltage supply. When we exit from the while loop we close the instrument and handle errors.

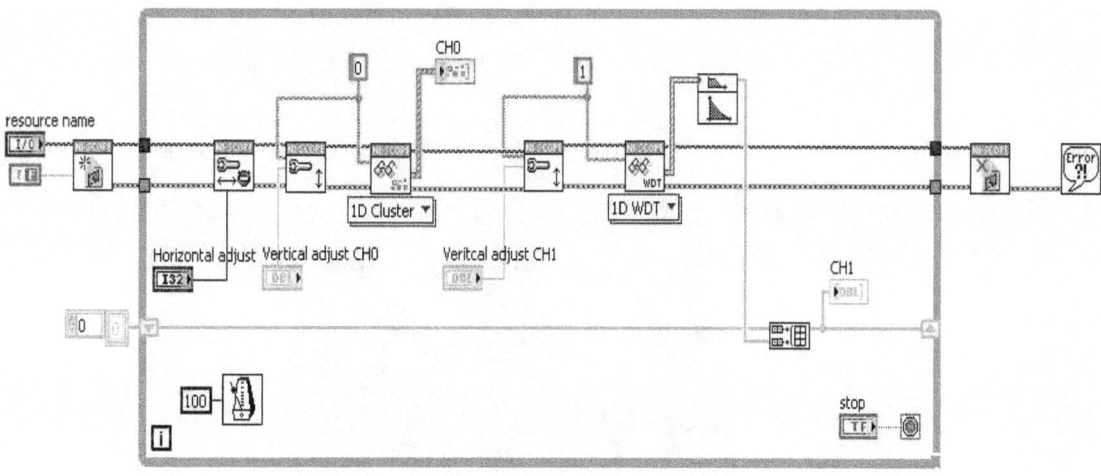

Figure 12-34 Block Diagram for the Power Supply

12.12 Virtual instruments for the Compact RIO chassis

12.12.1 Advanced motor control

In figure 12-35, we have used both the PXI and the Compact RIO chassis. The trapezoidal signal is generated with the NI PXI-4110 Power Supply. The signal then is amplified with the NI 9505 H – bridge. The Compact RIO is programmed to make a PID loop too. The motor is connected to NI 9505 H – bridge. The motor has a disk with 100 slots, which rotates between an Opto-coupler pair. The Opto-coupler is connected to the counter to read the frequency and calculate the RPM.

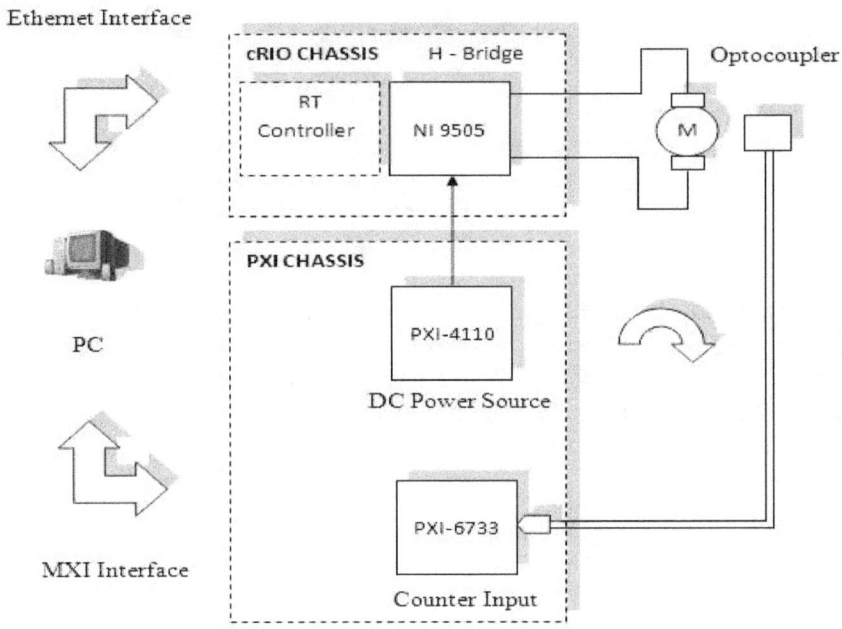

Figure 12-35 Block Schematics of the Advanced Motor Control

In figure 12-36, we can see the experimental setup with the motor and the Compact RIO.

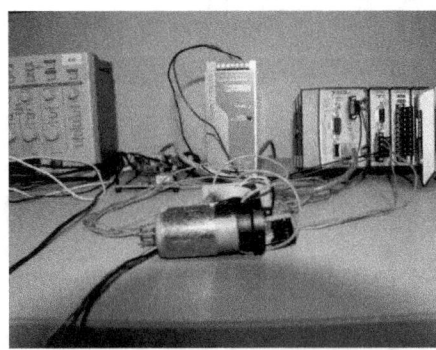

Figure12-36 Experimental Setup of the Advanced Motor Control

312

In figure 12-37,we can see the front panel of the signal generation program, which is similar to the program presented in paragraph 3.4. In the signal generation program the NI PXI-4110 power supply is used. In the Front Panel we have the resource ID, channel 1 is activated, the current limit is set to 100 mA and the voltage is measured at the output of the power supply.

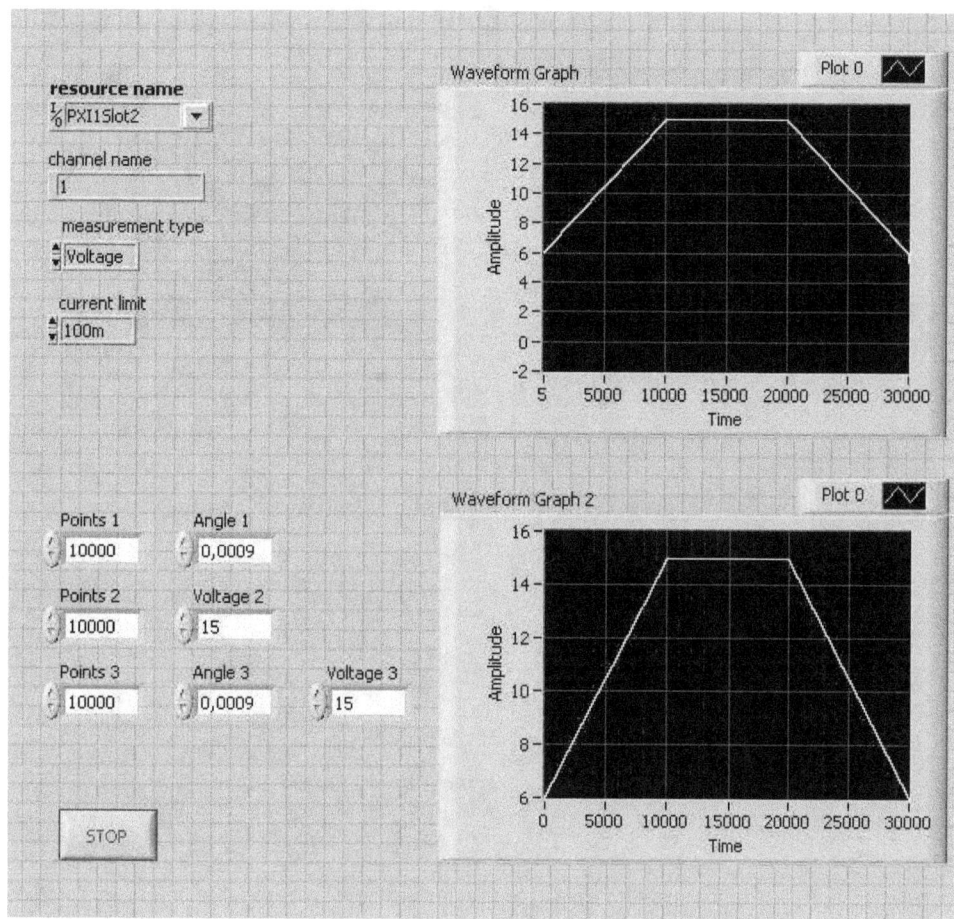

Figure 12-37 Front Panel of the Signal Generation

In figure 12-38, we have the RPM reading program. We have resource ID, rising edge, minimum frequency set to 100 kHz and maximum set to 1 MHz, 100 samples acquisitioned at the channel and the sampling divisor set to 5. We have put indicator for frequency and RPM and a Waveform Graph to represent the RPM. This graph should be similar to the generated signal.

In figure 12-39, we have the Front Panel of the FPGA programming. We have a lot of controls and indicators, which are use to control some functions of the NI 9505 H – bridge, which are also shown on the card with some LEDs. We have Enable Drive, Disable Drive and Enable Emergency-Stop.

We show the status of the Drive, the Drive Fault and the Over temperature Fault, the supply current (Vsup Present) present and the presents of the analog input's trigger (AI Trigger).

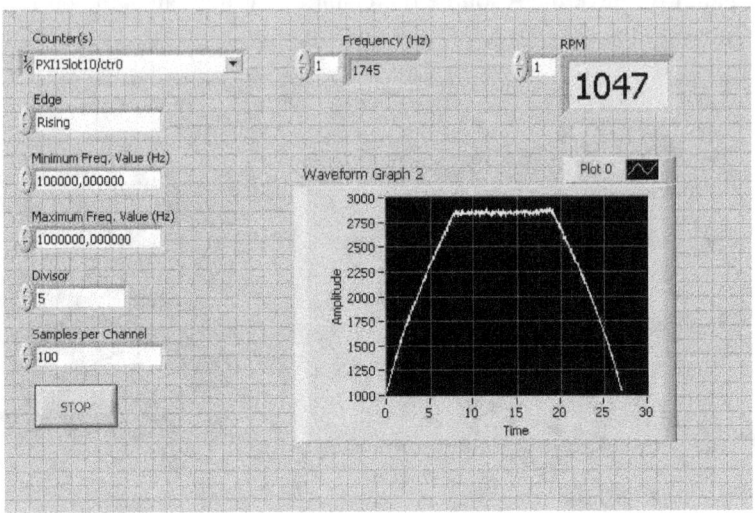

Figure 12-38 Front Panel of the RPM Reading

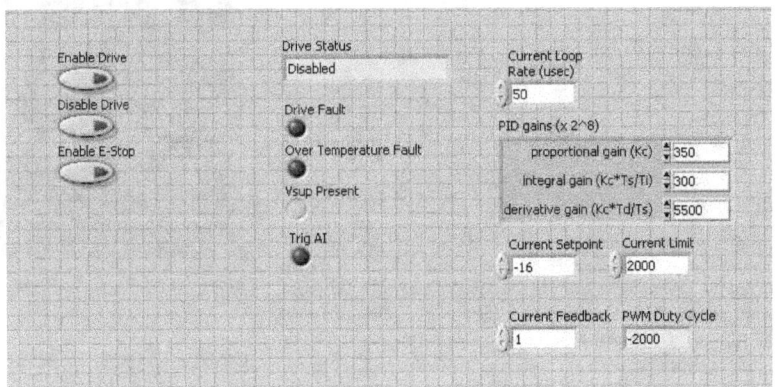

Figure 12-39 Front Panel of the FPGA Part

We have the Current Loop Rate set to 50 µs, this rate is the rate of the PID loop. We have the PID parameters: 350 for the proportional gain, 300 for the integral gain and 5500 for the derivative gain. We have a current set point at a negative value (in our case -16) this for rotating the motor in one direction, if it's positive the motor rotates in other direction. We limit the current at a certain value (in our case 2000) and we set current feedback to 1, after we read the PWM Duty Cycle.

The PWM Duty Cycle has the same sign as the Current Set point, this way to show the direction of the motor rotation. The Front Panel is almost the same with this one for all Block Diagrams which will be presented next.

In figure 12-40, we can see the Block Diagram of the trapeze generation program. This program is similar to the first While loop from figure 12-25, but it's made with a power supply, not a DAQ. The programming starts with channel creation. After entering in the While loop we set the voltage. Here is connected the trapezoidal signal.

314

After we set the current, we enable output and we measure the voltage at the power supply's ports. When we exit from the While loop, we disable output and close the instrument.

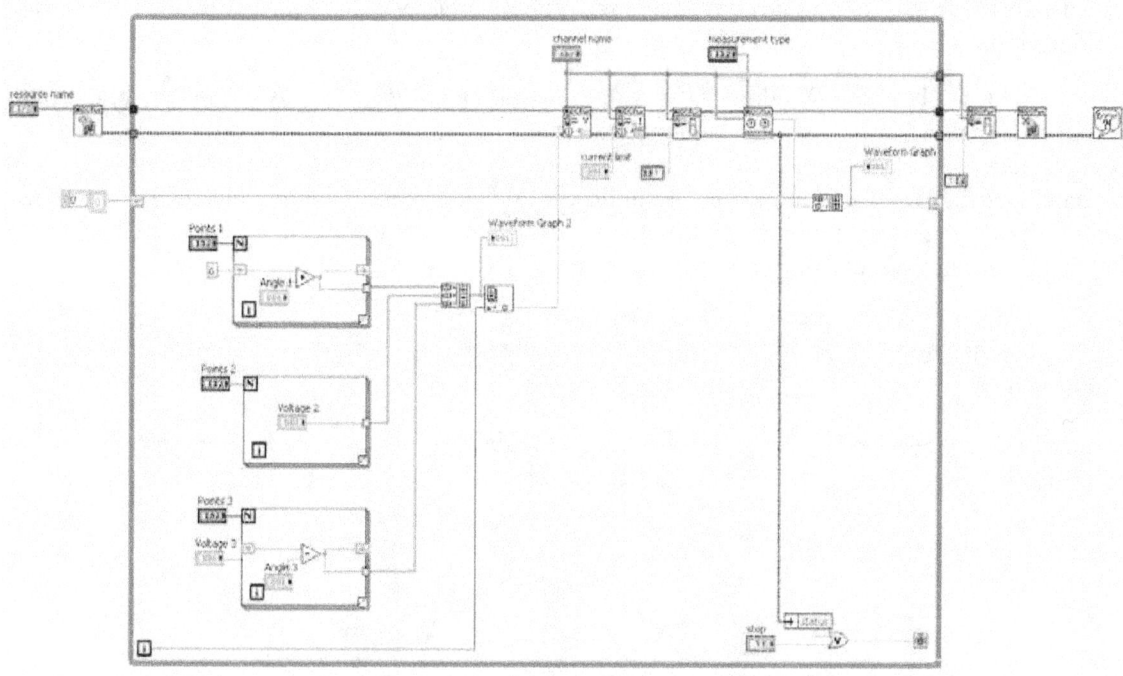

Figure 12-40 Block Diagram of Trapeze Generation with Power Supply

In figure 12-41, we have the RPM reading Block Diagram which is same with the second While loop from figure 12-25.

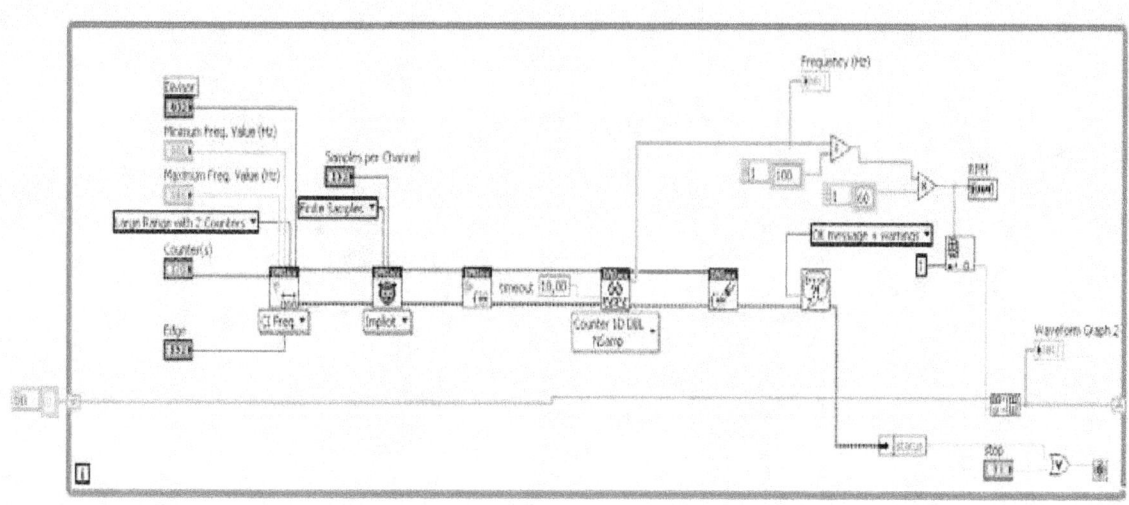

Figure 12-41 RPM Reading Block Diagram

In figure 12-42, we can see the we have a big While loop with all the controls used in the FPGA part and called by the Real-Time controller. At the left part we have loaded the FPGA part and at the output we closed it and we handle errors. In figure 12-43, we have the block Diagram of the networked real-time host. This VI is made copying the VI form figure 12-41, and adding global variables to send the data trough network to the Windows host. We have the global variables also before of the FPGA part loading for initialization.

In figure 12-44, we have the Windows host with the same global variables as in the networked real-time host, but mirrored. This means if in the networked real-time host we had a control in the windows host we have indicator and vice-versa. We can imagine an invisible line between the global variables with the same name starting from the control to the indicator, global variables communicates not only between different VIs on the same computer, but between VIs on different computers, connected via networks cables if they are in the same subnet.

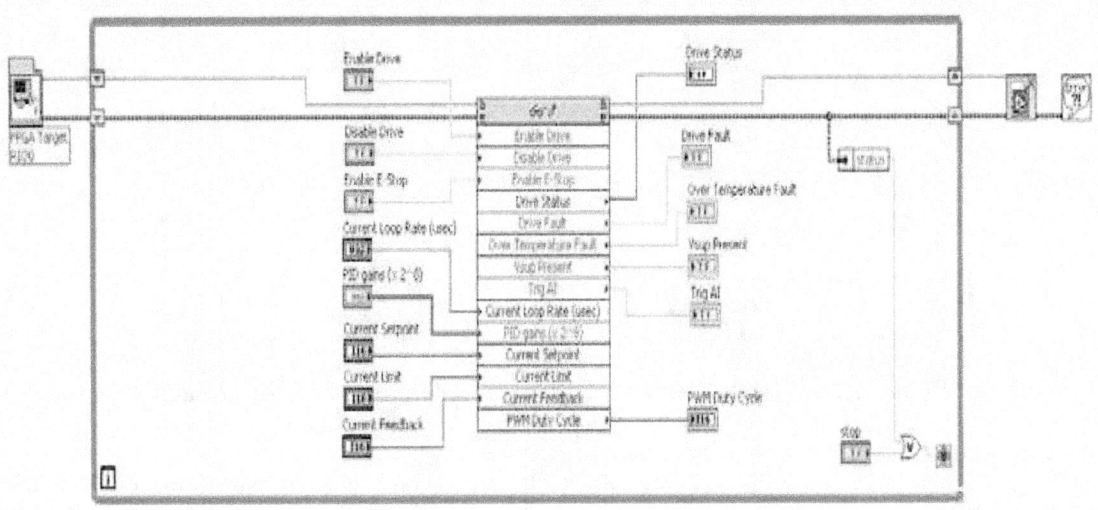

Figure 12-42 Block Diagram of the Real-Time Host

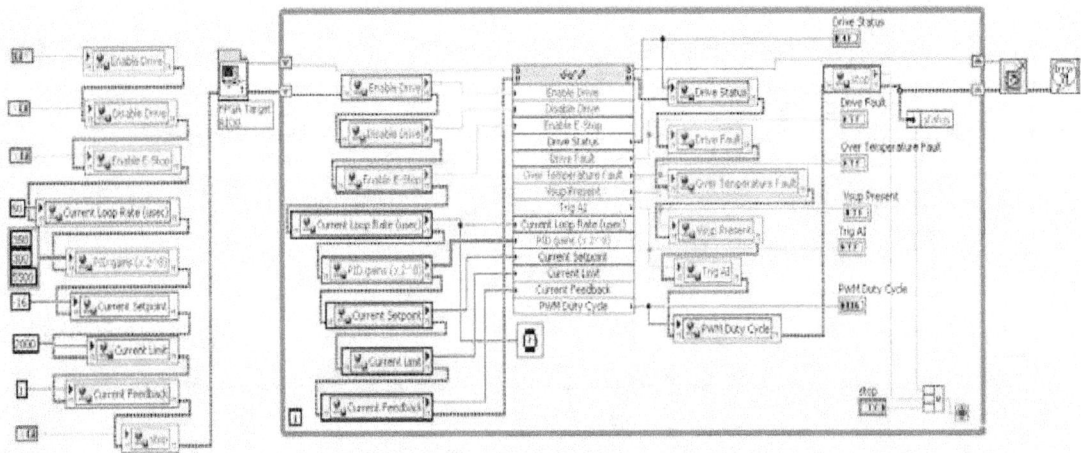

Figure 12-43 Block Diagram of the Networked Real-Time Host

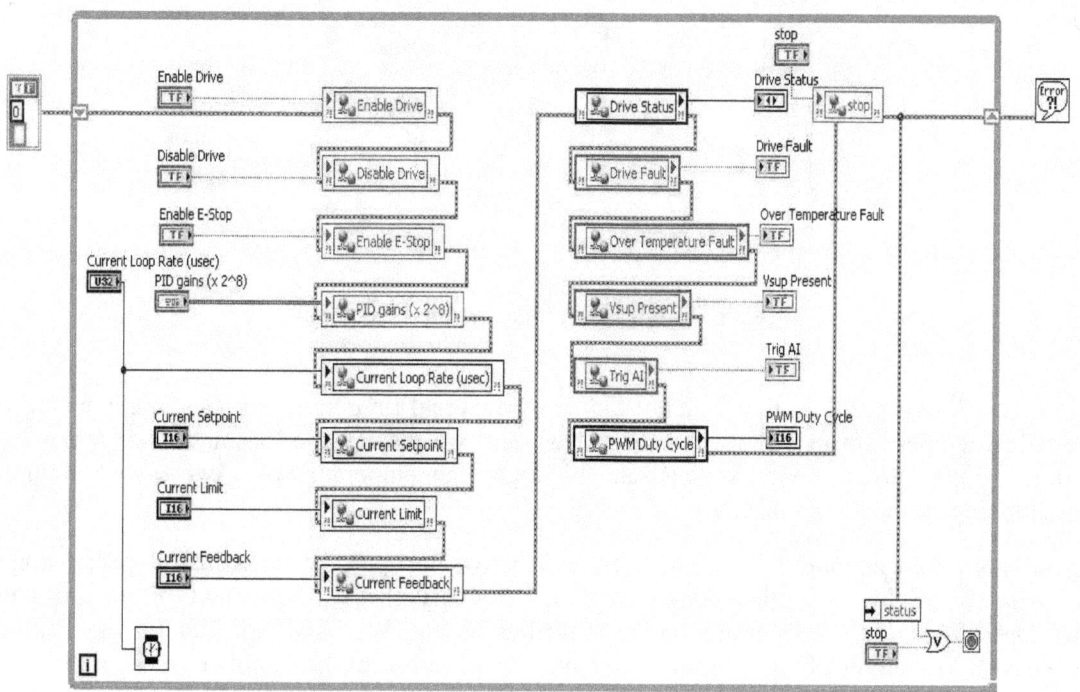

Figure 12-44 Block Diagram of the Windows Host

12.12.2 Advanced temperature measuring

In figure 12-45, we can see the block schematics of the experiment. We have the LM35 centigrade sensor connected to the NI 9201 analog input C series module which is connected to a CompactRIO. The CompactRIO is connected via Ethernet interface to the PC.

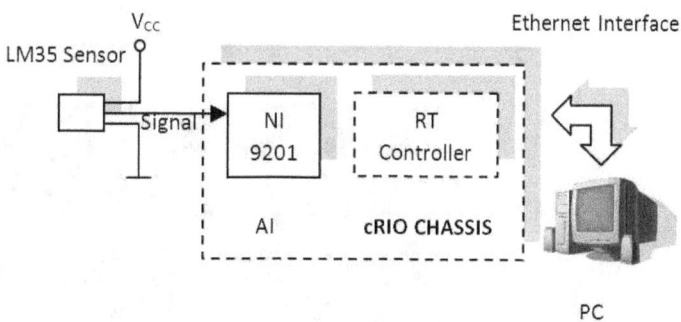

Figure 12-45 Block Schematics of the Advanced Temperature Measuring

In figure 12-46, we have the Front Panel of the FPGA part. We have only raw data collected here; we just measure the voltage in mV with the NI 9201 analog input module. We can set the loop timing and we visualize errors.

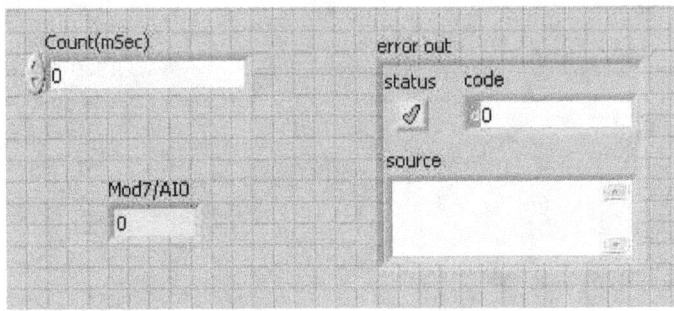

Figure 12-46 Front Panel of the FPGA part

In figure 12-47, we have the Front Panel of the networked real-time host part. Here we have the voltage converted in temperature in °C, we visualize errors and we display the log of the temperature o both Waveform Graph and Chart. Here we can see a big log of the temperature over one day when turning the air conditioning system OFF and ON.

In figure 12-48, we can see the Front Panel of the Windows host part. We have here a special stop button (Stop Windows GUI). With this button we can stop only the windows part of the program, the temperature acquisition will continue on the real-time system. We have indicator for the Temperature and we have a Waveform Chart for graph. We have c dial control for the sample interval in ms.

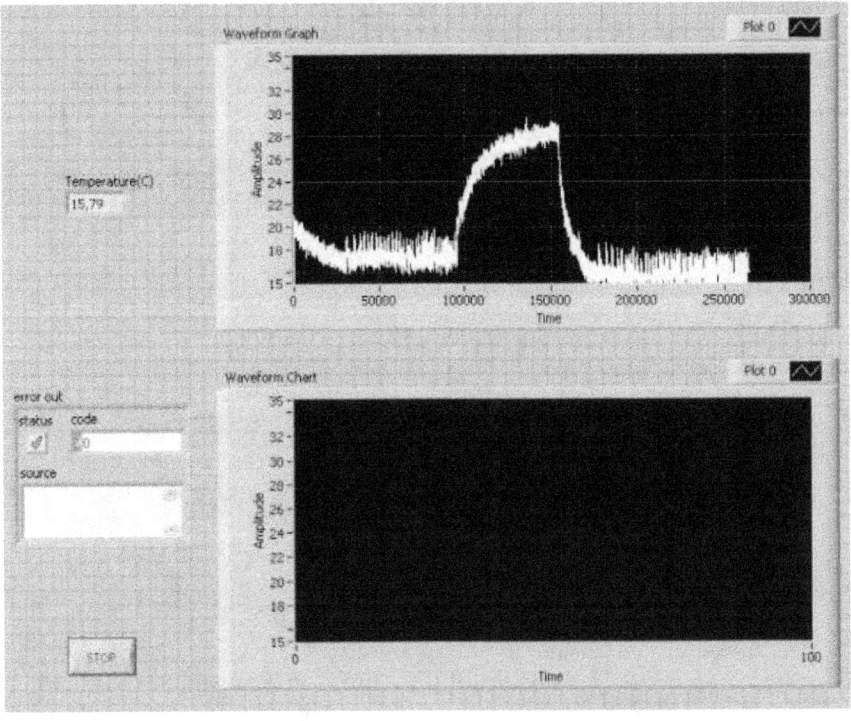

Figure 12-47 Front Panel of the Networked Real-Time Host

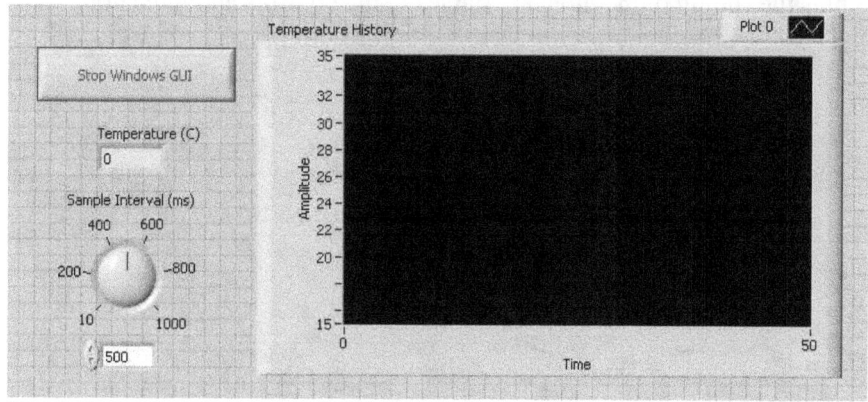

Figure 12-48 Front Panel of the Windows Host

In figure 12-49, we have the block diagram of the FPGA part. We have While loop and a Flat Sequence with some special FPGA timing and the acquisition with the FPGA node.

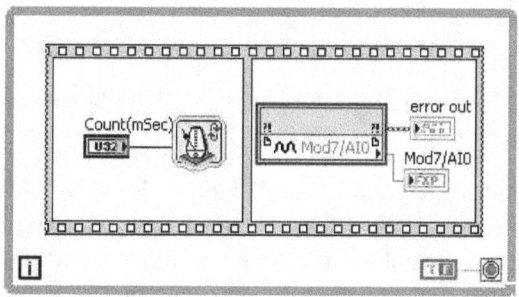

Figure 12-49 Block Diagram of the FPGA part

In figure 12-50, we can see the Block Diagram of the real-time host part. We have the FPGA program loaded in the real-time host (FPGA Target RIO0). We have a while loop with the used variables, the conversion from voltage to temperature in °C with multiplying with 100, a mean of 100 values. After the exit from the While loop we have the closing of the FPGA program loaded in the real-time host and the handling of errors.

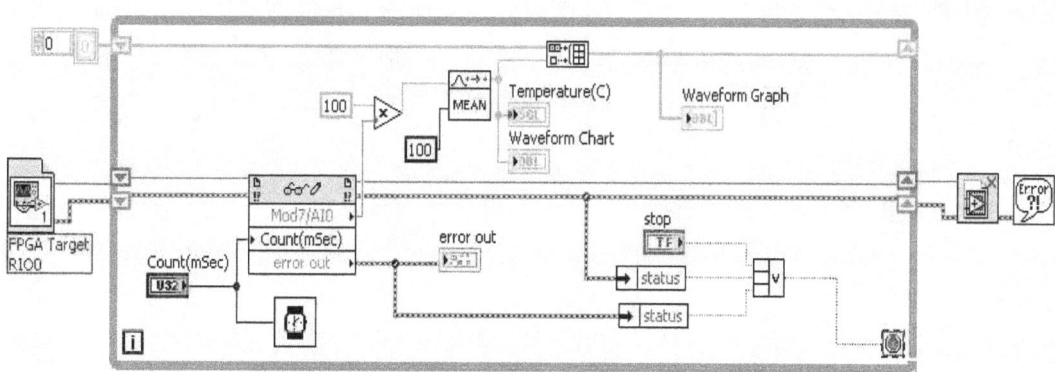

Figure 12-50 Block Diagram of the Real-Time Host

319

In figure 12-51, we have the Block Diagram of the networked real-time host, which is the copy of the real-time host plus the adding of the global variables to the controls and indicators.

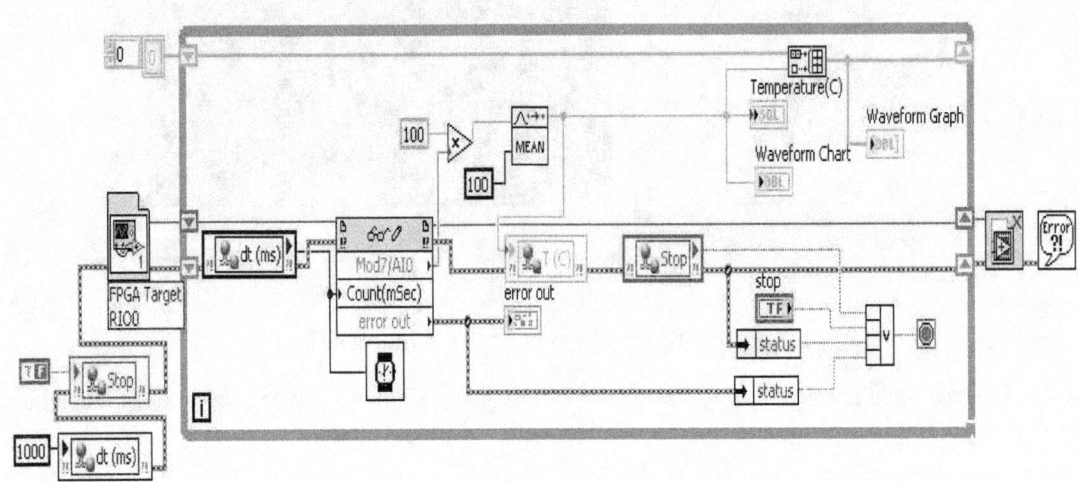

Figure 12-51 Block Diagram of the Networked Real-Time Host

In figure 12-52, we have the Block Diagram of the Windows host, which has the global variables as the mirror image of the networked real-time host. This means where we have control, we will have indicator and vice-versa. We can imagine an invisible line, connection between these global variables. These lines transport the data between the two VIs via Ethernet. The data is sent from the networked real-time host, but there is also feedback from the Widows host.

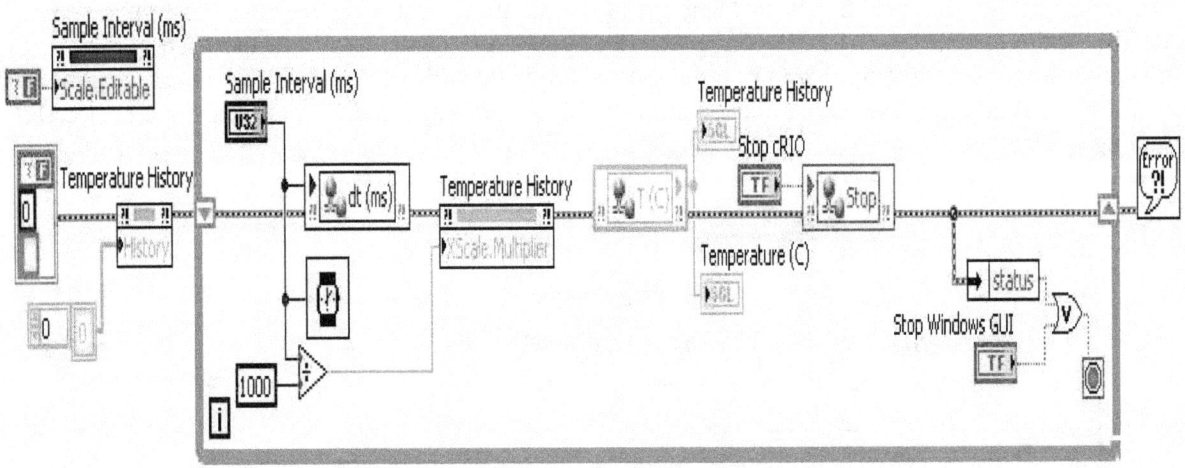

Figure 12-52 Block Diagram of the Windows Host

References

- **Lisa,K. W. and Travis,J.** (1997) , LabVIEW for Everyone: Graphical Programming Made Even Easier, New Jersey, Prentice Hall PTR

- **Essick, J.** (1998), Advanced LabVIEW Labs, New Jersey, Prentice Hall

- **Blume, P. A.** (2007), The LabVIEW style book, New Jersey, Prentice Hall

- **Relf, C.G.** (2003), Image Acquisition and Processing with LabVIEW, New York, CRC Press

- **Ertugrul, N.**(2002), LabVIEW for Electric Circuits, Machines, Drives, and Laboratories, Upper Saddle River, New Jersey, Prentice Hall PTR

- **Gupta, S. and John, J.** (2005), Virtual Instrumentation Using LabVIEW, New Delhi Tata McGraw Hill

- **Folea, S.,** (2011), LabVIEW: Practical Applications and Solutions, Croatia, InTech

- **King, R. H.** (2012), Introduction to Data Acquisition with LabVIEW, New York, McGraw-Hill

- **Bishop, R. H.** (2009), LabVIEW 2009 Student Edition, Prentice Hall-first Edition

- **Essick J.** (2012), Hands-On Introduction to LabVIEW for Scientists and Engineers, USA ,Oxford University Press

- **Mihura B.** (2001), LabVIEW for Data Acquisition, New Jersey, Prentice Hall

- **Stamps, D.** (2012),Learn LabVIEW 2010/2011 Fast, Mission KS, SDC Publications

- **Larson, R. W.** (2010), LabVIEW for Engineers, New Jersey, Prentice Hall

- **Tubbs, S. P.** (2011), LabVIEW for Electrical Engineers and Technologists, Stephen Philip Tubbs

- **Stamps, D.** (2013), Learn LabVIEW 2012 Fast, Mission KS, SDC Publications

- **Johnson, G. W.,** (1997), LabVIEW Graphical Programming: Practical Applications in Instrumentation and Control, New York, McGraw-Hill

- **Fairweather, I. and Brumfield, A.** (2011), LabVIEW: A Developer's Guide to Real World Integration, Seattle WA, Chapman and Hall/CRC

- **Sokoloff, L.** (1997), Basic Concepts of LabVIEW 4, New Jersey, Prentice Hall

- **Sokoloff, L.** (2004), Concepts of LabVIEW 6.0 : Applications, USA, Pearson,2004

- **Sumathi, S. and Surekha, P.,** (2007), LabVIEW based Advanced Instrumentation Systems, New York, Springer

- LabVIEW Student Edition (www.ni.com/labviewse)

- Web resources (ni.com)

  - NI Developer Zone (zone.ni.com)

  - Application Notes

  - Info-LabVIEW newsgroup (www.info-labview.org)

  - Instrument Driver Library (www.ni.com/idnet)

  - Hans Petter Halvorsen (Telemark University College) blog

- **Johnson,** G. and **Jennings,** R. (2006), LabVIEW graphical programming, New York, McGraw-Hill Professional

- **Eaton,** J.K. and **Eaton,** L. (1995), LabTutor: A Friendly Guide to Computer Interfacing and LabVIEW Programming, USA , Oxford University Press

Index

www.ingramcontent.com/pod-product-compliance
Lightning Source LLC
Chambersburg PA
CBHW080234180526
45167CB00006B/2276